The Six Sigma Black Belt Handbook

The Six Sigma Black Belt Handbook

Tom McCarty
Motorola University
Schaumburg, Illinois

Lorraine Daniels
Motorola University
Tempe, Arizona

Michael Bremer
The Cumberland Group
Hinsdale, Illinois

Praveen Gupta
Quality Technology Company
Schaumburg, Illinois

McGraw-Hill

New York Chicago San Francisco Lisbon London Madrid
Mexico City Milan New Delhi San Juan Seoul
Singapore Sydney Toronto

The McGraw-Hill Companies

Cataloging-in-Publication Data is on file with the Library of Congress.

3 4 5 6 7 8 9 0 DOC/DOC 0 1 0 9 8 7

ISBN 0-07-144329-0

The sponsoring editor for this book was Kenneth P. McCombs and the production supervisor was Pamela A. Pelton. The art director for the cover was Handel Low.

Printed and bound by RR Donnelley.

Contents

PREFACE

In the late 1980s, Bill Smith, a senior field engineer at Motorola coined the term "Six Sigma" as a more rigorous approach to reducing defects in Motorola business processes. His approach quickly demonstrated the power of establishing common metrics, setting outrageous goals, and applying analytical rigor to achieve those goals. In order to spread the skills and knowledge required to support this approach across the corporation, Motorola University was created. Their charter was to build employee capability globally, with a special emphasis on both technical skills and the leadership skills required to achieve the highest levels of business performance and customer satisfaction. Since those beginnings, instructors and coaches from Motorola University have been instrumental in driving the evolution of Six Sigma. As a result, Six Sigma has evolved from a metric used to express the ratio of defects, to opportunities for error in a business process to a disciplined and rigorous methodology for eliminating variability in all processes, to an overall management system that fosters customer focused execution across an organization.

The *Six Sigma Management System* has evolved to become an integration of business best practices that unleashes the power of the Six Sigma tools and methods in a way that fosters focused execution and breakthrough improvements.

Using the Six Sigma Management System as the overall context for driving business improvement, this book provides unique insight for Black Belts, Six Sigma leaders, and all Six Sigma practitioners for how to apply the wide variety of tools and methods that sit inside of the Six Sigma tool set.

Written by experienced instructors and coaches from Motorola University, this book was inspired by years of teaching and coaching focused on helping countless aspiring Six Sigma practitioners understand and apply Six Sigma skills and knowledge to their critical business issues. These instructors and coaches know from experience what Black Belts and Six Sigma leaders need to know to be successful in their Six Sigma implementations. As a result, the book avoids the theoretical and focuses instead on the practical insights that these instructors know will drive success at both the project level and in an organization-wide implementation.

The book is organized to provide insight into key areas that are all essential elements of a successful project or a successful implementation. The authors mean to send a message that Black Belt mastery of the technical tools is important but not sufficient. While the Black Belt practitioners will find a solid body of practical tips for using the most important tools and for successful application of the Define, Measure, Analyze, Improve, Control (DMAIC) methodology, they will also learn how to effectively organize and lead a variety of improvement teams, and how to become an effective coach to the leaders upon whom they rely for their success. Leaders, on the other hand, will learn enough about the tools and the DMAIC methodology that they will identify better projects, become better sponsors and coaches, and ultimately develop organizations where a Six Sigma Management System drives breakthrough success.

ACKNOWLEDGMENTS

Turning a vision into reality is an art form that is consistently demonstrated by only the best of leaders. Tom McCarty, Motorola University Director of Consulting Services, is one of those leaders. Nearly a year ago, Tom had a vision for developing a comprehensive Six Sigma Handbook that would capture the breadth and depth of Motorola's long-standing history and experience with Six Sigma and other process improvement techniques. And as all good leaders do, Tom enlisted the help and support of the best and brightest in the business. This book is the result of countless hours of hard work from several Motorola University Six Sigma professionals.

Tom led the way by capturing his years of experience in working with leadership teams within Motorola, as well as countless other organizations, to provide practical advice for leaders looking to bolster the success of their Six Sigma implementations.

Lorraine Daniels is truly a unique Six Sigma Master Black Belt. Lorraine can undoubtedly go head-to-head with any Six Sigma expert on the technical aspects of the methodology. But more importantly, Lorraine possesses a special talent for communicating, explaining, and teaching the concepts to just about anyone. And she put that talent to good use in Part Four of this handbook.

Michael Bremer brought 30 years of experience in continuous improvement and new insights into the integration of Lean and Six Sigma.

Praveen Gupta is already well-published in the area of Quality Systems and contributed the Six Sigma Measurement content and one of the supplements about the relationship of Six Sigma to Innovation.

John Heisey and Kathy Mills have more than 25 years of experience in Business Process Improvement. They did a great deal of research to support the book and in particular, developed the background and model for the Six Sigma Management System.

Jim Dovick has been working with Motorola for the last 15 years in assessing leadership skills and shared some new perspectives on the leadership roles and behaviors associated with Six Sigma.

Candace Medina is a Master Black Belt who contributed unique insights into the application of Six Sigma in non-manufacturing environments.

Anyone who has published a book before knows that the creative writing is where the fun is. But the real work is in the editing, proofing, and document management that must occur in order for a book to be published. Stacy Hanley has shown talent, tenacity, and drive in getting that work done on time, under budget while leading real collaboration across the team of writers. For all of that, we are truly grateful.

Part One:

The Six Sigma Management System

This first part of *The Six Sigma Black Belt Handbook* focuses on the extension of Six Sigma into a management system that encompasses all levels of an organization. Motorola University consultants have found that while implementing Six Sigma through individual projects has produced significant results in many organizations, sustainable, breakthrough improvements are realized by those organizations whose leadership has embraced Six Sigma and incorporated it into their vision, strategies, and business objectives - in short, adopted Six Sigma as the system for managing their organizations. The *Six Sigma Management System* enables a leadership team to align on their strategic objectives, establish their critical operational measures, and determine their organizational performance drivers and then use those to implement, drive, monitor, and sustain their Six Sigma effort.

The four chapters in this part of the book will:

- Introduce the *Six Sigma Management System*, and distinguish it from the Six Sigma metric and Six Sigma methodology
- Explain the background (Chapter 1), principles, and elements of the *Six Sigma Management System* (Chapter 2)
- Describe the Six Sigma leadership modes (Chapter 3)
- Provide insights into Six Sigma leadership (Chapter 4)
- Illustrate key tools used to implement the *Six Sigma Management System*

Chapter

1

Introduction to Six Sigma

Six Sigma has been labeled as a metric, a methodology, and now, a management system. While Green Belts, Black Belts, Master Black Belts, Champions and Sponsors have all had training on Six Sigma as a metric and as a methodology, few have had exposure to Six Sigma as an overall management system. Reviewing the metric and the methodology will help create a context for beginning to understand Six Sigma as a management system.

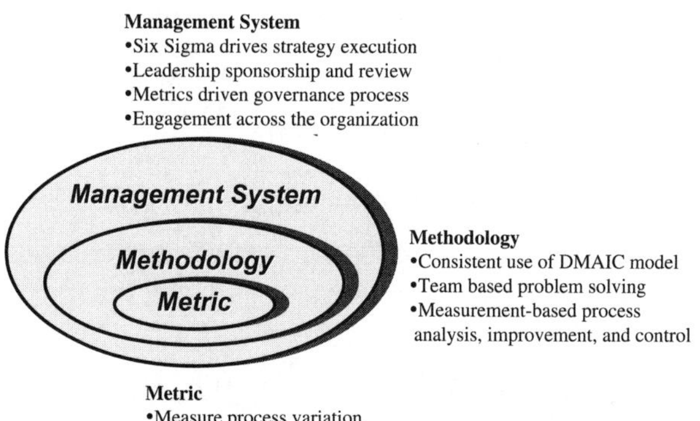
Management System
•Six Sigma drives strategy execution
•Leadership sponsorship and review
•Metrics driven governance process
•Engagement across the organization
Management System
Methodology
Metric
Methodology
•Consistent use of DMAIC model
•Team based problem solving
•Measurement-based process analysis, improvement, and control
Metric
•Measure process variation

Figure 1-1 Six Sigma as a Metric, Methodology, Management System

Six Sigma as a Metric

Sigma is the measurement used to assess process performance and the results of improvement efforts - a way to measure quality. Businesses use sigma to measure quality because it is a standard that reflects the degree of control over any process to meet the standard of performance established for that process.

Sigma is a universal scale. It is a scale like a yardstick measuring inches, a balance measuring ounces, or a thermometer measuring temperature. Universal scales like temperature, weight, and length allow us to compare very dissimilar objects. The sigma scale allows us to compare very different business processes in terms of the capability of the process to stay within the quality limits established for that process.

The Sigma scale measures Defects Per Million Opportunities (DPMO). Six Sigma equates to 3.4 defects per million opportunities. The Sigma metric allows dissimilar processes to be compared in terms of the number of defects generated by the process in one million opportunities.

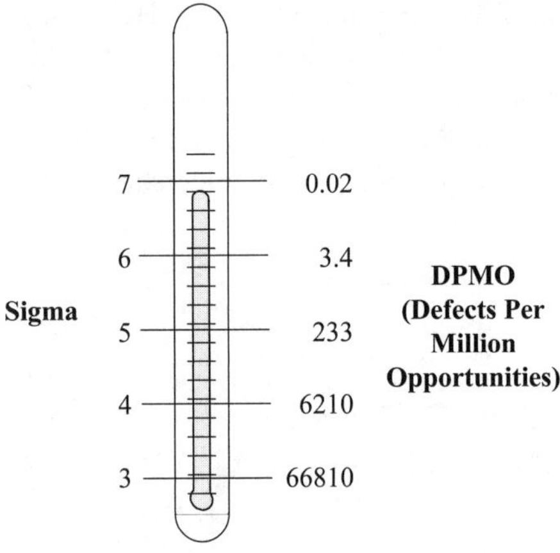

Figure 1-2 Sigma Scale

A process that operates at 4.6 Sigma is operating at 99.9% quality level. That means:

- **4000** wrong medical prescriptions each year
- More than **3000** newborns being dropped by doctors/nurses each year
- 2 long or short landings at American airports each **day**
- **400** lost letters per hour

A process that operates at the 6 Sigma level is operating at 99.9997% quality level. At 6 Sigma, these same processes would produce:

- **13** wrong drug prescriptions per year
- **10** newborns dropped by doctors/nurses each year
- 2 long or short landings at U.S. airports each **year**
- **1** lost letter per hour

Mikel J. Harry, one of the developers of Six Sigma at Motorola, has estimated that the average company in the Western world is at a 4 Sigma level, while 6 Sigma is not uncommon in Japan.[1] Dave Harrold, in *Control Engineering* [2] cites benchmark sigma levels broken down by industry and type of process:

- IRS phone-in tax advise - 2.2 σ
- Restaurant bills, doctors prescription writing, and payroll processing - 2.9 σ
- Average company - 3.0 σ
- Airline baggage handling - 3.2 σ
- Best in class companies - 5.7 σ
- U.S. Navy aircraft accidents - 5.7 σ
- Watch off by 2 seconds in 31 years - 6 σ
- Airline industry fatality rate - 6.2 σ

Clearly, the value of sigma is its universal application as a measuring stick for organizational and process quality. With sigma as the scale, measures of as-is process quality and standards for should-be process targets for quality improvement can be set and understood for any business process.

Six Sigma as a Methodology

The Six Sigma methodology builds on the Six Sigma metric. Six Sigma practitioners measure and assess process performance using DPMO and sigma. They apply the rigorous DMAIC (Define, Measure, Analyze, Improve, Control) methodology to analyze processes in order to root out sources of unacceptable variation, and develop alternatives to eliminate or reduce errors and variation. Once improvements are implemented, controls are put in place to ensure sustained results. Using this DMAIC methodology has netted many organizations significant improvements in product and service quality and profitability over the last several years.

The Six Sigma methodology is not limited to DMAIC. Other problem-

---

[1] Harry, Mikel. "Six Sigma: The Breakthrough Management Strategy Revolutionizing the World's Top Corporations." New York, N.Y. Random House Publishers, 2000.

[2] Harrold, Dave, "Designing for Six Sigma Capability", *Control Engineering*, January 1, 1999

solving techniques and methodologies are often used within the DMAIC framework to expand the tool set available to Six Sigma project teams. These include:

- Theory of Inventive Problem Solving (TRIZ)
- Lean
- Ford 8Ds (Disciplines)
- 5 Whys
- Is/Is Not Cause Analysis

Utilizing the sigma metric and marrying this variety of approaches with the DMAIC methodology, the Six Sigma methodology becomes a powerful problem-solving and continuous improvement methodology.

Clearly, the use of a consistent set of metrics can greatly aid an organization in understanding and controlling their key processes. So too, the various problem-solving methodologies significantly enhance an organization's ability to drive meaningful improvements and achieve solutions focused on root cause. Unfortunately, the experience of Motorola University consultants has demonstrated that good metrics and disciplined methodology are not sufficient for organizations that desire breakthrough improvements and results that are sustainable over time.

In fact, conversations with organizational leaders who report dissatisfaction with the results of their Six Sigma efforts have shown their Six Sigma teams have sufficient knowledge and skill related to good use of metrics and methodology. However, all too often, these teams have been applying the methodology to low level problems, and have been working with process metrics that don't link to the overall strategy of the organization.

It is this recurring theme that has driven Motorola University to develop the concept of Six Sigma as a management system, first introduced in the book "The New Six Sigma".

Six Sigma as a Management System

Six Sigma as a best practice is more than a set of metric-based problem solving and process improvement tools. At the highest level, Six Sigma has been developed into a practical management system for continuous business improvement that focuses management and the organization on four key areas:

- understanding and managing customer requirements
- aligning key processes to achieve those requirements
- utilizing rigorous data analysis to understand and minimize variation in key processes
- driving rapid and sustainable improvement to the business processes.

As such, the *Six Sigma Management System* encompasses both the Six Sigma metric and the Six Sigma methodology. It is when Six Sigma is implemented as a management system that organizations see the greatest impact.

These organizations are among those that have demonstrated that breakthrough improvements occur when senior leadership adopts Six Sigma as a management system paradigm.

- In 1999, ITT Industries implemented Value-Based Six Sigma (VBSS), the company's "overarching strategy for continuous improvement" (source: ITT Industries website). In the 2003 letter to the shareholders, the then-Chairman, President, and CEO, Louis J. Guiliano, wrote, "VBSS gives us the tools and discipline we need to make fact-based decisions, to solve problems and to find solutions in a systematic and measurable way. Now in its fourth year, the VBSS strategy is already making a huge difference to our customers. The thrust of our VBSS projects has shifted from simple initial short-term projects that drive out costs and waste, to more comprehensive projects that focus on making improvements that mean the most to our customers. VBSS is generating many new ways of growing our business and increasing our capacity, as well as saving millions of dollars. It is a key contributor to our robust cash flow performance. As our VBSS project leaders grow in numbers and in expertise - more than 10 percent of our 39,000 employees are now certified as Champions, Black Belts or Green Belts - they are increasingly focused on projects that are changing the way we do business in profound and enduring ways. Through the combined efforts of the Champions, Black Belts, Green Belts, and the teams they lead, we are seeing real progress in the quality of the products and services we are providing to customers. We are shortening cycle times, reducing lead times, and eliminating excess inventories; we are meeting or exceeding on-time delivery commitments, and dramatically reducing defect rates. Our customers have noticed these improvements, and we see this reflected in enhanced customer loyalty and market share position. More than that, our employees gain satisfaction from working on these teams."

- General Electric started its quality focus in the 1980s with Work-Out. Today, Six Sigma is providing the way "to meet our customers' needs and relentlessly look for new ways to exceed their expectations. Work-Out® in the 1980s defined how we behave. Today, **Six Sigma is the way we work**. Six Sigma is a vision we strive toward and a philosophy that is part of our business culture. It has changed the DNA of GE and has set the stage for making our customers feel Six Sigma." (source: General Electric website)
- Raytheon has used Six Sigma to cut billions of dollars in costs, improve cash flow and profits by millions, improve supplier and customer relationships,and build internal knowledge networks. "Raytheon Six Sigma™ is the philosophy of Raytheon management, embedded within the fabric of our business organizations as the vehicle for increasing productivity, growing the business, and building a new culture. Raytheon Six Sigma is the continuous process improvement effort designed to reduce costs." (source: Raytheon website)
- Honeywell views its Six Sigma initiative (called Six Sigma Plus) as the way to maintain its position with its customers as a premier company. "At Honeywell, Six Sigma refers to our overall strategy to improve growth and productivity as well as a measurement of quality. As a strategy, Six Sigma is a way for us to achieve performance breakthroughs." (source: Honeywell website)
- Valley Baptist Medical Center (Harlingen, TX) has incorporated Six Sigma Quality as one of its Seven Strategic Initiatives. The hospital has been recognized with a number of national awards, including the "Top Performer" in the country for the overall quality of physician care in the emergency room. (Award presented by independent marketing research firm Professional Research Consultants.) James G. Springfield, President and CEO of Valley Baptist Health System, says they have made Six Sigma the company's system for operations. (source: Valley Baptist Health System website).

These and other organizations have discovered that successful practice of Six Sigma requires the adoption of a management system to strategically guide their Six Sigma programs. The next chapter will explore the principles behind the *Six Sigma Management System*, as developed by Motorola University consultants to guide their clients to build strategic management systems and achieve breakthrough results.

Chapter

2

Foundations of the Six Sigma Management System

The *Six Sigma Management System* is robust, and designed to guide any organization's performance improvement initiatives at all levels. This management system:

- is built on the business process model of organization structure
- uses a data-driven management approach based on an unique operational measurement system
- is centered on a model of a high-performing, ethical leadership team
- applies a team-based model as its fundamental work unit.

Six Sigma's Business Process Management Model

Six Sigma as a management system incorporates the business process management (BPM) model. The *Six Sigma Management System* treats the business process as its fundamental organizational building block. The business process is the operational unit that is measured, managed, and continuously improved through the *Six Sigma Management System.*

The BPM model is best understood when contrasted with the classic functional model of management. In this classic model, the building block of an organizational unit is the functional department. Before 1990, most American companies operated their businesses in functional silos and basically ignored the ideas of business process management. In the minds of functional management, process design involved writing policy and procedure manuals for functional departments to follow.

Motorola invented and pioneered Six Sigma in the late 1980s. In the 1990s, leading businesses, including Motorola, turned to process reengineering to compete in markets that were exploding with improvement in the variety and quality of customer choices. Markets had changed from supplier-driven, push-controlled to customer demand-controlled dynamics. As these leading companies experimented and developed their understanding of process management and redesign, their leaders' vision evolved from a focus on managing specialized functional divisions of labor to the focus on managing business processes. BPM became their fundamental operating model. Through years of effort and work with BPM, they invented *continuous process improvement* as an operational strategy that combined the strengths of the Six Sigma improvement methodology and the BPM model.

Motorola and a few other global technology giants led a dramatic change in the fundamentals of how goal-driven organizations are designed and operated. They fought against a 150-year practice of designing organizations exclusively with the hallowed building blocks of the discrete functional departments - accounting, manufacturing, marketing and sales, etc. Instead, these leaders chose the "business process" as their organizational building block. Through trial and success, these companies have demonstrated the supremacy of the business process as the fundamental building block and the management unit to measure and control in high performance companies.

The process was a more natural unit to manage in the manufacturing companies that led the revolution than in the service businesses and government agencies that have since adopted the *continuous process improvement* strategy. Motorola, GE, Raytheon, and others began their process improvement efforts in their manufacturing operations with a goal of improving the quality and reducing the cost of their products. In manufacturing, these companies all achieved a very impressive, breakthrough level of success that proved the viability of BPM and continuous process improvement using Six Sigma methods. Billions of dollars were saved and customers were delighted with the quality and value of the products they received. AlliedSignal's Raymond C. Stark, Vice President of Six Sigma & Productivity, attributed Six Sigma practices with saving the company $1.5 billion between 1994 and 1998. [1]

---

[1] Harrold, Dave, "Designing for Six Sigma Capability", *Control Engineering*, January 1, 1999

But, at Motorola and GE, the application of BPM and Six Sigma continuous process improvement was not limited to the manufacturing arena. GE launched a corporate wide quality improvement strategy in 1995 when Jack Welch, Chairman and CEO, committed GE's empire to reach Six Sigma quality by the year 2000. Welch was quoted in 1997 that he expected his managers to be "committed zealots" of Six Sigma. The following year the company credited Six Sigma with adding $300 million to 1997 operating income. [2]

In that time period, Motorola's leadership worldwide embarked on the redesign, in fact, the redefinition of the total set of core business processes that had to be managed and improved to compete and survive in the changing global marketplace.

Motorola's core business process redesign experience enhanced and extended the definition of a process from a "manufacturing process" to a true "business process". In the broadest sense, a process is a structure for action to achieve predetermined goals. The classic definition, from Thomas Davenport, in *Process Innovation,* [3] states that a process is:

> "A structured, measured set of activities designed to produce a specified output for a particular customer or market. ... A process is a specific ordering of work activities across time and place, with a beginning, an end, and clearly identified inputs and outputs."

The key elements in this definition include structured and measured activities done in a specific ordering. A process must be bounded by a beginning and an end with clearly identified inputs and outputs. Those elements of the definition of a process remain fundamental to all process improvement work, especially Six Sigma. However, from the Motorola experience, a "business process" has come to have even richer meaning and greater utility as a conceptual tool for process management and continuous process improvement.

The most fundamental characteristic of a "business process" is not the individual structured measured activity or its inputs and outputs. It is the synchronization and coordination of structured, measured activities that tie them into business processes. That synchronization and coordination is typically accomplished through managing the flow of information through the "business process".

---

[2] Bylinsky, Gene, "How to Bring Out Products Faster", *Fortune*, November 23, 1998

[3] Davenport, Thomas H., "Process Innovation, Reengineering Work Through Information Technology", Harvard Business School Press, 1993

This enhanced understanding of the "business process" allows organizations today to manage and improve core business processes and business service processes. A "service process" is seen as coordinated set of collaborative, transactional activities that deliver value to customers.

A "core business process" is typically strategic to the survival of an organization and is:

- large, complex, and long-running. A single instance of a process such as "order fulfillment" or "design and develop new product" may run for months, or even years.
- multi-dimensional, with end-to-end flows involving materials, information, and even internal and external business commitments.
- widely distributed across traditional organizational boundaries both within and even between organizations.

When Motorola, GE, and other leaders began their efforts to redesign and redefine their businesses in terms of processes, their businesses were made up of processes in an organic or unmanaged state. These same organic process conditions continue to be encountered by every organization that is beginning the Six Sigma journey into process management and continuous process improvement.

Organic processes in their unmanaged state share many characteristics which make them difficult to deal with, at least initially.

- Cross-functional organic processes exist inside all large organizations, even those that are functionally managed. These processes are implicit, accepted, and mostly unmeasured, having evolved within the history of the organization.
- Organic processes are functional, producing some successful output units, but are uncontrolled and unreliable in terms of the quality and productivity they generate.
- Organic processes fiercely resist efforts at managed change, due to the threatening nature of moving from a trusted order to an unknown new order.
- Organic processes are difficult to see inside any organization that has not consciously designed or explicitly documented their processes.

- Organic processes interact with other processes. They divide and combine with one another as their undefined boundaries change.
- Organic processes evolve through:
  - unplanned changes and series of small adjustments in their internal activities, and
  - the acquisition or loss of process participants and their capabilities.
- Organic processes are often partially automated. For the sake of speed and reliability, routine or mundane activities are performed by computers wherever possible. Automated components of organic processes are normally the result of the one-to-one conversion of original manual activities into automated activities.
- In organic processes, people perform the tasks that are too unstructured to delegate to a computer or that require personal interaction with customers.
- Quality and productivity are often dependent on the intelligence, judgment, and efforts of individuals.
- People interpret formal and informal information flowing though the process, make judgments, and act to solve perceived internal or customer related problems.
- People modify processes to adapt to varying requirements. This makes processes dynamic and adaptive to demands from customers and unstable with variable output quality and quantity.

Through Six Sigma process improvement methodologies, organic processes are restructured, made explicit and visible, and ultimately, are brought under control. Motorola's people and leadership went through great struggles to accomplish this. Every Motorolan felt the stress of this enormous Six Sigma process redesign challenge, as safe traditional roles in organic processes were peeled away to build the new order of core process management. Today, as part of the *Six Sigma Management System*, leaders who are in charge of segments of Motorola's business operations are given the title of "Process Owner" and tasked with the continued maintenance and improvement of the processes they own.

The *Six Sigma Management System* has adopted BPM as the model for creating and deploying processes as fundamental organizational business units. Today, organizations that practice Six Sigma management treat their processes with great care and combine continuous process improvement with planned life-cycle process management. Managed processes have become critical proprietary intellectual property. Some say that managed processes are the business today and continuous process improvement will be the future of the business.

Six Sigma's Data-Driven Management Approach

A second cornerstone of the *Six Sigma Management System* is application of empirical scientific principles to manage business processes. The influence of scientific thinking, the belief that scientific methods can reliably solve business problems, and the acceptance of the need to base business decisions on factual data have all contributed to evolutionary development of the Six Sigma measurement system called the Dashboard system. The Dashboard system is the management information system that connects business strategy with the day-to-day ground floor operations of a business.

Belief in Root Causes

Six Sigma leaders take the position that while business processes are very variable in the organic state, they can always be controlled by finding and removing the sources of unwanted variability. In Six Sigma's business vocabulary, these unseen sources of unwanted variability are called the "root causes". There are basic tenets of science that underpin the belief that root causes can be fixed to improve processes. These are:

- Every process event has one or more root causes.
- Process events in business organizations are not random or chaotic.
- Analytical tools and reason will succeed in pulling complex events apart to uncover the contributing root causes.
- Most, if not all, underlying root causes can be controlled in order to manage the nature and occurrence of process events to improve process performance.

So, Six Sigma is a very optimistic body of management theory and practice. Every individual who practices Six Sigma must be comfortable with these tenets about root causation. Six Sigma's problem solving methodology, DMAIC, is built on these optimistic beliefs. If these beliefs were not true, neither Six Sigma nor any similar management discipline could reliably succeed in improving processes. In practice, if a Six Sigma project does temporarily fail to find the root cause of a process problem, no one even thinks that a root cause does not exist. The root is still there waiting and will be found and fixed.

Management Based on Actual Process Measurements

Six Sigma leaders take the position that empirical data obtained from measuring actual process events is the primary source of knowledge required to manage and improve that process. Organizations that adopt the *Six Sigma Management* System commit to creating a Six Sigma measurement system, called a Dashboard system, to collect data and to use that data as the basis for strategic decision-making.

It seems apparent that this type of empirical data-driven approach to management should be universally applied in the technology-rich world of the 21st century. But the experience of Motorola Six Sigma consultants argues otherwise. Too many private and public organizations are still guided strategically by untested assumptions and historical "truths". In these companies and agencies, organizational strategy and day-to-day operations are disconnected or, at best, loosely tethered.

Efforts towards building data-driven management systems have been going on for more than 100 years. In American industry, the evolution of management theories has left a legacy of bureaucratic organizational structures and pre-conceived notions about measurement systems that must be recognized and systematically rebuilt in order to implement Six Sigma successfully.

Impact of Scientific Management Theory on the Six Sigma Management System

U.S. industry has a long history of using detailed specifications of work tasks and quantitative measures of results as the basis for managing ground level operations. At the turn of the 20th century, Frederick Taylor developed the original Scientific Management Theory, which championed the idea of using standards for and measures of operational job tasks and work products. The approach was well designed for organizations with assembly lines or any repetitive work activities. Scientific managers endeavored to improve production efficiency through work studies, better tool design, and even economic incentives tied to productivity. The American automobile industry was the success model of scientific management going well into the 1950s.

Because it worked at the ground level (on the factory floors), American industries typically kept various versions of data-based performance measurement systems in place and continue to evolve them today. Also, there was and still is a wide acceptance across industries that day-to-day ground level operations should be managed by data-driven decision making. This environment that has survived 100 years was quite friendly to Six Sigma tenets, and was in place where Six Sigma was born, in the 1980s on a Motorola factory floor.

The Impact of Bureaucratic Management Practices

During the 20th century, another development in the thinking of American managers interacted with scientific management theory. This school of management attempted to address some of the same issues that the *Six Sigma Management System* handles today - how to centrally, strategically manage large-scale, diverse, ground-level operations. At the turn of the 20th century, a management thinker named Max Weber overlaid his concept of layers of management with hierarchies of organizational authority on scientific management theory. Weber's "bureaucratic organization structure" allowed a small number of top managers to "control" large numbers of dispersed factories with diverse assembly lines.

The Role of Financial Management Systems (Measurement in Dollars)

Today's layered management organization charts and the hierarchies of decision-making within these charts reflect Max Weber's legacy. The bureaucratic model worked well in large organizations when combined with financial management practices. Basically, the detail-rich data from ground level operations was converted to the lowest common denominator of business - dollars. Financial reporting systems converted operating results into the costs incurred and the value of goods and services produced. Financial planning systems required that operational plans be formulated as budgeted costs and forecast revenue from goods and services produced. As this financial information from each operational unit was reported up to each higher level in the management hierarchy, there were rollups to show the cumulative financial condition and prospects of the organization. These financial reporting systems did - and still do - provide the primary information needed by financial managers and by capital markets. And, for many decades, the combination of the bureaucratic decision-making model and financial management practices worked well, allowing large organizations to operate in stable markets that used almost unchanging technologies to produce their products.

The Impact of Rough Seas in the Evolution of Six Sigma Management Thinking

But markets grew more diverse and turbulent. Disruptive technologies became the norm of competition. Under these conditions, exercising operational management control through bureaucratic financial management systems became problematic. The long control lines of the bureaucratic organization prevented management from seeing financial results in time to order financial adjustments. To make matters worse, financial measurement systems rarely reported useful information about why operational processes were consuming too much cash and failing to generate enough valuable goods and services to survive. Management teams that did not understand this new world of the increasing rate-of-change in everything were blindsided and their businesses foundered. The metaphor of turning the great ship too slowly to avoid disaster found its way into common business parlance.

Improving on Operational Measurement Systems

In an effort to redesign their operational management systems to improve control and flexibility, "scientific" management theory made a series of adaptations and improvements starting in the late 1950s. Among the earliest move was the adoption of the cybernetic model, and the idea of feedback control loops. In the 1960s, ideas about short cycle feedback, control loops, and flexible contingency planning were borrowed from the fields of Systems Theory and Operations Research. These theories were very popular with the Pentagon for planning and executing military logistical operations in Vietnam.

From the mid 1970s well into the 1980s, the first generation of "production control" systems were installed in leading edge factories across the U.S. Johnson & Johnson and Brown & Williamson Tobacco Company both invested heavily and successfully in automated real-time control of production lines. Companies like Honeywell and Emerson Fisher invented whole new product lines of production control equipment based on cybernetic principles.

Motorola's Contributions

Throughout the 1980s, Motorola engineers implemented computer integrated manufacturing lines that that used custom designed robots and real time statistical process control. Product quality and productivity on these lines, ranging from Motorola pagers to Motorola computer chips, jumped by orders of magnitude. This effort led to winning the first Malcolm Baldrige U.S. National Quality Award in 1988. Statistical process control (SPC), long sought as a "holy grail" for manufacturing, had become a real-time operational reality. SPC is the mother concept for Six Sigma. The breakthrough thinking that came with Six Sigma is that the principles of SPC were applied beyond as-is manufacturing process control to truly proactive business process improvement efforts.

Management Measurement and Information Systems Play Catch-Up

Throughout the 1980s and 1990s in large organizations, top-level strategists and decision-makers remained far from their factory floors and their other strategic business operations. The "data" that these leaders saw were still converted, manipulated, summarized, and finally homogenized to provide an after-the-fact brief, mostly about the size and value of the results produced by their various operating units. Business strategy and operations remained very loosely tethered.

Motorola's top leadership clearly recognized this problem in the mid 1980s. The company's IT organization attempted to solve the problem through a revolutionary information systems design. At that time, a worldwide IBM mainframe network provided Motorola leaders with, basically, financial rollups. Motorola's top leaders who were very manufacturing savvy wanted more. The company had explicitly recognized the key to competitive survival was the ability to adjust to increasingly rapid change in markets and technologies.

Motorola IT leaders were tasked with designing and building an internal, but worldwide, network to connect their factories to their executive suite with the right kind of hard operational data and information. Leadership knew that constant changes and improvements were needed in their ground level operations, and that the leadership had to strategically manage the direction those changes were taking. Motorola invented the beginnings of a new management information model that used new kinds of measures including internal leading indicator metrics. But computer networking technology in the 1980s was not up to the task of fully implementing this model. Most of the required hardware and software did not exist as off-the-shelf products, and the build-from-scratch effort proved too complex and costly to fully implement at that time.

These new ideas were not lost, however. Motorola continued to build and implement successive generations of integrated measurement systems that did progressively better in connecting ground level operations to the executive suite. When Motorola won the Baldrige award again in 2002, Robert L. Barnett, Motorola corporate executive vice president said,

> "Six Sigma [management system] helps you gauge quality in several different areas, not just in manufacturing. There are many different processes related to leadership, strategy and customers, and Six Sigma can be used in any of these categories to assess how well the process is working and what the business results are. We have an ongoing valuation and improvement of processes to allow us to get better results."

The Six Sigma Dashboard System

What Motorola has learned from decades of experience is captured today as the Six Sigma model for balanced, integrated measurement systems - or the Dashboard system. The mechanics of the Dashboard system will be described in the next chapter of this book. What is important to note here is that the Dashboard system is an integrated operational and financial measurement system. It is a balanced measurement system intended to connect ground level operations to all levels of any organization's management hierarchy. The Dashboard system still summarizes the massive details of ground level operations. Designed to report selected sets of performance indicators to the appropriate process owners, business sector owners, and corporate executives, the genius of the Dashboard system derives from the selection of what raw data to measure and summarize.

Designing a working Dashboard system requires a profound knowledge of the operational details of the business processes of an organization. That is the kind of knowledge that Six Sigma teams generate.

The *Six Sigma Management System* incorporates the Dashboard measurement system that was grown through two decades of Motorola experience with process improvement and Six Sigma. Motorola's leaders' belief in the need for operational data to manage their business at all levels led to the development of an integrated measurement system that connects executive suite strategy with ground level operations and day-to-day results. Leaders make data-driven decisions and formulate strategies that enable continuous process improvement and rapid adaptation to change. And this system allows Six Sigma initiatives to be deployed and directed with maximum strategic impact.

The Foundation of Six Sigma's High Performance Leadership Model

Six Sigma is a powerful methodology that will produce breakthrough results when it is deployed by a visionary leadership team that is totally committed to *organizational* success. The *Six Sigma Management System* produces an organization where the adage, "Lead, follow or get out of the way of continuous improvement!" operates 24/7.

The creative, visionary role of leadership is the central tenet. The *Six Sigma Management System* is a top down, leadership-driven business improvement system. To align expectations and focus work efforts, leaders at the top of an organization create the strategic direction and then clearly and enthusiastically communicate it to everyone else. The thinking and energy of the leadership team must be the force that connects the organization's raison d'être to its energy source - its people. In a Six Sigma organization, no management structure or set of business tools can substitute for the constant, energetic, crisply focused advocacy of an aligned, committed leadership team. After the leadership's vision is effectively communicated and the organization is empowered to create the vision, then new ideas for change will bubble up from the ground levels of a Six Sigma organization.

The role of leadership and the implementation of the *Six Sigma*

Management System are the central focal points of this handbook. Other chapters will expand on the topic of leadership. Here, the foundations of the current thinking about Six Sigma leadership roles and practices are explored.

When implementing the Six Sigma Management System, the threefold challenge to leadership is to:

- envision the future,
- lay out a roadmap to get there, and
- mold the organization into a shape that can follow the map.

This is not a newly minted challenge. Is it safe to say that, at their birth, great organizations have almost always been the product of the creative, breakthrough visions of their leaders.

> "Every thoughtful man has an idea of what ought to be; but what the world is waiting for is a social and economic blueprint. ... We want those [leaders] who can mold the political, social, industrial, and moral mass into a sound and shapely whole." - Henry Ford [4]

Henry Ford intuitively saw the key roles of leadership - creating the vision of what ought to be, drawing the blueprint to follow to build the vision, and molding the organization to support the vision. He did not have - or need - the benefit of the following 80 years of evolution of management theory and practice. Much of that thinking was drawn from the experience of large industrial businesses that operated comfortably in stable environments, and used bureaucratic organizational structures for command and control decision-making. This experience relegated management to a very conservative stewardship role. "Just don't rock the boat." and "If it's not broke, don't fix it." management thinking dominated for 50 years.

Descriptions of top management activities like "annual strategic planning" covered the mundane tasks of projecting revenues from sales for the next several years, approving R&D efforts to do incremental product/service changes, and determining if it would be absolutely necessary to risk the introduction of a new line. Leaders by virtue of their many years of inbred experience in the business acquired a strong belief in the "truths" and

---

[4] Henry Ford, Ford Ideals: Being a Selection from Mr. Ford's Page in the Dearborn Independent (1922) Kessinger Publishing Company 2003, ISBN: 0766160343

dogmas of their industry. Using the purse strings, the top of the organization protected the clock-like status quo against unnecessary and risky innovations.

Along the way, many great organizations like Johnson & Johnson, Hewlett-Packard, 3M, and Microsoft were founded by leaders who had personal visions along with a set of strong core values and a relentless drive for organization building. But, for a long time, the innovators who founded these businesses were considered special people.

By the 1990s, Hamel and other leading management theorists were strongly encouraging top management in large organizations to escape the conservative constraints of their history, and to reinvent their businesses in innovative, visionary ways.

> "Where are you likely to find people with the least diversity of experience, the largest investment in the past, and the greatest reverence for industry dogma? At the top!" - Gary Hamel [5]

In the interim between Ford and Hamel, a whole body of literature about management's roles, responsibilities, and styles was written. A commonly accepted distinction had been established between leaders and managers. *Managers supervise people and do the ground-level work of organizations. Leaders change and improve the way organizations do their ground-level work.* In the new world of accelerating technological and market changes, top management had to become the change leaders in the adaptive organization.

> "The job of the people with the most formal authority, the "chiefs" -- chief executive officer, chief operating officer, chief financial officer -- is to create an environment in which change insurgency can flourish." - Bob Reich [6]

The introduction of Six Sigma as a business strategy at Motorola in the 1980s was itself an act of strategic, visionary reinvention. Motorola's top-level management team, led by Bob Galvin and George Fisher, recognized that there was greater strategic potential in Six Sigma ideas. The original Six Sigma process improvement methodology was being invented and applied at the factory level to improve Motorola product quality. Galvin envisioned the Six Sigma company. Through the direct advocacy of every

[5] Gary Hamel, "Strategy as Revolution" (Jul-Aug 1996). Harvard Business Review, pp. 69-82.

[6] Robert B. Reich, "Your Job is Change", *Fast Company*, October 2000

key leader, Motorola communicated this vision of a defect-free organization. They remolded their worldwide organization into alignment with the vision, and empowered every Motorolan to take action to support the "cause" of winning the global battle for superior quality and value.

A bias toward action is another core leadership tenet of the Six Sigma Management System.

> "Rule #1. A Bias for Action!" - Tom Peters [7]

The action bias is based on the long-standing success of the Motorola results-oriented business culture. Since the 1970s, at Motorola, every plan at every level has resulted in "to-dos" and assigned accountability for each "to-do". The culture explicitly recognizes that imperfect plans that are implemented and adjusted successfully on the fly are far more valuable than the ultimate perfect strategic document. Strategic planning, in the *Six Sigma Management System*, is a sub-activity of the greater leadership responsibility of "Aligning the Organization".

There is a creative tension within the *Six Sigma Management System*. The Six Sigma DMAIC methodology aims for continuous improvement of the organization's processes. In many cases, this improvement is "incremental" improvement. In stark contrast, the Six Sigma leadership model aims to guide the organization using innovative and discontinuous agile thinking about how and where to apply the DMAIC process improvement methodology. Since the publication of Being Digital, his seminal book on the Internet Age, [8] Professor Nicholas Negroponte of MIT has repeatedly warned about the conflict between the disciplined mindset that generates incremental progress and the stimulation of creativity. Because of this, he argues that big companies with their stable routines aren't - and can't be - good at innovation.

In the *Six Sigma Management System*, both innovation and incremental change are critical to achieving breakthrough results that can be sustained by the organization. It is the role of leadership to create visions and then to foster innovation by encouraging risk-taking by those who identify opportunities for improvement and set "stretch goals" for Six Sigma improvement initiatives. A leadership that accepts nothing less than breakthrough innovation achieved through carefully managed incremen-

---

[7] Tom Peters, "Re-Imagine!", Presentation to Motorola Leaders, July 14, 2004, tompetersnew.com

[8] Nicholas Negroponte, Being Digital, Random House Inc. 1996

tal change processes is a signal characteristic of successful Six Sigma organizations like GE, Allied Signal and Motorola.

In the *Six Sigma Management System*, fostering innovation and aggressive goal setting are key leadership responsibilities necessary to "Mobilize the Organization" to implement change. Motorola leaders lead. They communicate. They dramatize the need to achieve the vision. Every member of every team is challenged and empowered to do their very best. And every project is justified by its potential to impact strategic objectives and achieve the vision.

> "Do you have awesome Talent ... everywhere? Do you push that Talent to pursue audacious Quests?" - Tom Peters [9]

As a Six Sigma company, Motorola is totally committed to being a magnet for talent and a developer of leaders. One of key accomplishments has been the creation of the Motorola Global Leadership Supply Process. Motorola's 2002 Summary Annual Report [10] states,

> "We completed a rigorous ranking of 1,200 executives as part of our new Leadership Supply process, which is designed to ensure that we can locate - internally or externally - talented, trained, respected and motivated people always ready to accept new assignments."

Guided by a Six Sigma trained leadership team of Sponsors and Process Owners, Motorola has certified a vast cadre of Master Black Belts, Black Belts, and Green Belts who are the living infrastructure of the Six Sigma Management System.

The roots of the last central Six Sigma leadership tenet have been fed by Motorola's ethical business values. The Six Sigma Management System, in practice, relies on the trust of and respects the rights of all stakeholders. These values are based on Motorola's key beliefs of ***Uncompromising Integrity*** and ***Constant Respect for People*** that have been in existence for decades. Since its establishment in the 1970s, a formal code of business conduct has provided Motorola employees guidance for their business activities, placing a priority on establishing trust with stakeholders. That ethical tradition continues. In 2003, Matt Barney, Motorola's Director of Six Sigma Business Improvement wrote,

---

[9] ibid., page 3

[10] http://www.motorola.com/General/Financial/Annual_Report/2002/letter02.html

> "CEO Chris Galvin has consistently emphasized ethics, both in terms of role modeling expected behaviors and managing employee performance. ... he has led the effort to establish ethics as a regular metric in Motorola's performance appraisal system; all leaders must score high to continue their employment." [11]

Today, it is expected that every successful Six Sigma company will be led by a high performance leadership team. Six Sigma has evolved from a ground level quality improvement methodology to a business management system. Along the way, successful experiences have shown that visionary leadership drives Six Sigma to achieve significant, big, innovative organizational performance improvements. Visionary leaders are those who are action-oriented and trustworthy; they are champions of innovation and change, dynamic communicators with stories of compelling quests, and they develop and empower and enable their people.

Foundations of the Six Sigma Work Ethos

At its best, a visitor to a Motorola Six Sigma facility will meet many bright, committed Motorolans whose jobs are to operate Six Sigma processes. These people will be able to describe each of their processes and the impact of the outputs of the process on their customers. They will demonstrate pride in the sigma quality levels attained and a sense of ownership and responsibility in maintaining and participating in the continuous improvement of the process. From the ground-level up through the entire organizational structure, people who work at Motorola and at other Six Sigma companies are educated in process thinking and highly aware of their important roles in process improvement and adapting to change.

This is an ideal work ethic that Six Sigma companies strive to engender and sustain. It is these ground-level, hands-on people who must - and do - sustain the big performance improvements created by the DMAIC performance improvement projects. In many cases, it is also the profound knowledge of ground-level, hands-on people that provides key insights into the root causes of process performance variation, thus enabling the development of innovative and significant solutions.

---

[11] Matt Barney and Tom McCarty, The New Six Sigma, Prentice Hall PTR, 2003

Deming's 14 Points for Management

This Six Sigma work ethos can easily be traced back to the influence of W. Edwards Deming whose seminal work on process control and quality and productivity improvement began in the late 1940s and continued forward into the 1990s. Deming repeatedly delivered his now famous 14 points for management. Twelve of his 14 points are abstracted below, because these points reflect his view of the strategic need for leadership to encourage a work ethic of commitment to continuous improvement. [12]

- "Create constancy of purpose toward improvement of product and service, with the aim to become competitive, stay in business, and provide jobs."
- "... We are in a new economic age. Western management must awaken to the challenge, must learn their responsibilities, and take n leadership for change."
- "Improve constantly and forever the system of production and service, to improve quality and productivity, and thus constantly decrease costs."
- "Institute training on the job."
- "Institute leadership. The aim of supervision should be to help people and machines and gadgets to do a better job..."
- "Drive out fear, so that everyone may work effectively for the company."
- "Break down barriers between departments. People ... must work as a team, to foresee problems of production and in use that may be encountered with the product or service."
- "Eliminate slogans, exhortations, and targets for the workforce asking for zero defects and new levels of productivity. Such exhortations only create adversarial relationships, as the bulk of the causes of low quality and low productivity belong to the system and thus lie beyond the power of the work force."
- "Eliminate work standards (quotas) on the factory floor. Substitute leadership."
- "Remove barriers that rob the hourly paid worker of his right to pride in workmanship. The responsibility of supervisors must be changed from sheer numbers to quality."
- "Institute a vigorous program of education and self-improvement."
- "Put everybody in the company to work to accomplish the transformation.

---

[12] Deming, W. E., Quality, Productivity and Competitive Position, M.I.T. Center for Advanced Engineering Study, 1982.

How to Develop and Sustain the Six Sigma Culture and Ethos

Deming's points for leaders apply equally well today, decades later, to developing and sustaining the culture and work ethos in which Six Sigma can thrive. The following are 12 updated restatements of the Deming message as applied in the *Six Sigma Management System*:The transformation is everybody's job."

- People at work are rational and economically motivated. They will respond with their own enlightened self interest when leadership clearly and openly shares a strategic view of the value chain of the organization. Performance improvement produces greater quality and value, which in turn produces customer satisfaction and loyalty, which in turn produces profits and jobs in a highly competitive environment.
- The first job of leadership is to recognize the need for and then initiate strategic changes in direction.
- The term of the commitment to improve processes is "forever". The improvement task will never be finished.
- Learning by doing is the most effective method.
- Leaders, including supervisors, enable their people to improve the process and thereby achieve gains in performance.
- Continuing respect for people builds trust. Innovative thinking, including disruptive ideas and risk taking, are encouraged and rewarded.
- Everything is team-based. Teams are a force multiplier. Cooperation across traditional organizational boundaries is a requirement for survival.
- All people are inherently complex and flexible and, therefore, error-prone when given the wrong type of task in a process. Incentives and motivational techniques are at best irrelevant in such circumstances. The only solution to human errors is to design them out of the work process.
- Arbitrary work standards and quotas are disincentives to performance.
- Quality and productivity are driven by personal pride. Leaders empower people to be the best they can.
- The primary organizational value of the human resource is the intellectual property people can create. People must be developed and educated to maximize their potential to participate in change and improvement.

- Everyone has a critical role in achieving the vision in the Six Sigma organization.

Contributions of *Kaizen* and Japanese Work Methods and Culture

As it evolved, Six Sigma also absorbed many principles of *Kaizen*, the productivity movement introduced in the U.S. in the mid 1980s. The practice of *Kaizen* reflected thirty years of learning experiences with quality improvement efforts in Japanese industry beginning in the 1950s. The Japanese were the early adopters of the statistical methods of Deming and Joseph M. Juran, and they heeded their advice to refocus quality control efforts from making technical improvements in factories to an overall concern for the entire business organization. By the 1980s, organization-wide quality practices had become a strategic tool of Japanese business management and a competitive weapon in global competition. Many of these practices were brought back to the U. S. as the *Kaizen* movement.

The Japanese word *Kaizen* was Americanized to mean "continuous improvement", but a more literal definition of the Japanese meaning is "to take apart and put back together in a better way". Thirty years of development in the unique business culture of Japan made Kaizen a movement that strongly depended on the powerful culturally-based personal commitment of employees to their company and to their role in working continuously to build a <u>better</u> way.

Along with *Kaizen*, closely related Japanese methods were adopted by American companies including the use of quality circles and participatory work teams as instruments of improvement and change. The *Six Sigma Management* System has been directly influenced by 25 years of American learning experiences with teams as work units for implementing change as well as operating processes. When teams are empowered to take care of and improve their processes, individual team members tend to develop more personal commitments to the team, and that attaches to the mission of the team as pride of accomplishment. The *Six Sigma Management System* explicitly recognizes and deploys a variety of types of teams, including *Kaizen* teams, to take on different kinds of performance improvement challenges. The chapters about Lean and *Kaizen*

Process Improvement Teams in this handbook discuss the how different kinds of teams can work effectively within the Six Sigma DMAIC model.

The *Six Sigma Management System* Model of Team Development

Over many years of experience of deploying process improvement teams, Motorola's Six Sigma leaders have informally adopted Dr. Bruce Tuckman's 1965 model of the four stages of team development [13]. Every team is expected to go through the stages of forming, storming, norming, and performing. Six Sigma leadership watched teams go through these stages many times and began to understand the impact of these stages on the ability of a team to do its job of process performance improvement.

The Forming Team

In the forming stage, Six Sigma teams show a high dependence on leadership for guidance and direction. The team's purpose, objectives, and scope of work must be clearly communicated by the leader. A true team does not yet exist. In their own minds, team members are clarifying what the task assignment is and what it means to them personally and individually. Team roles and responsibilities are unclear and members may participate reluctantly. Leaders must be prepared to work closely with the team to get the elements of the Six Sigma Team Charter in place. Leaders should expect to do a lot of explaining about the strategic "why's" of the team's assignment.

The Storming Team

In the storming stage, the Six Sigma team's internal structure is born. Individuals try out team roles and responsibilities. In the storming process, they struggle to adjust to working together as teammates, and clash with each other's personal habits and working styles. Storming is an emotional adjustment and various teams will experience this stage with very different intensity levels. The primary issue is how the team will function to accomplish its objectives. Among team members, there may be significant disagreement and conflict about how everything should be accomplished. Cliques and factions can form, and there may be power

---

[13] Bruce W. Tuckman, "Developmental Sequence in Small Groups", *Psychological Bulletin*,1965 Volume 63, Number 6, Pages 384 99.

struggles. Members may ignore internal team procedures and test the tolerance of their leaders. Decisions do not come easily within the storming group. The team needs to be focused on its assigned objectives to balance the distraction of its internal relationship struggles. Leaders take on a coaching role, generating ideas for the team to debate, and suggesting and encouraging compromise decision-making to enable progress toward the team's primary performance improvement objective.

The Norming Team

In the norming stage, group identity and cohesiveness among members is established. Members achieve a sense of belonging and become comfortable sharing ideas and feelings, and giving and receiving feedback. The Six Sigma team builds a shared commitment to their assignment and their internally developed goals. Team roles and responsibilities are clear and accepted. The team discusses and develops its internal processes and working style. Big decisions are made by group agreement. Win-win agreements and consensus displaces compromise decision-making. Smaller decisions may be delegated to individuals or small teams within group. The sense of team unity is strong, and the team may engage in fun and social activities. The team leader becomes a facilitator and enabler. The team shares some of the leadership.

The Performing Team

In the performing stage, the Six Sigma team becomes an interpersonal force of interlocked roles and shared commitments that will support rather than hinder task performance. Performing teams have the *esprit* to deal with obstacles, setbacks, and complex problem-solving challenges. The focus of the team shifts to achieving its process performance improvement objectives. There is a shared commitment to over-achieving strategic objectives. The team makes most of its decisions independently while considering the criteria agreed to with their sponsor and the leadership. The team attains a high degree of autonomy in pursuing their DMAIC tasks and is able to function without direct participation of the sponsor. Disagreements occur but can be positively resolved within the team. The team is able to adjust its internal processes for itself and redefine roles as needed. The performing team takes on more delegated tasks and assignments from the team Black Belt leader. Team members recognize when to seek assistance or instruction from their internal leader or outside resources. The sponsor delegates and governs through regular progress reviews.

The *Six Sigma Management System* Transactional Leadership Styles

In practice, Six Sigma companies expect their performance improvement teams to go through these four stages of development. Experience has shown that when teams achieve the performing stage, they are also more often successful in accomplishing their assignments. Therefore, the *Six Sigma Management System* looks to leaders, team leaders, sponsors, process owners, and champions to practice leadership behaviors that promote the development of performing (stage four) teams. The team development practices for leaders in Six Sigma companies also apply to developing the same Six Sigma work ethos in the many operational teams that run, maintain, and continuously improve existing business processes.

Roots of the Six Sigma Leadership Styles Model

The School of Behavioral Management Theory has influenced thinking about leadership practices for the *Six Sigma Management System*. These theories prescribe behavioral styles for leaders based on correlated patterns of work performance that various leadership behaviors engender in individuals and teams. Behavioral Management Theories - including the Managerial Grid of Blake and Mouton, the Situational Leadership® Model of Hershey and Blanchard, Fiedler's Leader-Follower Contingency Model, and House's Path-Goal Theory of Leadership - have all influenced Six Sigma leadership practices.

Leader-Follower Contingency Model

In one of the earliest theories that tried to match varieties of leadership behavior with people and situation, Fred Fiedler proposed the Leader-Follower Contingency Model. There are three conditional elements of Fiedler's model: leader-member relations, task structure, and position power of the leader. Depending on these three conditions, Fiedler predicted either task-motivated or relationship-motivated leadership behaviors would be more or less successful. [14]

---

[14] F. E. Fiedler, "The contingency model and the dynamics of the leadership process", In L. A. Berkowitch (Ed.) Advances in Experimental Social Psychology (Vol. 11), Academic Press, 1978

The Managerial Grid

Blake and Mouton originally identified two metrics for measuring manager behavior: a scale rating the manager's concern for people, and a scale rating the manager's concern for production. Combining these two scales into a 10 by 10 grid, they coined names for each five "styles" of management behavior based on a manager's ratings on the two metrics. [15]

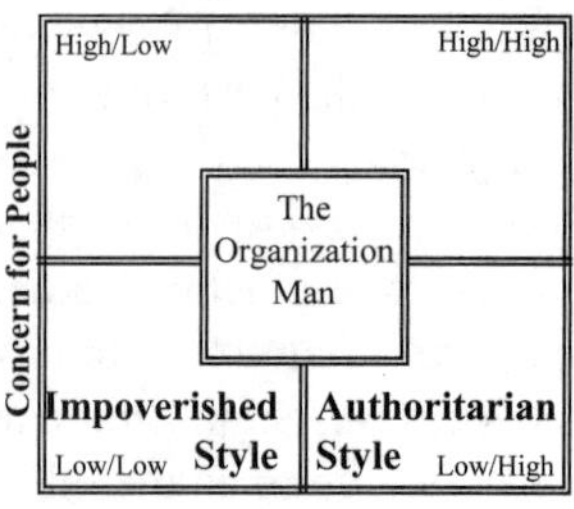

Figure 2-1 Managerial Grid

The Situational Leadership® Model

Paul Hersey and Blanchard developed their Situational Leadership Model that suggested that leadership styles built on the two dimensions of task focus and people focus should be matched to the situation. As Paul Hersey describes it, "Situational Leadership is about being effective as a leader. This involves matching your leader behaviors (those behaviors you use when attempting to influence someone else) with the needs of the individual or group that you are working with. It is adapting the combination of directive behaviors and supportive behaviors appropriately to the readiness of others to perform specific tasks or functions." [16]

The Path-Goal Model of Leadership

Robert House, another behavioral management theorist states that, "The reformulated [Path-Goal] theory…addresses the effects of leaders on the motivation and performance of immediate subordinates …[L]eaders, to be effective, engage in behaviors that compliment subordinates environments and abilities in a manner that compensates for deficiencies and is

[15] Robert R. Blake & Jane S. Mouton, The New Managerial Grid, Gulf Pub. Co., Book Division, 1978

[16] Paul Hersey, " Situational Leadership®: Conversations with Paul Hersey", The Center for Leadership Studies, Inc., 2001

instrumental to subordinate satisfaction and individual and work unit [team] performance." [17]

The Six Sigma Transactional Leadership Styles Applied

Evolving from this history, there are four styles of leadership behavior - Directing, Coaching, Supporting, and Delegating - that are commonly expected of Six Sigma leaders.

The Directive Leader

When leading DMAIC or other process improvement teams, the Directive leader is needed to get the stage one (forming) team to understand and accept their assignments and build their Team Charter. A Directive leader is entirely in charge of what is happening now and dictates where the team is going. This leader provides clear, action-oriented directions for people to follow and explains why.

The Leader as Coach

For the stage two process improvement team in the storming mode, a leader who is a strong and stable coach is most effective. A Coaching leader is a recognized expert who is involved with the details of team activity but acts as a teacher and constructive critic. The coach leads with knowledge and encourages team members to develop their own understanding and perspective. Leadership coaching behavior is also called "selling" by some behavioral management theorists. At the storming stage in Six Sigma team development, the "selling" leader encourages people to adjust some of their individual expectations, to learn about and accept other team members' perspectives, and to discover the mutual interests that they share around the team assignment and charter. This Coach helps team members develop their own emotional attachment to the team itself, to the charter of the team, and to the quest of the campaign.

The Participative Leader

In the forming stage, the Six Sigma leader is expected to gradually move away from the modes of setting the team direction and active coaching toward a "Participative" style of leadership. When the team is able to move ahead and

---

[17] Robert J. House, Path-Goal Theory of Leadership, Lessons, Legacy and a Reformulated Theory", *Leadership Quarterly* 7 (3), 1996

make decisions and complete their tasks, the leader plays the role of team member. When the team struggles, the leader "participates" with ideas and suggestions for the team to discuss and make decisions. The leader also participates as the team's external representative to get needed resources or outside help.

The Delegating Leader

Finally, in the last stage, the performing team works best with a Delegating leader. The role of the leader becomes one of providing high level direction about what should be accomplished strategically, and then reviewing progress and providing feedback to the team on its accomplishments.

Norms and Variability in Six Sigma Team and Leadership Behavior

Six Sigma is a statistical methodology at its base and all Belt-level practitioners are trained in the statistical concepts of averages and standard deviations. This makes it possible for Six Sigma practitioners to understand the four-stage team development model as a statistical abstraction of variable team behavior. At each stage of team development, the "modal" or typical behavior is surrounded by variation in team behavior that falls into the other stages. The same statistical view of the four Six Sigma leadership styles also applies. At any point in time in managing teams, leaders will use a mix of behavioral styles appropriate to the varying complexity and clarity of the task and problem at hand, and modes of behavior of the team. One can gather data and create a metric shown as histogram distribution of leader behaviors in a time period or stage of team development. The leader behaviors can be classified into four histogram categories, one for each of the four styles of leadership behavior. In the forming stage, for example, one would expect a distribution of leader behavior to look something like that in the figure below.

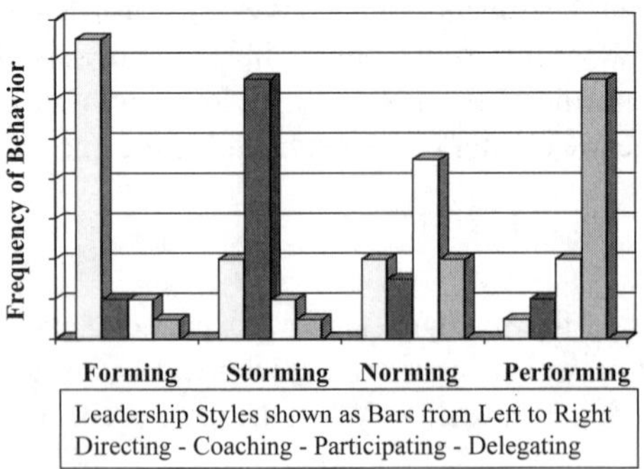

Figure 2-2 Leadership Styles in each Teaming Stage

Some critics of all behavioral management theories argue that many, if not most, people lack the personality flexibility to adopt radically different styles of leadership in varying circumstances. That position is arguable since no hard data exist to support or contradict it. On the other hand, Six Sigma is based on the belief that people, leaders, and team members are able to do exceptional things. Six Sigma companies recruit and develop talent as a top priority and assign their most talented performers into their Six Sigma programs.

Transactional vs. Transformational Leadership in Six Sigma

Six Sigma thinking about the four leadership styles is only one dimension of the leadership role needed from campaign leaders. Leaders must create the ideal cultural environment where the process improvement work ethos can thrive and generate breakthrough changes. The "Direct-Coach-Participate-Delegate" styles of leadership have been called a "transactional" model of leadership. Transactional models focus on leading people through the performance of routine tasks and activities that are the stuff of everyday work. Contrasted with transactional models, there are "transformational" models of leadership. These are the models that describe and analyze the need for, and the impact of, charismatic leadership, and change leadership. The *Six Sigma Management System* looks at leadership as a multi-dimensional complex, where both transactional and transformational leadership are critical dimensions of leadership performance. Chapter 5 (Leadership Behaviors) looks extensively at the transformational dimension of leadership that is needed to engage an organization of people in the quest to create, drive, and direct a breakthrough campaign and then sustain Six Sigma results in the world of constant change.

There are two integral parts to the Six Sigma work ethos. The first is the ongoing commitment of people to making their processes work and work better. The following quote attributed to Deming gives a feeling for the cultural that nurtures this commitment. " Don't blame the singers if the song is written poorly; instead, rewrite the music." The second part of the ethos is the team spirit. There is great power in the team experience to engage individuals in a shared group identity and a commitment to a quest.

Summary of Six Sigma Management System Foundations

There is a growing cohesive body of knowledge surrounding Six Sigma companies and the methods and processes their leaders are using (and continuously improving) to run these companies. This chapter has looked back at the roots of this emerging *Six Sigma Management System.* The chapter has attempted to establish a perspective on how Six Sigma leaders operate and why they got this way. Six Sigma as a management system evolved as a blend of applied science and engineering, practical business experience gained in more than fifty years in both Japanese and American industry, established theories of leadership, and proven team-based methods of organizing into productive work units. The *Six Sigma Management System* is a unique way of thinking, combined with a set of management tools and practices that can be very valuable to companies engaged in Six Sigma and other process performance improvement initiatives.

Chapter

3

The Leadership Modes

This chapter is about the implementation of the *Six Sigma Management System* that creates the leadership structure and focus needed to run a Six Sigma campaign. The Six Sigma campaign is an initiative involving multiple team-based process improvement projects that is directed and organized from the top-down. As part of the campaign, each process improvement team will apply the most appropriate approach to its project - that may be the Six Sigma DMAIC approach, or it may be another proven team-based problem-solving approach such as *Kaizen* or Lean. The *Six Sigma Management System* is the vehicle for directing and organizing the work of all process improvement teams.

The key tenet of the Six Sigma Management System is that breakthrough improvement and rapid execution occurs most effectively in those environments where the leadership team is able to achieve definitive purpose and a clear, laser sharp focus for the improvement efforts. Many leadership teams use the equation Y_= f(x1+x2+x3......) as a way of symbolizing their need to focus. At Motorola, leadership teams use the following framework to express the relationship between the "Big Y" and the "vital X's":

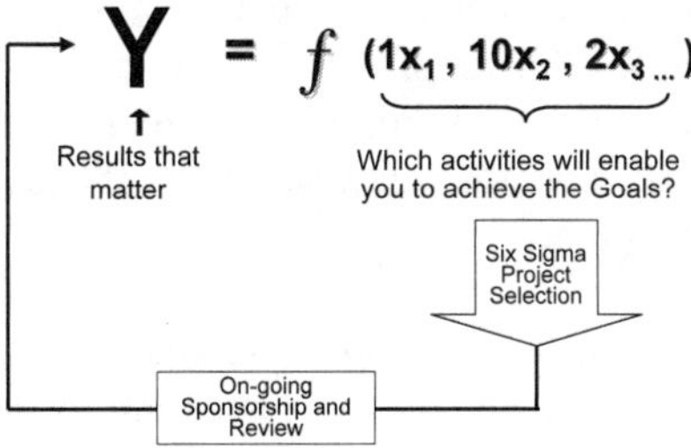

Figure 3-1 Relationship between Big Y's and Vital X's

Leadership teams that are able to align on a single "Big Y" (or a small set of "Big Y's") achieve a high level of focus, can communicate the focus areas more readily, and often create unique competitive advantage by concentrating on a unique set of "Big Y's."
Big Y's are the most important metrics that represent the results that a leadership team is driving in their organization; and Vital X's are the key activities that support those "Y's". For example, if a team decided that their key competitive weapon could be speed to market for new products, then their Big Y could be Product Introduction Cycle Times. The Vital X's would be projects that drive improvement in the New Product Introduction process such as improved requirements gathering or better handoffs between marketing, engineering and manufacturing.

The Big Y and Vital X symbols become shorthand for leadership teams to express their alignment in knowing the results that they are trying to achieve, selecting projects that support those results, and then staying engaged with the projects to ensure that results are achieved.

The *Six Sigma Management System* outlines four modes of leadership behavior that energize Six Sigma initiatives and enable the Big Y and Vital X symbolism to become real management practice.. Based on the experience of Motorola Consultants, this chapter will first provide an overview of those four modes, and then describe the activities involved in each one.

Overview of Leadership Modes

Implementing the *Six Sigma Management System* is not a discrete or one-time activity. A Six Sigma organization continuously evaluates itself and looks for opportunities for improvement, guided by the *Six Sigma Management System* tenet of continuous process improvement. Thus, implementing the *Six Sigma Management System* is a journey that lasts for the life of the organization.

As leadership implements the *Six Sigma Management System*, and continuously evaluates its organizational processes and results, it will move among four modes of operation:

- Aligning (and re-aligning) the organization to the right targets
- Mobilizing (and re-mobilizing) the organization to commit to reach these targets through Six Sigma continuous process improvement
- Accelerating the rate of change that is achieved through Six Sigma initiatives
- Governing the campaign and the organization for sustained business process improvement

Align Mode

In the Align Mode, leadership is focused on aligning the organization to a clear, actionable set of initiatives that directly support strategic objectives. A variety of management activities and tools are applied in the align mode, all with the intention of achieving the same purpose. Without leadership, organizational groups and individuals in them tend to be drawn inward toward a focus on their own work performance and their own view of the organization's priorities. Organizational groups and individuals in them may even have diverse agendas and priorities, and, without leadership, might initiate overlapping or even conflicting Six Sigma projects. It is essential that Six Sigma campaigns be guided from the top of the organization, following top management's true vision for the organization.

Leadership may chose to initiate Align Mode activities with a baseline audit such as the Six Sigma Implementation Rapid Assessment. This baseline assessment will help leadership determine how far it will need to move the organization in order to progress toward implementing the *Six Sigma Management System*, and can help make the case for the need for change.

In the Align Mode, leadership also reviews and refreshes the organization's overall mission, strategies, and goals with a renewed customer focus. (In Six Sigma terminology, this emphasis on customer focus is called listening to the Voice of the Customer (VOC).) Strategy is reviewed and mapped to ensure a balanced focus among customers, internal processes, organizational growth and development, and financial accountability. A Dashboard lays out metrics and stretch goals for key strategies. Leadership then identifies points in the organization where performance improvement should have measurable impact on goals, and generates companion Six Sigma process improvement project candidates. Leadership summarizes their alignment activities in the Six Sigma Alignment Roll Up, to be used as a communications tool to get the rest of the organization, down to individuals, aligned and then mobilized.

Mobilize Mode

In the Mobilize Mode, leadership focuses on people, and obtaining their commitment. Leadership selects its initial cadre of change leaders - known as Sponsors, Champions, Master Black Belts, Black Belts, and Green Belts in Six Sigma terminology. These individuals, along with process owners, start their training. Initial project teams are formed and empowered to work as agents of change within the organization using the Six Sigma DMAIC approach along with other process improvement methodologies.

Using this cadre of change leaders and agents, leadership begins to communicate its customer-focused vision, Six Sigma strategy, and stretch goals to the rest of the organization. Leadership also spearheads the effort to change the culture of the organization and to break down barriers to adoption of Six Sigma. Leadership must be passionate advocates of Six Sigma, and communicate their enthusiasm and motivation for change. As Jack Welch, chairman and CEO of General Electric said in 1997, "You can't behave in a calm, rational manner. You've got to be out there on the lunatic fringe."

Accelerate Mode

In the Accelerate Mode, leadership focuses the organization on action with a mind toward getting results. Leadership assists project teams in completing Six Sigma team charters, encouraging aggressive schedules and short cycle progress and results milestones. Project Champions provide guidance and coaching, and ensure teams get any additional training as required, in order to accelerate their progress. Teams' progress and results are publicized to the rest of the organization, to encourage and promote future change.

Govern Mode

Leadership enters the Govern Mode the moment it kicks off the first Six Sigma project. In the Govern Mode, leadership ensures that the Six Sigma teams and the entire organization become and remain committed to supporting strategic goals. Leadership must clearly communicate the strategic impact of each project in the Six Sigma campaign. Champions continue to guide and coach, and work with process owners and other stakeholders to gain and retain support. The Six Sigma project teams implement the DMAIC methodology and regularly report progress to their Champions. Dashboards are built to show selected critical performance changes for target processes. Process owners monitor improvements in their operations using their own dashboards. The results shown on these critical process dashboards are rolled up into a strategic organizational dashboard. Leadership monitors project progress and dashboard results. They publicize the campaign's progress, and make adjustments are required. Vigilant governance is the key to successful Six Sigma campaigns.

The next part of this chapter looks at the activities involved in each leadership mode in more detail.

Align Mode Activities

Organizations generally begin implementing their *Six Sigma Management System* in the Align Mode. Align Mode activities identify and bring organizational targets into clear focus to direct all ensuing implementation activities. Align mode activities create the structural foundation for the *Six Sigma Management System.*

Align Mode Activity - Conduct Six Sigma Implementation Rapid Assessment

Some organizations, particularly those that believe their past management efforts have been fragmented or varied by division, location, temporal space, etc., find that entering the Align Mode with a baseline audit is helpful to make the case for accelerated change.

Organizations have used the Baldrige National Quality Program Criteria for Performance Excellence for their baseline audits. This is an in-depth, comprehensive audit of an organization in these seven categories of criteria:

1. Leadership
2. Strategic Planning
3. Customer and Market Focus
4. Measurement, Analysis, and Knowledge Management
5. Human Resource Focus
6. Process Management
7. Business Results

Motorola Consultants have developed a more time-efficient baseline audit called the Six Sigma Implementation Rapid Assessment to provide leadership with baseline measures of the organization's current condition on key business and process quality criteria. The Six Sigma Implementation Rapid Assessment consists of two levels of questions:

- Level One questioning is designed to uncover the state of readiness (or willingness) to focus on the right things, apply a structured approach with analytical rigor, emphasize data-driven decision-making, and quantify improvements. *(Figure 3-1)*
- Level Two questioning is designed to focus on efforts related to key business processes. *(Figure 3-2)*

A Rapid Assessment Snapshot is used to capture and display an assessment of the organization versus best practices *(Figure 3-3)*. This establishes the performance gap that leadership will work to overcome through its implementation of a Six Sigma campaign. It also provides a vehicle that leadership can use to communicate the need for implementing process improvement efforts to the rest of the organization.

Level One: Management System and Data Driven Decision-Making

- Do the company's senior executives demonstrate an understanding and commitment to the Six Sigma approach and/or accelerated business improvement?
- How is the leadership team identifying, prioritizing, and driving the results that matter?
- How is senior management monitoring performance, selecting improvement efforts, and reviewing projects?

- How are resources identified and allocated to drive improvement projects to full and timely execution?
- What methodology/approach is being used to attack improvement efforts?
- How does the organization use data to drive customer and business critical
- Are appropriate data used for participative and preventative problem solving methods?
- How are potential improvement solutions for implementation evaluated and quantified?
- How is impact determined after implementation?

Figure 3-2 Level One Questioning

Level Two: Key Business Processes

New Product/Service Development Processes

- Does the organization demonstrate a disciplined product/service development process, supported by a control and review process with appropriate metrics?
- Is there a customer-driven product/service development roadmap with appropriate customer review processes?

Supplier Management

- What is the leadership-driven strategy regarding the procurement of supplied products and services?
- What processes are in place to govern the selection and management of suppliers?

Management of Operational Processes

- How are operational processes routinely reviewed?
- How are plans put in place to drive the required rate of sustainable improvement?

Management of Customer-Facing Processes

- How are the customer-facing processes - like customer service, sales, marketing and field services - routinely reviewed?
- How are plans put in place to drive the required rate of sustainable improvement?

Figure 3-3 Level Two Questioning

Six Sigma Implementation Rapid Assessment Snapshot		
Category	Best Practices	
Six Sigma Management System	Clear leadership alignment to the "vital few" metrics; visible and active selection and review of the critical improvement projects; trained and committed resources supporting the projects.	4 - Excellent ☐ 3 - Very Good ☐ 2 - Fair ☐ 1 - Poor ☐ 0 - Not Doing ☐

New Product/ Service Development Process	A disciplined, customer-driven process exists for identifying and prioritizing new product/service requirements. All product/service development projects follow a stage/gate review process to ensure resource optimization and timely delivery. Long-term product/service development strategies are synchronized through a roadmap development process.	4 - Excellent ☐ 3 - Very Good ☐ 2 - Fair ☐ 1 - Poor ☐ 0 - Not Doing ☐
Supplier Management	Procurement and supplier management processes are synchronized across the organization. Supplier performance management is proactive and data driven. Suppliers are actively engaged early in the product design and development process.	4 - Excellent ☐ 3 - Very Good ☐ 2 - Fair ☐ 1 - Poor ☐ 0 - Not Doing ☐
Management of Operational Processes	Operational processes are routinely reviewed and plans are in place to drive the required rate of improvement. Process characterizations, control plans, quality checks, and problem prevention measures are acted upon and updated. Pertinent methods of statistical quality and process control effectively and efficiently used.	4 - Excellent ☐ 3 - Very Good ☐ 2 - Fair ☐ 1 - Poor ☐ 0 - Not Doing ☐
Management of Customer- Facing Processes	Customer service, sales, marketing and field service processes are routinely reviewed and plans are in place to drive the required rate of improvement. Process characterizations, control plans, and quality checks are maintained and acted upon. Pertinent methods of statistical quality and process control effectively and efficiently used.	4 - Excellent ☐ 3 - Very Good ☐ 2 - Fair ☐ 1 - Poor ☐ 0 - Not Doing ☐
Data Driven analysis for prevention, problem solving and decision making	Organization uses data driven problem-solving methods across the spectrum, from executive decision-making to front line root cause analysis. Data collection processes are systematic and efficient. Teams demonstrate the ability to apply appropriate tools to the problem that is being solved.	4 - Excellent ☐ 3 - Very Good ☐ 2 - Fair ☐ 1 - Poor ☐ 0 - Not Doing ☐

Figure 3-4 Level Three Questioning

Align Mode Activity - Refresh and Refocus the Organization's Vision, Mission, and Core Values

The Vision

Leadership's vision is what it aspires to achieve in the longer term. It is a picture of what the organization will be like in the future. The vision answers the question, "If a Martian visited your organization in 10 or 20 or 30 years, what would they see - or, better yet, sense?" Having a vision is so essential to the life of an organization that some have said leadership equals vision.

> "Leadership is lifting a person's vision to higher sights, the raising of a person's performance to a higher standard, the building of a personality beyond its normal limitations." - Peter Drucker, Consultant, Author

> "The very essence of leadership is that you have to have vision. You can't blow an uncertain trumpet." - Theodore Hesburgh, President, University of Notre Dame

Having a vision implies that leadership has embraced far-reaching aims for the organization, and has conceived what the organization will be like when it achieves its aims.

Any far-reaching aim that leadership envisions is what Jim Collins (former member of the faculty at Stanford University's Graduate School of Business) calls a "Big Hairy Audacious Goal", or BHAG (pronounced "bee hag") [1]. A BHAG is a kind of long-term stretch goal. BHAGs must be set carefully. BHAGs must present a compelling challenge, yet be attainable by the organization. They must be relevant to the organization and its core values. BHAGs must also be clear and understandable by the organization. BHAGs draw from leadership's passions (what we firmly believe we are and can do), the organization's core competencies (what we are good at), and the value the organization provides (what our stakeholders pay us to do).

Complimentary to the BHAGs is leadership's vision of what the organization will be like when it achieves its BHAGs. What milestone products/services will the organization provide? How will customers regard the organization? What will the organization's public image be?

---

[1] www.jimcollins.com

What markets will the organization be doing business in? What will be the core competencies? What major social strides will the organization have made?

Visioning is much more than aiming to improve what the organization is doing now. Organization building is incremental. Visioning has to be revolutionary. Vision implies imagination and foresight and a "big picture" view. As Nicholas Negroponte of MIT wrote, "Incrementalism is innovation's worst enemy." [2]

Revisiting the vision at the very start of the Align Mode enables leadership to answer the question, "Where can our organization be in the future, as a result of implementing Six Sigma as our management system and our business process improvement methodology?" It allows leadership to get excited about the potential radical long-term advances the organization can make. It opens the possibilities of true breakthrough achievements.

Once the vision has been refreshed, leadership must communicate it clearly to the rest of the organization. Every individual within the organization must understand and share the vision. It is most desirable for leadership to exude its vision in all its actions and words. As Jack Welch, Chairman and CEO of General Electric said, "Good business leaders create a vision, articulate the vision, passionately own the vision, and relentlessly drive it to completion."

Articulating the vision most commonly means crafting a vision statement. A vision statement does not have to be lengthy or wordy. In fact, many times leadership spends too much time crafting its vision statement but not enough time communicating its vision to the organization.

Consider this vision statement of the spirits company Whyte and MacKay:

> "We will be the most agile, exciting spirits business. Our competitors will fear us. They will watch us build valuable allegiances with our most trusted customers, excite our consumers, and create an environment where our employees can thrive. We summarise all of this in one phrase, 'Fighting Spirit'." [3]

---

[2] Negroponte, Nicholas, "The Balance of Trade of Ideas", *WIReD*, Issue 3.04, April 1995. Also at http://web.media.mit.edu/~nicholas/Wired/WIRED3-04.html

[3] http://www.whyteandmackay.co.uk/dev/new_content/company-vision.asp

This is the type of statement designed to motivate and energize an organization. The language is inspirational. The statement conveys an infectious enthusiasm for the business and its possibilities. It is a statement of vision.

The Mission

An organization's Mission statement articulates what the organization is and does. It states the reason for the organization's existence. A Mission statement needs to be more than the vague, "We're in the insurance business" or "We make consumer electronics". The Mission statement provides concrete direction for the organization by concisely stating:

- who the organization is
- what types of products and/or services it provides
- its key processes and/or technologies
- what markets/customers it serves
- what (competitive) advantages it offers its markets/customers.

Mission statements that contain these elements are sometimes called "hard" mission statements. The following mission statement from Northwestern University Hospital is a hard mission statement.

> "Our Mission:
> Northwestern Memorial HealthCare is an academic medical center where the patient comes first. We are an organization of caregivers who aspire to consistently high standards of quality, cost-effectiveness and patient satisfaction. We seek to improve the health of the communities we serve by delivering a broad range of services with sensitivity to the individual needs of our patients and their families.
>
> We are bonded in an essential academic and service relationship with the Feinberg School of Medicine of Northwestern University. The quality of our services is enhanced through their integration with education and research in an environment that encourages excellence of practice, critical inquiry and learning." [4]

---

[4] http://www.nmh.org/about_nmh/mission.html

By contrast, this mission statement from Pfizer Inc. is more of a "soft" mission statement.

> "Our Mission
> We will become the world's most valued company to patients, customers, colleagues, investors, business partners, and the communities where we work and live.
>
> Our Purpose
> We dedicate ourselves to humanity's quest for longer, healthier, happier lives through innovation in pharmaceutical, consumer, and animal health products." [5]

It emphasizes philosophical and social issues and the organization's image, not its specific products, processes, markets, or customers. Some "soft" mission statements emphasize the value the organization provides to its shareholders. A "soft" mission statement is more like a vision or values statement than a statement of the business the organization is in. It is important for the Mission statement to concretely spell out the organization's identity, focus, and direction.

Organizations historically have expanded through acquisition and merger and movement into new markets and have contracted through divestiture. Leadership's agreed-to Mission statement can guide the growth or contraction of its organization. Leadership must evaluate whether each acquisition or divestiture makes sense in terms of the organization's mission.

Leadership's renewal of the organization's mission statement can lead to re-thinking what business the organization is in, and what business it should be in. For example, in 1996 PepsiCo, the manufacturer of Pepsi Cola, rethought its mission and, as a result, decided to "spin off its restaurant businesses, sell its food distribution company, and focus on its core beverage and snack food businesses". [6] As another example, the drinks company Diageo, which was formed in December 1997 through the merger of GrandMet and Guinness, was at the time of the merger a broad-based consumer goods company, with core business being food and drinks. In mid 2000, Diageo leadership has rethought its mission and realigned its business behind premium drinks. [7]

---

[5] http://www.pfizer.com/are/mn_about_mission.html

[6] http://www.pepsico.com/company/history.shtml

[7] http://www.diageo.com/pageengine.asp?status_id=3000&page_id=22&site_id=3§ion_id=24

So a Mission statement guides leadership's actions in the organization's overall direction. It also provides a focus for the rest of the organization in knowing where the business is headed.

Core Values

The core values of an organization are the ethical principles underlying the organization's actions with its customers, suppliers, employees, and the community. They are the higher principles, ideals, and beliefs that guide the delivery of a product or performance of a service. Customers innately support an organization's principles, ideals, and belief just by doing business with an organization. An organization sometimes calls its Core Values its Credo.

These, for example, are the core values of Anheuser-Busch Inc.:

- Quality in everything we do
- Exceeding customer expectations
- Trust, respect and integrity in all of our relationships
- Continuous improvement, innovation and embracing change
- Teamwork and open, honest communication
- Each employee's responsibility for contributing to the company's success
- Creating a safe, productive and rewarding work environment
- Building a high-performing, diverse workforce
- Promoting the responsible consumption of our products
- Preserving and protecting the environment and supporting communities where we do business [8]

An organization's statement of core values guides the behavior of every employee and sets the expectations of the organization's customers and suppliers as well as those of the community. Core values also state, in general, how an organization will deal with outside regulation such as environmental controls, diversity requirements, or employee safety regulations. In short, they convey how the organization will behave as a citizen.

Align Mode Activity - Gather Voice of the Customer

The organization's vision, mission, and values define the groups of target customers to be served and satisfied. In order to be able to satisfy

[8] http://www.anheuser-busch.com/misc/vision.html

customers, the organization must know what its customers want. Six Sigma demands that organizations pay very close attention to what their customers are saying. That is what the term "Voice of the Customer" intends to convey.

The Voice of the Customer tells an organization:

- What kinds or types of products/services the customer is willing to "pay" the organization to provide - whether payment comes directly as in the private sector, or payment comes in the form of monetary support through taxation as in the public sector
- Which products/services customers judge as "more is better" and "less is worse" - those are the products/services that are most important to them
- The absolute minimum requirements that any product/service must meet - these become "dissatisfiers" if the organization does not meet them
- What delights customers when the organization's products/services improve - these are called "delighters" in Six Sigma terminology
- What results customers want the organization to achieve - those are the aspects of organizational performance that matter most to them
- What values customers want the organization to demonstrate - these affect the organization's core values statement

Because gathering this type of customer information is generally not part of the typical day-to-day organizational processes, Six Sigma promotes it as a separate activity. Leadership must segment the organization's customer base into homogenous subgroups or segments - customers who buy the same products and use the same services - and then study their expectations related to the specific products and services that leadership views as vision-critical, mission-critical, and value-critical. These are a few sources of customer information:

- History of customer behavior - their buying/service access habits and trends
- Management's experience with and judgment about customers/markets
- Judgment of individual employees who have customer contact
- Customer surveys

- Direct customer feedback (e.g., comment cards, interviews, focus groups)
- Meetings of customer groups (e.g., town meetings and other public forums, shareholder meetings)
- Organizations that represent customers (professional associations, civic associations)
- Published research by government and private groups

Public sector and non-profit organizations also frequently tap the experience of other organizations with similar visions, missions, and values for customer information.

Customers being interviewed or surveyed commonly ask for things that are whims or relatively unimportant to them. It is important that leadership determines the true importance of each requirement to each customer group.

And, because Six Sigma is a data-driven management system, it is important to measure customer expectations. How fast is "faster"? What does "more reliable" translate into? How often is "frequent"? Does "accurate" mean 100% error-free? Organizations implementing the Six Sigma Management System must be able to <u>measure</u> whether they are truly meeting customer expectations.

Align Mode Activity - Develop/Validate Strategy

In implementing the *Six Sigma Management System*, leadership must review the current strategy and strategic initiatives and adjust strategy as needed to ensure it is aligned with the refreshed vision, mission, and core values. A leadership review of strategy in light of new or refreshed information about customer expectations helps leadership ensure that key customer expectations are addressed. Leadership then "maps" all of their separate strategies together, to ensure the composite strategic direction of the organization is balanced across all aspects of organizational performance. This concept of strategy mapping was developed by Dr. Robert Kaplan and Dr. David Norton [9], and was based on Drs. Kaplan and Norton earlier work, the Balanced Scorecard. [10]

---

[9] Robert S. Kaplan and David P Norton, *Strategy Maps: Converting Intangible Assets into Tangible Outcomes* (Boston: Harvard Business School Press, 2004)

[10] Robert S. Kaplan and David P. Norton, *The Balanced Scorecard: Translating Strategy into Action* (Boston: Harvard Business School Press, 1996)

The Balanced Scorecard

The Balanced Scorecard was developed to move organizations away from measuring organizational performance using solely traditional financial metrics. Taking four views of the organization, instead of one view, leadership creates the balanced overall strategic direction the supports their vision.

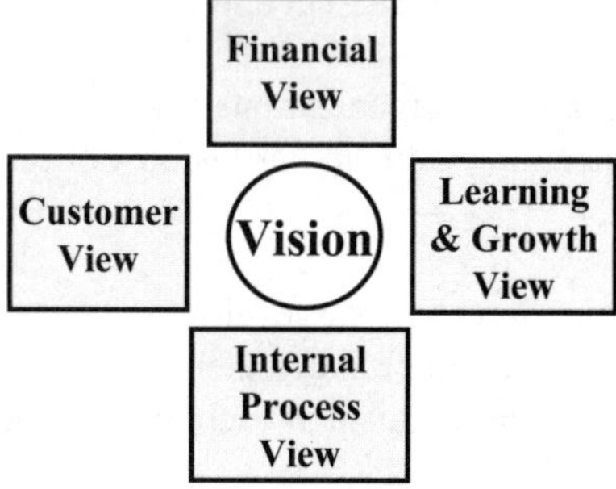

Figure 3-5 Four Views Supporting the Vision

- Customer View - The organization as seen by customers as a producer of valuable, beneficial services and products. The composite strategic direction must include customer-focused strategies. These strategies must focus on the factors that customers care about and judge the organization on.
- Internal Process View - The organization as a set of processes that must generate added value and produce beneficial products and services for customers. Strategies must focus on the quantity and quality of the outputs of business processes and improving the capabilities of those processes to produce consistently high quality outputs.
- Learning and Growth View - The organization as a living entity that needs to grow and change to remain successful. Strategies must focus on improving the capabilities and skills of people in the organization to allow growth and meet future changing demands.
- Financial View - The organization seen as a fiscal entity using measures of revenue and expense. Measurements tell the story of past activities and financial objectives, including cost control and revenue generation. Selected financial objectives must focus on areas where improvements in operations are expected to have measurable impacts.

When leaders apply the Balanced Scorecard method, they review all their strategies, and then select key strategies, keeping in mind that they need to have some key strategic initiatives in each of the four perspectives. The Balanced Scorecard graphic that is created shows the strategic objectives to pursue in order to accomplish the organization's vision and mission, while living by its values. The graphic helps focus the organization's vision, mission, and values into specific actionable strategic objectives.

- Customer focused objectives help the organization achieve its vision of superior service, responsiveness, customer loyalty, etc.
- Internal Process focused objectives help achieve the vision of superior operations (even when the customer is not immediately or directly impacted).
- Learning and Growth focused objectives help achieve the requirement of a strong, flexible organization built on skilled, dedicated people who have the necessary physical resources to do their jobs.
- Financial Results focused objectives help guide the organization in areas where cost versus benefits is an important leadership decision factor. The private sector sets profit and revenue goals. For the Public Sector Organization (PSO), the pure financial view of private sector profit and loss is not applicable. PSOs are organizations that use public resources to create benefits and value for the communities they serve. PSOs cost taxpayers money and sometimes constrain freedoms to do their jobs. PSOs depend on due process or other authorities to fund and legitimize their activities. Therefore many PSOs use the category called *Accountable Resource Management* (ARM) to replace the pure financial view. ARM covers the view of the organization that includes objectives about:
  - Costs incurred to deliver benefits
  - Operational efficiency
  - Social and economic burdens on citizens
  - Building support from authorities and agencies that provide funding and legitimacy to the organization. PSOs strive to meet the objectives of donor institutions. PSOs strive to be open and accountable in their use of resources to generate value to benefit citizens.

The Strategy Map

The Strategy Map begins with the graphic display of a Balanced Scorecard that shows all key strategies sorted into four box matrix. Leadership creates its Strategy Map by simply drawing arrows between strategies showing how each strategy relates to and supports other strategies. The Strategy Map shows the value chain of the organization. When an organization displays its strategic objectives on the map, leadership has a graphic display of where the objectives fit in the value chain.

- Happy customers are loyal and "pay" for services and support financial objectives.
- Improved processes deliver better products/services and support customer objectives.
- Learning and growth objectives strengthen the organization.
- Financial objectives provide funds to achieve other objectives.

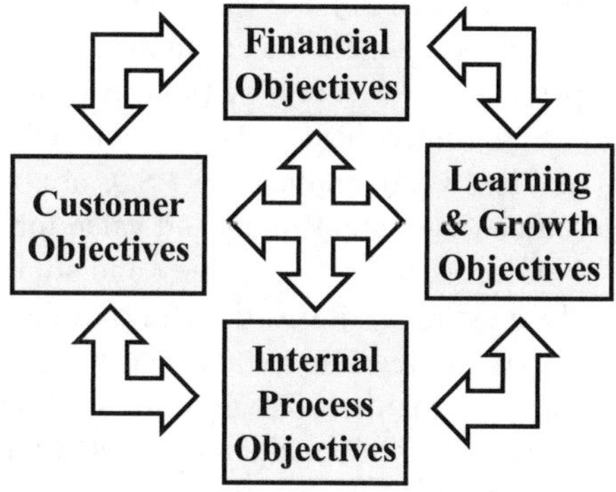

Figure 3-6 Relationships in a Strategy Map

The exact relationships among strategies are usually intricate and sometimes arguable. It is important, however, for leadership to align on what they believe are the most important strategic objectives and the relationships among them.

Develop Dashboard

The tool called a Dashboard operationalizes the Strategy Map. The Dashboard is a measurement system that allows leadership to track the progress of the organization toward meeting strategic objectives laid out on the Strategy Map.

The unique aspect of dashboards as measurement systems is their intentional simplicity and visual nature of the way they display organizational performance. Dashboards provide leaders with graphic displays of critical indicators of their organization's current performance - as a car dashboard displays speed, engine temp, oil pressure, trip odometer, etc. Dashboards show the indicators that must be tracked to operate the vehicle successfully.

The planners of every vehicle select what goes on the dashboard display. Once dashboard items are selected, the information displayed is the at-a-glance guide that shows operators the current status of their vehicle.

Dashboards for organizations are "at-a-glance" displays of the current status of metrics for their selected key strategic objectives. Once dashboards are in use, they make it very clear when the organization is moving toward its goals and when it is not.

A Dashboard shows:

- Selected strategic objectives
- One or more metrics for each objective
- A "stretch goal" for each metric
- The current actual status of each metric

<u>Building the Organizational Dashboard</u>

To monitor the organizational impact of Six Sigma campaigns, the leadership team must select objectives and related metrics to display on the top-level organizational Dashboard. There are different opinions about the best number of objectives and metrics to display on a Dashboard. The purpose of the Dashboard is to focus instant and constant attention to the status of the selected objectives. The items chosen for display on the Dashboard must represent the consensus of the leadership team relative to the strategic focus for the Six Sigma campaign.

With the Dashboard, the leadership team operationalizes the Balanced Scorecard strategy to serve customers and stakeholders, and to improve the business itself by tying strategies to key objectives and the ways these objectives will be measured (metrics).

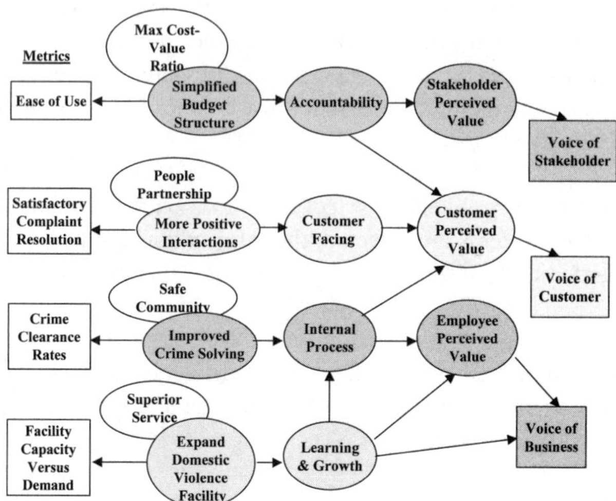

Figure 3-7 Operationalizing the Strategic Scorecard

The graphic shown represents selected performance improvement initiatives from a large police agency. The police agency tracks four key metrics. These metrics and their matching objectives are tied back to the customer view, the financial view/accountability view, the internal business process view, and the learning and growth view. It shows the strategic importance leadership in this police agency places on moving these metrics in a positive direction.

Strategic Objectives and Dashboard Metrics

Many important strategic objectives are soft. For example, "Improve our customers' perception of product quality" could be a soft objective of any manufacturing business. Organizations use soft objectives because, even though they are hard to measure directly, normal people all agree that they are "real" and very important to achieve.

Dashboard metrics, on the other hand, are concrete measures with observable scales. For example, one metric associated with the perception of product quality is number of returned units of the product.

Metrics must be:

- Valid - Typically, multiple metrics must be selected to track the achievement of soft strategic objectives. Leadership asserts that, summed together, each set of supporting metrics shows a reasonable picture of the status of the strategic goal and related supporting objectives. This is called metric validity.
- Reliable - Metrics themselves cannot be soft. The metric measurement must be something concrete and observable so that everyone can agree on the same measured value for each metric. That is called a reliable metric. Metrics selected must be reliable or they are useless.
- Measured on an accurate, differentiating scale - Metrics must give accurate data that show significant changes in the thing measured.
- Cost-effective, technically and practically doable, and collectible in a timely manner to support decision-making - The cost, the technical challenges, and the practical difficulty of collecting data for a metric are always considerations.

It is the challenge of leadership to find metrics that are both practical to implement and that provide valid, reliable, and actionable data.

Stretch Goals for Dashboards

In the *Six Sigma Management System*, the Dashboard provides leadership the data for making decisions about the Six Sigma campaign. The goals laid out on the Dashboard are the goals for the Six Sigma Campaign. Actual results will be tracked and evaluated against these goals. Dashboards display the status of each metric versus the goal set for that metric. Following Six Sigma practice, leadership sets stretch goals for each metric to challenge the organization.

> "The greatest danger for most of us is not that our aim is too high and we miss it, but that it is too low and we reach it" - Michelangelo

Stretch goals should force "outside the box" thinking. A goal that can be achieved by an obvious or known solution is not a stretch goal. But there is a fine line between an impossible goal and a stretch goal. Leadership must believe that the organization's best people can achieve stretch goals.

Once leadership has built the Strategy Map and Dashboard at the highest level of the organization (with strategic objectives, metrics, and stretch goals), the process cascades down through the organization.

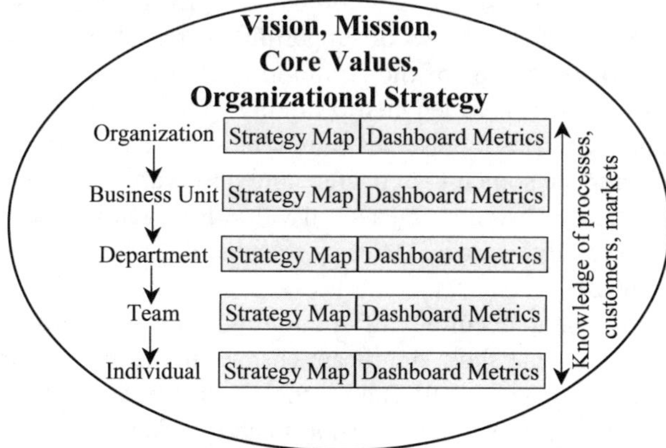

Figure 3-8 Cascading the Strategy Map and Dashboard

At leadership's discretion, each business unit, department, process improvement team, and individual in the organization builds a Strategy Map and Dashboard that, ultimately, supports those of the leadership. Building Strategy Maps and Dashboards at all levels of the organization is an important part of aligning the rest of the organization to the same top-level strategies. And Dashboards are the measurement systems for the Governance Mode of implementing the Six Sigma campaigns. When Motorola's Commercial, Government and Industrial Solutions Sector (CGISS) won a 2002 Malcolm Baldrige National (U.S.) Quality Award, one of its achievements cited was "full organizational alignment through the use of *Performance Excellence Scorecards*, cascading strategic objectives down from the sector level into all businesses and support functions. In turn, these cascade into the *Personal Commitment* plans of all employees worldwide." [11]

Align Mode Activity - Identify and Analyze Performance Drivers

Layers of Dashboards reaching from ground-level operations to the top of the organization become the basis for leaderships' nomination of candidates for Six Sigma performance improvement projects. Projects are nominated because they are expected to significantly impact one or more

[11] Motorola CGISS Best Practices, available at http://www.motorola.com/cgiss/docs/CGISS_PE_Best_Practices.pdf

metrics on the top level Dashboard. Potential projects involve three primary ingredients: a business process, a recognized need to improve that process in some way, and a clear link that shows how improving that process will positively impact the achievement of a top level objective in a way that will be measured by its Dashboard metrics.

The leadership team finds project candidates by identifying and analyzing "performance drivers". A performance driver is a factor that will cause organizational performance to change, and thus move one or more key metrics on a Dashboard. A fundamental belief of Six Sigma is that there are causes for everything, including changes in performance. Some causes or performance drivers are direct and obvious; some drivers are indirect and their relationship to metrics is very subtle and hard to detect. It is the job of leadership to get their best minds to identify performance drivers for the organization's key metrics. Then they must figure out how to manipulate these drivers to cause improved performance in the targeted metrics. Performance driver "handles" are things that can intentionally, systematically be changed or manipulated to control performance drivers.

Suppose the strategic goal of a baking company is to make the cracker that customers rate best over all others. The metrics for this include the taste, saltiness, color, and crispness of the cracker. Currently, customers tell the company that their crackers need to be crispier. Suppose that oven temperature is a performance driver of the crispness of the company's crackers. Then the handles for oven temperature will include the amount and evenness of heat energy applied, the thermostat used, the insulation of the oven, etc. The process that bakes the crackers and controls crispiness and contains the handles that control baking temperature then becomes a candidate for a Six Sigma project.

As another example, suppose a police agency has a strategic objective of "Improve the perception of safe schools and neighborhoods". The metrics chosen are:

- gang activity
- number of residential burglaries
- drug activity
- violent crimes

as measured by numbers of reports for each neighborhood.

Leadership chooses to analyze number of residential burglaries, and determines that the drivers include:

- investigative success at solving residential burglaries
- truancy rates
- home security

Leadership believes that if truancy is reduced, residential burglary will also be reduced. Leadership then nominates one candidate Six Sigma project the "Truant Reporting/Handling Process" which includes the handles of:

- timeliness and accuracy with which schools report truants
- effectiveness of truant officers in handling truants
- effectiveness of protocols for involving parents

The goal of the project would be to improve the performance of the Driver (i.e., reduce truancy) by changing the values of the Handles.

To be a "good" candidate for a Six Sigma project, leadership must believe that a significant improvement in measured process performance can be achieved. For this example, leadership would have to believe that significant improvement in the truancy handling process is possible, and that will significantly reduce residential burglaries, and that will in turn have a significant positive improvement in the public's perception of safe schools and neighborhoods. Leadership formulates this cause-effect relationship using all the objective evidence and expertise available to them. This relationship will be further defined, measured, and analyzed when the project is handed off to a Six Sigma DMAIC team.

Identifying performance drivers (and handles) related to the organization's key metrics is the way to identify Six Sigma projects that are highly probable to have a large impact on achieving the organization's strategic objectives, mission, and vision. It ensures that Six Sigma projects are aligned with the objectives, mission, and vision. This is where Six Sigma campaign success is created. This is the first step in creating the potential for breakthrough results.

Align Mode Activity - Create the Six Sigma Alignment Roll Up

The next step in creating the potential for breakthrough results is to summarize leadership's alignment activities in the Six Sigma Alignment Roll Up document. That document is the summary record of the leadership's strategic planning and is used as a communication tool to get the rest of the organization, down to individuals, aligned and then mobilized. The Six Sigma Alignment Roll Up tool will be covered in the next chapter.

Mobilize Mode Activities

In the Mobilize mode of the Six Sigma Management System, leadership communicates the strategic alignment message to the entire organization. Part of this is accomplished by the development of Dashboard objectives and metrics at all levels throughout the organization. Part of this is accomplished by leaders who, as a group and individually, describe the need for change and improvement and their support for Six Sigma and process improvement methods. Leaders gain the commitment of their people to the objectives and stretch goals of the campaign. Another part involves educating the entire organization to think in terms of business processes with customer focus. At the same time leadership, must build the Six Sigma process improvement infrastructure of trained people and dedicated resources required for success. Once teams are trained, they are assigned projects and empowered to start immediately to create productive change.

Mobilize Mode Activity - Select and Train Six Sigma Change Leaders

The *Six Sigma Management System* requires not only committed, aligned leadership, but also trained Six Sigma professionals and practitioners dedicated to continuous process improvement. While there is no prescribed organizational structure, the typical Six Sigma organization includes:

- Sponsors
- Champions
- Process Owners
- Green Belts (GBs)
- Black Belts (BBs)

- Master Black Belts (MBBs)
- Project Teams

Sponsors

Sponsors are members of the top leadership team. Many organizations have one, high-level executive as the formal Six Sigma Sponsor. The Sponsor directs the Six Sigma Campaign on a day-to-day basis, much as a Marketing Vice President would oversee the activities of the Marketing Department. The Sponsor spearheads the Six Sigma campaign, managing it both as an aligned set of projects and as the more traditional ongoing political campaign with the objectives of keeping current supporters and gaining new ones. Informal Six Sigma Sponsors are individuals who communicate the potential of Six Sigma by using Six Sigma in their work and by actively advocating Six Sigma to the rest of the organization.

Note: In some organizations, Sponsors and Champions can be one in the same - the terms and roles are applied interchangeably.

Champions

Champions are executives who take responsibility for one or more individual Six Sigma projects. Champions are often the managers of the business divisions and sectors that run the processes their teams are assigned to improve. Champions approve project plans. They authorize the resources for Six Sigma projects. Champions work closely with teams to ensure that teams understand the strategic objectives of the Six Sigma Campaign. They advise their teams when developing team charters. Champions also meet regularly with teams, reviewing their results and providing guidance and recommendations. Champions are a conduit for feedback to the leadership group.

Note that in some organizations, Sponsors and Champions can be one in the same - the terms and roles are applied interchangeably.

Process Owners

Process owners are responsible for everything between the beginning and end points of a process. This may include activities that cross functional boundaries. Highlighting Process Owners in the organization emphasizes the Six Sigma Management System's process focus.

A Process Owner may also be a Champion. The Process Owner owns many of the resources that a project team will need to do its job. The Process Owner, for example, can arrange for the time and cooperation of process performers. Process Owners must support the team and its goals because ultimately, the Process Owner must implement any substantive changes made to improve the process. Process Owners must be advocates of process improvement and Six Sigma, and must be managers and champions of change.

Green Belts (GBs)

A Green Belt is a Six Sigma practitioner, usually part-time, who has been trained in the Six Sigma DMAIC problem-solving methodology and basic statistical tools. Green Belts are most commonly members of project teams. In some organizations or on some projects, Green Belts are project leaders.

Black Belts (BBs)

A Black Belt is a full-time Six Sigma practitioner who has had rigorous training in the statistical methods used to gather and analyze data in a Six Sigma project. Good Black Belt candidates are technically oriented individuals. Black Belts must also have the skills and knowledge to lead projects across an organization, to act as mentors and coaches to Green Belts, and to identify good candidates for Six Sigma projects.

Master Black Belts (MBBs)

Master Black Belts are highly experienced Black Belts. An MBB is a full-time position. MBBs are the technical leaders of a Six Sigma campaign. MBBS train BBs and GBs in statistical methods. MBBs assist, mentor, and coach BBs in the correct application of the statistical methods. MBBs are highly technical, but must also be able to communicate with and teach BBs and GBs.

Project Teams

Project Teams apply the DMAIC methodology to the assigned process to gather as-is process data, analyze the data, determine root causes of variation, generate solutions, and then implement process improvements and monitor results. The Champion usually selects the Project Team Leader. The make-up of the Project Team is then determined jointly. Project Teams are made up of one or more Green Belts plus individuals who have expertise in and/or responsibility for the assigned process. The Champion meets regularly with the Project Team, reviewing progress, providing feedback, assistance, and direction, and obtaining resources when needed.

Training

Everyone involved in the Six Sigma campaign - from leadership down through individuals on the project team - should receive some training. Leadership, Sponsors, and Champions generally go through two to three days of Six Sigma leadership training that orients them to what Six Sigma is, what the possibilities are, and the structure of a Six Sigma campaign. Motorola University also utilizes a hands-on Jumpstart workshop for leadership, where leadership performs the Align Mode activities to get a fast start on implementing a Six Sigma campaign.

Most Green Belt training is one or two weeks long, followed by completion of a project. BBs should help GBs select a project prior to attending classroom training or beginning self-administered GB training so the GB candidate can think about how to apply the knowledge and skills acquired in the training.

Most BB training is four weeks long. Motorola University Black Belt training spreads the four weeks over a four-month period to allow for intermediary application of skills and knowledge to an actual project. BB candidates must have a project selected before beginning training so they can begin to immediately apply what they learn in the classroom.

MBB candidates will have completed Black Belt training. MBB candidates can benefit from additional courseware relevant to their expanded role as the organization's Six Sigma trainers and mentors such as training delivery, advanced statistics, and team building.

All other individuals in the organization should receive at least Six Sigma orientation or awareness training, or what Motorola University refers to as Foundations Training, to help them understand the rationale and methodology of the Six Sigma continuous process improvement campaign.

Senior Leadership Team	***Leadership Jumpstart Event (2 days)*** - align with business strategy - secure commitment, prioritize and resource projects
Champions	***Six Sigma Management Training (2 days)*** - equip owners to manage in Six Sigma environments - ensure the advancement of the improvement projects
Black Belts	***Six Sigma Black Belt Training (20 days)*** - prepare Black Belts s to lead Six Sigma improvement teams - equip Black Belts with SPC tools and techniques
Green Belts	***Six Sigma Green Belt Training (5–10 days)*** - prepare team members to execute project activities - provide foundation of improvement processes and tools
Foundations Training – Organization Wide	

Figure 3-9 Six Sigma Training by Population

Mobilize Mode Activity - Plan the Six Sigma Campaign

In the Align Mode, leadership identifies candidates for Six Sigma projects that are aligned with the organization's vision, mission, core values, strategic objectives, and goals. Leadership develops a "hopper" of projects. In the Mobilize Mode, leadership plans how to implement those projects.

In most organizations, the amount of resources needed to work on all the projects in the hopper far exceeds the resources available. Leadership needs to prioritize and make project selections, to make the best use of available resources.

There are a number of criteria for leadership to qualify candidate projects for assignment to teams:

- How available is the data required to study the problem? Data

collection can be expensive, and Six Sigma is a data-centric problem-solving methodology.
- What are the expected benefits? How much benefit is expected? Candidate projects that have the potential to benefit customers, stakeholders, and the business are prime Six Sigma projects.
- Will the human and physical resources required to complete the project be available? How do the expected benefits measure up against required resources?
- How long will the project take to complete? Projects should be do able in six months or less.
- What is the Sponsor's/Champion's level of commitment to the project? Up-front commitment is a requisite for Six Sigma projects.

Planning the campaign also requires leadership to evaluate the potential synergy and soft impact of project ideas. Which projects will complement each other? In what order does it make sense to kick off the set of projects? Which projects will send the right message to the organization? To achieve breakthrough results, the set of selected projects should be a true campaign, not just a hodge-podge of isolated efforts.

Mobilize Mode Activity - Select and Empower Project Teams

Once a project has been selected for implementation, the project Champion selects the team leader. Together, they identify the team members. Team members should be selected for their expertise in the process and technologies that are used. Individuals who are strong communicators should also be included. Selection to be part of a Six Sigma team should indicate a recognition that the individual is a superior performer and a leader. It is also important to match the expected difficulty and complexity of each project to the level of Six Sigma certification and prior project experience of the team members.

As project Champions are recruiting team leaders and team members, it is imperative that leadership sends the message of its strategic purpose and alignment to the teams. Teams must feel the challenge and understand the rationale for the improvement challenge. Leadership must clearly communicate the reason for engaging in improvement efforts. "Great Groups coalesce around a genuine challenge, a problem perceived as worthy of a gifted person's best efforts. Much of the joy typical of Great Groups

[12] Warren G. Bennis and Patricia Ward Biederman, *Organizing Genius* (Perseus Books, 1997)

seems to reflect the profound pleasure humans take in solving difficult problems." [12]

Teams must be empowered with the authority to investigate and act, but must clearly understand their boundaries. Teams must be free to take calculated risks in order to achieve stretch goals. Leadership must set clear boundaries, then let the teams go. "Groups become great only when everyone in them, leaders and members alike, is free to do his or her absolute best." [13]

The Team Charter is the official document of a team's empowerment. Leadership (particularly the project Champion) will assist the team in beginning to develop the Team Charter and review and approve the Charter when finished. The Team Charter guides the work of the team. It states the problem to be solved, the business case for solving the problem, the project's goals, the boundaries for the team's work, the project plan and schedules, and the roles and responsibilities of the project Champion, team leader, and individual team members. The Team Charter sets the expectations for what the team will do and accomplish. Leadership involvement in developing the Team Charter assures that the Charter aligns with the vision, mission, core values, strategic objectives, and goals that leadership has set out for the organization.

Mobilize Mode Activity - Align the Rest of the Organization

Along with empowering project teams, leadership must begin to align the rest of the organization to the Six Sigma campaign, and to the organization's vision, mission, core values, strategic objectives, and goals for the campaign. Leadership must clearly communicate the need for change, and the value of using the *Six Sigma Management System* to achieve that change. The organization must come to view Six Sigma not as "the wildest chimera of a moonstruck mind," [14] but as a way to "make a dent in the universe." [15]

It requires a culture change for Six Sigma to flourish in any organization. Many organizations have a cultural resistance to change, and many individuals in an organization have a personal resistance to change. Leadership must overcome this resistance by being true change advocates and instigators. Motorola University consultants have found that the

[13] ibid

[14] *The Federalist* on President Thomas Jefferson's Louisiana Purchase

[15] Steve Jobs, co-founder of Apple Computer

results of the baseline audit are often helpful in making the case for accelerated change. The Rapid Assessment boils the story down to a set of a few understandable, critical dimensions where the organization must show improvement. Any story that leadership can tell to dramatize the need for change can help win hearts and minds.
And there are hundreds of Six Sigma success stories leadership can use to demonstrate results. Motorola, for example, credits $1.6 billion in savings since it instituted its Six Sigma program. The Idaho Emergency Medical Services Bureau reduced the time to process certification applications and issue certification cards from 22.8 business days (on average in the 1999-2001 time period) to 2.2 days (average for 2003), using Six Sigma. [16] The city of Fort Wayne, Indiana used Six Sigma to reduce the response time to pothole complaints from an average of 21 hours to 3 hours. [17] Commonwealth Health Corporation in Kentucky launched Six Sigma in 1998. It realized savings of over $3 million in 2001, and cumulative savings through early March 2002 were over $7 million. [18] The list goes on, demonstrating Six Sigma's effectiveness in dollar savings and process improvements.

In addition to aligning the organization on the need for accelerated change through Six Sigma, leadership needs to convey the organization's direction and goals. The Six Sigma Alignment Roll Up also becomes a communication tool for telling the organization where it is headed, and why.

The Deployment Plan

The keys to mobilizing the rest of the organization are effective communication, and enthusiasm on the part of leadership. These activities are so important that leadership will normally have a formal, written Six Sigma Deployment Plan prepared. This plan will begin with the leaderships' S*ix Sigma Alignment Roll Up*, which includes the vision, mission, values, strategies and initiatives along with key Six Sigma projects and the business results expected. Following that will be the Communication Plan that covers who is responsible and how the messages will be delivered. All types of media - from launch events to videos and posters - should be included.

---

[16] Dia Gainor, "Improvement of Cycle Time for State Issuance of Emergency Medical Services Personnel Certification". *EMS Management Journal* 2004 Jan-Mar 1(1):7-19.

[17] http://www.cityoffortwayne.org/6sigma.htm

[18] http://www.gehealthcare.com/prod_sol/hcare/pdf/chc_pioneer_six_sigma.pdf

Accelerate Mode Activities

The overall purpose of Accelerate mode activities is to move the Six Sigma organization into action with a focus on getting results quickly.

Project Champions should provide teams with guidance and coaching, and ensure they get any additional training as required, in order to accelerate their progress. Motorola University consultants have found the *action learning* methodology to be most effective. *Action learning* combines structured education with real-time project work and coaching to quickly bridge from "learning" to "doing". Expert support is provided to teams on a just-in-time basis. Specific, as-needed coaching adds to team training.

Implementing *action learning* builds skills and knowledge in teams on an as-needed, as-applied basis. A strong support structure of Six Sigma trainers, team coaches, and experts in Six Sigma tools must be available and involved with teams. Teams should be closely monitored during all phases of DMAIC, and encouraged to seek expert help and training. There is too much in the Six Sigma tool kit to learn in the classroom. The DMAIC methodology is a broad framework that follows many variations depending on the specific process and its performance problems. *Action learning* is similar to the hands-on internship required for many professionals. Exposure to real process conditions and real people struggling to run broken processes stimulates Six Sigma practitioners to seek new knowledge and build new skills in process improvement. When experts are closely involved, they can recognize the need for advanced Six Sigma tools and teach teams to apply these tools through their example. *Action learning* is a good vehicle for adding Lean tools and other process improvement approaches and thinking to the project team's arsenal. The *action learning* cycle is depicted below:

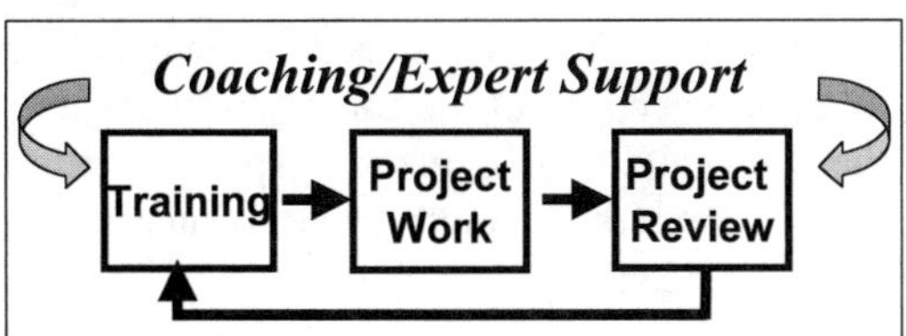

Figure 3-10 Action Learning Methodology

In order to accelerate results, leadership should be actively involved in "campaign management." This includes publishing schedules that tie project completions to strategic objectives, tracking progress, and publicizing success and on-schedule performance. Champions should publicize their teams' progress and results to the rest of the organization, to keep the organization informed and to encourage and promote future change. "Great Groups are contagious. They inspire other Great Groups." [19]

Champions should have project teams write their Team Charters with aggressive schedules and plans to deliver short cycle progress reports and reach concrete results milestones quickly. Changes happen in sprints, not marathons. Leadership needs to drive projects to achieve timely results. Success requires tight clock management. In the days of Total Quality Management (TQM) at Motorola, teams were given 11 months to operate. Motorola University consultants found that teams lost interest within 90 days! The lesson learned: Leadership must be aggressive and use a four to six month time frame for projects to be successful. If a project goes beyond six months, the team is lost.

Govern Mode Activities

In the Govern Mode, leadership focuses on managing, reviewing, and driving the completion of projects, and ensuring project work remains aligned.

Govern Mode Activity - Demonstrate Sponsorship

The first key activity in the Govern mode is the active sponsorship of the Six Sigma Campaign overall and the individual projects. Champions take ownership and responsibility for the success of the Campaign. Other members of the top leadership team also demonstrate their commitment to achieving the overall objectives of the campaign. Leaders must also visibly sponsor the teams and their projects. A sponsoring leader will publicly take ownership of a project and commit to success in achieving the stretch goals that have been set.

[19] Warren G. Bennis and Patricia Ward Biederman, *Organizing Genius* (Perseus Books, 1997)

Govern Mode Activity - Establish Communications Network

The second key activity in the Govern Mode is establishing a "knowledge sharing" program and driving a proactive communications network that includes all the Six Sigma teams, the people who own and operate the processes involved in the campaign, the business owners, general managers, and everyone whose work may impact (or be impacted) by others. Insights by one team may help another. Changes proposed in one process may require adjustments in other processes. Success stories will encourage everyone. Only the leadership has the capability to insure that this network is set up and continues to function. Some information will flow through the team Champions to process owners and other members of the leadership team. So the leadership team must consciously take on the responsibility to re-distribute knowledge gained down to other teams and operations. Also, the leadership team must require that peer-to-peer and team-to-team communications links are formally established and people are accountable for knowledge sharing.

The last and most important activity in the Govern Mode is conducting rigorous reviews of Six Sigma projects along with tracking the impact of process improvements on the key strategies that the leadership selected.

Tollgate Review Process

There is a formal process for Sponsors and Champions to review each team's progress and interim results. It is called the tollgate review process and takes place at the conclusion of each DMAIC phase. The tollgate review process includes a detailed list of questions to be answered at each phase. All the tollgate questions for each phase are presented later in this book

Project Status Reviews

In addition to tollgate reviews, normally conduct a weekly status review for each project. Conducting frequent reviews of projects allows teams to make mid-course corrections, and allows project sponsors and the leadership to see unforeseen barriers, and provide support or resources. Sometimes teams, early on in the DMAIC process, uncover "quick wins" - obvious or small improvements or changes that are easy, fast, and inexpensive to implement that the team itself can put in place with little risk. Sometimes, quick wins are enough to move the project toward its goal, and the Champion can then re-allocate some resources to other projects. Sponsors and Champions may also see an opportunity to apply quick wins implemented to other processes in the organization. Regular reviews help drive proactive dialogue and knowledge sharing among team members and throughout the organization

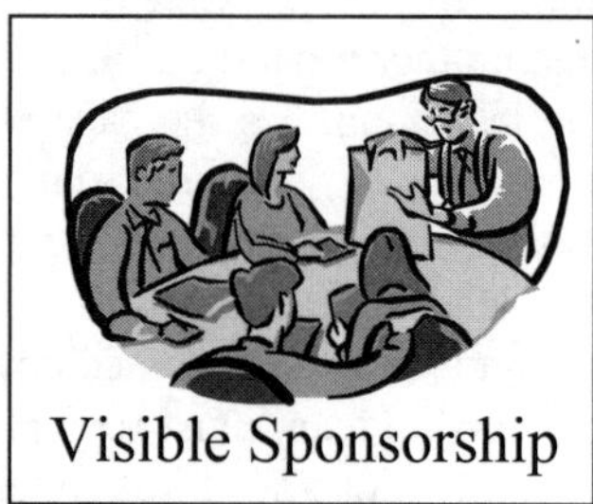

Figure 3-11 Six Sigma Weekly Project Status Report

Six Sigma Measurement System

Status reports and tollgate reviews of projects are consolidated for the leadership in a measurement system designed specifically for the governance of Six Sigma Campaigns. Information about projects is consolidated around the core processes of the organization. Core processes are super-processes that leadership sees as critical to the achievement of the vision and strategic to the survival of the business. The business will normally track the performance of those processes using metrics selected by the leadership.

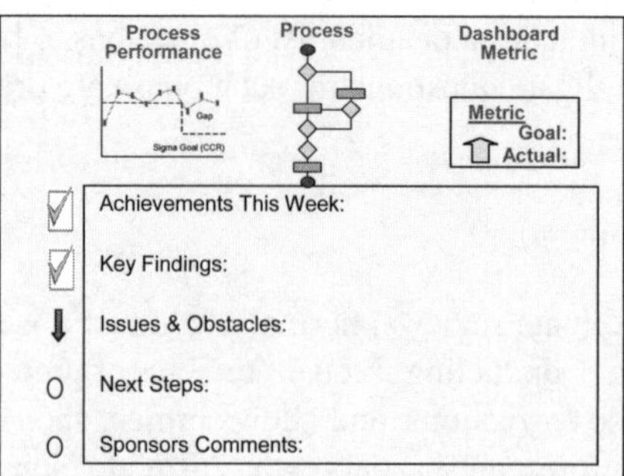

Figure 3-12 Typical Core Process Performance Metrics and Consolidated Project Reports

Dashboards

The Corporate Level Dashboard is the display for top leadership of the key metrics from core processes that should be moving as a result of process performance improvement. The leadership may have other business-unit level dashboards developed to assist in governance as well. These unit-level dashboards will provide displays of the current status of metrics selected for each of those business units.

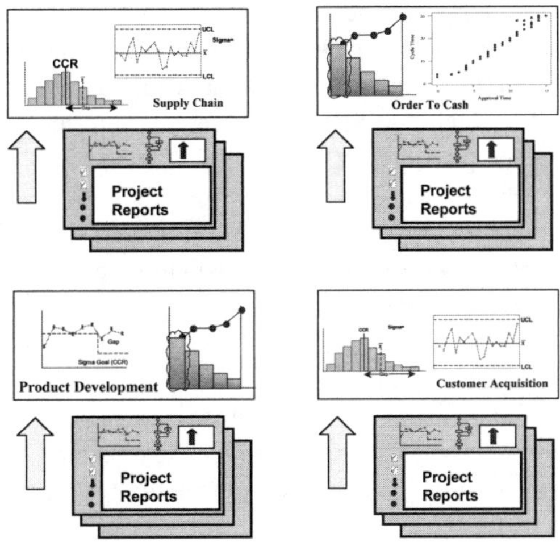

Figure 3-13 Corporate Dashboard

Structured review processes and measurement systems that support Dashboard displays are the tools leadership uses to govern Six Sigma Campaigns. What is critical in the Govern Mode is that leaders are involved in the campaign, constantly comparing results to their expectations, giving support and feedback to team participants, and making adjustments.

Leadership Modes Summary

Using the new *Six Sigma Management System*, the implementation of a campaign begins with the Align Mode activities. Leaders revisit and align on the vision, mission, values, and strategic objectives of the organization. To kick off the campaign, leadership uses Mobilize and Accelerate Mode activities and tools to get teams set up and empowered and the entire organization involved in the quest for significant process improvement. As teams begin their actual Six Sigma DMAIC activities, leadership Governs the Six Sigma campaign.

As teams complete projects, leadership returns to the Align Mode and adds new projects to the campaign, and then moves back into the Mobilize Mode to form and empower new teams. Leadership also regularly revisits the Voice of the Customer, and may revise its strategy and goals to more closely align with the VOC. This may cause a shift in project priorities or emphasis, which will impact implementation (mobilize and accelerate activities) and governance.

The leadership modes are expressed in the mindsets of the leaders of Six Sigma campaigns. It is important for leaders to recognize that the keys to leadership effectiveness are:

- alignment to clear objectives,
- mobilization of all elements of the organization,
- a focus on clock Acceleration to produce results, and
- a structured Governance system implemented by involved leaders.

Breakthrough results come from a unified management focus on continuous process improvement that guides the organization to achieve campaign goals through the application of team-based problem-solving approaches. The *Six Sigma Management System* is a key element in realizing such breakthrough organizational results.

Chapter

4

Insights into Six Sigma Leadership

Introduction

The notion of leadership as expressed in the context of a successful Six Sigma campaign cannot reside in just one person. It must take root in the very fabric of an organization. It must extend from top executives, who shape and frame the vision of the campaign, to those front-line managers, who inspire and lead teams through the various stages of execution. Any break in this chain of active leadership will almost certainly handicap even the best conceived and most promising of initiatives.

The demands of a strong Six Sigma program will stretch the aggregate capabilities of the leadership team. In play are the big picture issues - "Why are we doing this and what do we hope to achieve?" - plus a host of process and detail considerations - "How do we get there, how do we measure our progress along the way, and what do we gain in the end?"

And above all is the concerted effort to lead people. Only people can bring a Six Sigma initiative to life. The capacity of leaders to engage, motivate, and build collaboration throughout the organization will make the difference between a wildly successful effort and a mediocre outcome that may accomplish little more than creating some new process steps.

Studies and experience clearly show that critical leadership dimensions must be very active in order for this kind of program to really make a difference. The most brilliant corps of subject matter experts who have little skill for the people dynamic will have only limited impact on bringing about genuine change. And the "Great Communicators" who have just a surface grasp of the critical practices may get buy-in and enthusiasm from the organization, but miss the actual content and substance that will ultimately spell campaign success.

The value of searching out the key capabilities and competencies that are the foundation of effective Six Sigma leadership cannot be overstated. Some of these factors are the end product of extensive training and skill development. But others link to the natural core actions of individual leaders - the requisite behaviors that must be in place for the Six Sigma campaign to be successful. Those behaviors make up the Six Sigma Leadership Model that is the subject of this chapter.

The remainder of this chapter breaks down this Leadership Model into a series of leadership insights and describes exactly how each of these elements has a bearing on Six Sigma effectiveness. The chapter discusses the key capabilities individuals must possess to become great Six Sigma leaders. It also provides Target Conversations that organizations can use as tools to begin to look for and cultivate the strengths that are needed in great Six Sigma leaders.

Leadership Insight One: Passion For Delivering Customer Value

Top Six Sigma leaders are ardent believers in Six Sigma and its possibilities. They are convinced that, if properly done, the effort will lead to vastly improved business approaches that will advance the organization and its practices. In the end, they believe what happens in their organization will have great value and benefit to the customers they serve. It is this passion for delivering value to the customer that motivates leaders to undertake a Six Sigma effort.

Six Sigma leaders recognize that no two organizations are identical. Every set of customers is different, and what they perceive as a value proposition is unique to the individual organization and its business. A Six Sigma leader must bring much more than a paint-by-numbers kit to the effort. It would be a serious mistake for this leader to attempt to bend and twist an existing organization into a preconceived shape that someone decided represents "best practices" for one and all. To be sure, the templates and the process steps that drive Six Sigma are vital and proven over the years to work. But they are models, not cookie cutters. The exact implementation must be local, customized, and tailored to fit the exact requirements of the individual organization, its business, and the customers it serves.

The great Six Sigma leader thinks of each person touched by the campaign as a customer, and must retain this customer focus first, last, and at every step along the way. This requires listening and discovery to gain a solid understanding of the current status of an organization. And it requires insight into the attitudes of the people who are becoming immersed in an effort that will ultimately change the way they operate.

Dedicated leaders of Six Sigma campaigns have something of a missionary quality. They bring to the role a clear vision of value and of improvement. But they recognize that it will take more than passing out a new operating manual to gain buy-in and belief from a team. The impact of Six Sigma leadership is gauged by just how embedded the customer-focused practices become in the minds, hearts, and culture of the organization and its people. The ultimate success is realized when customers notice the difference and value the change.

Key Capabilities

A leader who has a passion for delivering customer value will possess these key capabilities:

- Creating Value: The leader is personally motivated by contributing to all customers in a way that brings about benefit and betterment.
- Active Listening: Discovery-skills through which information is surfaced about current practices, ideas for improvement, attitudes toward change - what it is that customers want.
- Partnership Communications: Skills for leading an interactive and responsive dialogue that establishes trust and belief among both external customers and all internal program partners.

Target Conversation Centered on Leadership Insight One

These questions will help uncover the underlying motivations of a potential Six Sigma leader related to customer focus:

- What do you think makes the difference between someone who is very strong at leading this kind of program versus a person who struggles and who is not that effective?
- Let's say you are meeting with a brand-new project team for the first time. What would be your opening message?
- Suppose you move to individual meetings with a number of key stakeholders to learn about the current operation. What approach would you take in these meetings?
- What would you find most satisfying about assuming a leadership role for a Six Sigma program?
- How do you expect that our customers will benefit from this effort?

Positive Evidence

Individuals who exhibit these qualities rate positively on a passion for delivering customer value:

- Includes the people factor as a critical element
- Addresses attitude and motivations as well as content
- Engaging and inviting approach with all partners
- Listens as much as directs and instructs
- Seeks to influence people and enable their potential
- Focus on outcomes and the positive values gained by the organization and its people
- Names specific results and tangible effect on customers

Counter Evidence

By contrast, individuals who exhibit these behaviors are questionable when it comes to passion for delivering customer value:

- Emphasis on process steps above all else
- Teaches a methodology rather than a mind-set
- Communications features instruction, not dialogue
- Seeks data and information, not attitude and emotions
- Strong personal organization sole element of effectiveness
- On-time completion of the program primary satisfaction
- Internal focus only; no link to outside customers

Leadership Insight Two: Focus On Execution

Like any meaningful initiative, a strong Six Sigma initiative requires pace and momentum, and a leader who from start to finish ensures a consistent pursuit of the steps that will lead to the ultimate goal. This focus on execution is the second element of a great Six Sigma leader.

The entire team needs to understand the exact size and shape of the goal and the critical pathway that must be followed in order to reach that destination. Good leadership here will spark both understanding and buy-in. The team will quickly recognize that they are embarking on much more than a "hope for the best" journey. Team members come to develop a spirit of "we know where we want to go, and we have a roadmap of steps and events that is sure to get us there".

The team's efforts cannot be a self-propelled adventure. In order to keep the team on track, the leader must have a continuing grasp of the master plan and all its components. This element of leadership involves somewhat emulating the behaviors of a chess player who must build a plan of attack, marshal all the resources on the board, and factor in the timing of all the moves that will guarantee the desired results. In Six Sigma terms, this implies managing a substantial number of people and capabilities into an integrated and coordinated whole.

To manage all of the multiple factors that affect execution, the leader must be comfortable operating from a system that charts every project step, tracks all reports and documentation, records meetings and decisions, etc. Good personal organization is paramount, and a sense of process is indispensable.

All the requirements of a Six Sigma program can appear to be a rather formidable challenge to a team that is experiencing it for the first time. The organized and in-control leader will be a powerful asset to reassure the team that they are on the right road to a successful outcome.

Key Capabilities

A leader who possesses the requisite focus on execution will exhibit these behaviors:

- Drive To Results: Keeping the end line always in view, defined in exact terms as to steps, due dates, key milestones, and checkpoints.
- Managing Resources: Developing and implementing the execution plan that integrates people, departments, cross-functional contacts, etc.
- Managing Process: Operating through a system that ensures central information, clear next-step actions, readily available indicators of current status, and a data/audit trail.

Target Conversation Centered on Leadership Insight Two

These questions will help evaluate the execution skills of a potential Six Sigma leader:

- How do you prefer to launch a new project that will involve several people and a fairly substantial number of execution steps?
- What steps do you take to make sure that you will bring this project in on time and within budget?
- In your experience, what are some of the pitfalls that slow down a project or prevent it from a successful outcome?
- Tell me about one of the best projects or programs you have managed and what made it so good.
- By contrast, what was the least satisfying project or program that you were part of and what went wrong?
- What steps do you feel are necessary to ensure that the outcomes we produce are appreciated and value by our customers?

Positive Evidence

Individuals who exhibit these qualities rate positively on execution skills:

- Detailed discussion on organizational approaches
- Works from a master plan from the very start
- Very hands-on with monitoring and tracking all project events
- Has alternative and back-up plans to adjust to new factors
- Keeps all partners aware of current project status
- Provides milestones and mid-course checkpoints to allow close monitoring of timeline and momentum
- Seeks customer input and validation

Counter Evidence

In contrast, the execution skills of individuals who exhibit these behaviors are questionable:

- No evidence of working to a master plan
- Captive to circumstances and unanticipated events
- Limited evidence of "preventive maintenance" provisions
- Blames poor project performance on uncontrollable factors
- Limited evidence of defined checkpoints along the way
- Equates good luck and hard work with project success

Leadership Insight Three: Fact-Based Decision Making

Leading - if it were easy, anyone could do it. The reality is that leading is usually not easy. There are countless times when effective leaders must put themselves on the line and make decisions that can spell the difference between success and disaster. Having the capabilities that surround strong decision-making is almost always an indispensable component of a solid leader.

It starts with the diligence of sound data-gathering. Simple intuition and quick knee-jerk reaction result in decisions that are a matter of guesswork and luck, which is not a combination that inspires a lot of confidence or generates reliable decisions. The capable leader searches out current information, local conditions, the views and opinions of others, and every other data point that might be applicable to the here-and-now decision.

There is an analytical component to this, as information is sorted and factored as part of this "critical thinking" discipline. "What-if?" scenarios are played out and alternatives are weighed. The end product is a decision that is informed, with foundation, and complete with fallback plans if necessary. Pure guesswork is taken out of the equation.

And this must happen in a timely manner. Analysis that becomes so exhaustive as to forever delay action is a silent enemy to pace and momentum. With appropriate facts in hand, the effective leader makes the call, clearly and without ambiguity.

Once the decision is put in place, the leader takes ownership of it. Should resistance be encountered, the great leader will present a ready defense of the position that is grounded in the fact-finding and logic that preceded it. And that ownership continues through the final outcomes. "I never expected that twist" or "certain things happened that were beyond my control" are statements rarely heard from the lips of the strong decision-maker.

Key Capabilities

A Six Sigma initiative is a fluid program, with a host of variables generated by the people and the local circumstances it addresses. The leader has no "bible" at hand with a foolproof answer for every situation. There are always decisions to be made, hopefully in the hands of someone who has these capabilities:

- Critical Thinking: Strong analytic steps will be key, including comprehensive data-search, weighing options and alternatives, best-case results, etc.
- Decisiveness: Timeliness and clarity are paramount and a reversal of decision is rare.
- Accountability: Full personal ownership accompanies the bottom-line decision and its outcomes.

Target Conversation Centered on Insight Three

These questions will help evaluate the decision-making skills of a potential Six Sigma leader:

- In your mind what makes the difference between a good decision-maker and one who is not so good?
- Tell me of a time when you made an important decision that was something of a personal risk for you, but turned out very well.
- Think of a decision you made in the past that didn't turn out well. What did you learn from that experience?
- What is your best approach for convincing people who initially are very much opposed to some action you are proposing?
- Let's say your own boss doesn't agree with one of your decisions but is willing to go along with you if you insist. What will you do?
- What part of decision-making do you do best?

Positive Evidence

Individuals who exhibit these qualities rate positively on decision-making ability:

- Stresses a thorough search for good information
- Seeks to diagnose current situation and here-and-now facts
- Readily produces the case and logic that underpin a decision
- Inquisitive about the viewpoints and perspectives of others
- Weighs alternative decisions and the likely outcomes of each
- Comfortable with standing behind decisions and ready to live with the outcomes that result

Counter Evidence

By contrast, the decision-making abilities of individuals who exhibit these behaviors are questionable:

- Espouses the notion that "any decision is better than no decision"
- Relies mainly on experience and solutions that worked in the past
- Indicates "gut and intuition" as a primary factor
- Works in a personal vacuum without seeking other views
- Needs sign-off of a superior to feel secure in decisions
- Disclaims ownership over bad outcomes - "not my fault"

Leadership Insight Four: Emphasis On Performance Metrics

It certainly would make no sense at all for a team to leap into a Six Sigma program with little more than a vague hope that it will all work out in the end and actually produce some benefit for the organization.

The great Six Sigma leader is on center stage to ensure that the very opposite takes place. To start, there must be a purpose and a point to every Six Sigma initiative. Exactly what is the destination point of every project and how will the team know that it has arrived? Executive leadership sets the strategic goals by first identifying which areas of organizational and business practice offer the greatest potential for improvement. Then each Six Sigma team defines the objectives of its project that align with the organization's strategic goals.

The strong Six Sigma leader will keep the team on task until those specific objectives are defined. Project objectives always include a yardstick: What is the gap between the organization's current practice and the desired outcomes, and how can that be expressed in a concrete and tangible way?

Even when the team has defined a goal and some high-bar metrics that will spell out the successful result of its efforts, there is still much more required to stay on course. The team will need a series of gauges on its dashboard to indicate progress (or the lack thereof!). This real-time grasp of project health is a vital part of a Six Sigma leader's role. Essential here is the capability to set norms and monitor progress through every phase of the project.

This, again, points to the need for excellent analytic skills and attention to detail. It can be deadly to rely on intuition and on a general impression that things are on track. Last minute discovery of previously unrecognized glitches can delay or even derail a team's entire effort. Effective leadership here means maintaining close proximity to information, data, and detail.

A strong Six Sigma program is a healthy challenge that will stretch out the best of teams. But it is not meant to be a blind incursion into previously unexplored lands with no idea of route or destination. The effective Six Sigma leader will assume the navigator role to keep all on a path that is sure to yield the desired results.

Key Capabilities

An individual who recognizes the criticality of performance metrics will exhibit these behaviors:

- Setting Goals: Skills for leading the process that results in a set of clear and crisp targets to serve as measurement guidelines through--out the program.
- Tracking Progress: Establishing the real-time system that will monitor projects and serve as an early-warning signal for potential problems or roadblocks.

- Detail Management: Underlying these practices are superior personal skills for organization and detail. There is recognition of the necessity to quantify all steps and to monitor progress-to-plan on a real-time basis.

Target Conversation Centered on Insight Four

These questions will help to evaluate the performance management capabilities of a potential Six Sigma leader:

- What are most important goals you have in front of you in your job right now?
- How will you know that you have successfully accomplished those goals?
- Is there some way you can tell if you are on track with hitting those goals completely and within the timeline you have set for yourself?
- Tell me of a time when a project that you were working on fell seriously behind schedule and what you did to get back on track.
- What do you regard as the best project you've ever worked on? Why? How did you personally contribute to that outcome?
- If you rate your own performance for the past six months on a 1-10 scale, ten being high, how would you rate yourself? Why?
- Is there any way that you think of yourself as a perfectionist in your job? Explain.

Positive Evidence

Individuals who exhibit these qualities rate positively on performance management ability:

- Strong emphasis on the tangible and quantifiable
- Overall targets are highly defined and measurable
- Clear system for measuring progress throughout projects
- Equates strong performance with specific and concrete results
- Signals of personal organization and attention to detail
- Rejects any aspect of "loose ends" or unfinished work

Counter Evidence

Whereas individuals whose behavior matches any of these criteria have questionable ability to set and manage to performance metrics:

- Description of goals is largely soft and intangible
- Project results lack definition and measurement criteria
- No evidence of using project indicators (gauges and metrics)
- Any indication of a reactive style; responds to negative events rather than anticipates them
- Disclaims high degree of personal organization or detail management

Leadership Insight Five: Visible Advocacy For Breakthrough Improvements

As with any leadership role, there is something of an art form to running a Six Sigma project. Every bit of the DMAIC process may be faithfully observed and the systems and metrics can be fine-tuned to perfection. But there is more to it than just good process management to bring a program through to complete success.

In almost all cases, an effective Six Sigma program will involve significant change as an organization moves from one place to another in the way it does its business and impacts customers. Six Sigma leaders commonly encounter resistance to that change. People can be stubborn when it comes to relinquishing old habits in favor of something entirely new. And often the embedded organizational culture itself represents a major barrier to this change.

It is the job of all Six Sigma leaders to assume the role of missionaries to shepherd an organization through this transition. All leadership - from the most senior executive to the most junior team leader - must breathe life and soul into the Six Sigma initiative, to ensure that it become far more than a minor stab at upgrading a few systems and procedures.

Often Six Sigma leaders will be required to be the advocates and the rainmakers for the cause. They will need to speak up and directly confront the barriers and the resistance that threatens the positive outcomes of the program. Logjams must be broken and adverse conditions must be reversed. Clearly, the mission of a Six Sigma leader is not one for the faint of heart.

On the other hard, this advocate-leader must be capable of exercising a very persuasive style, motivating and encouraging and nudging people into fully embracing the power of the program. And since there will certainly be some conflict along the way, there will be the need to unravel contrary viewpoints and to negotiate solutions that everyone can buy into. The ability to creating win-win agreements is an especially invaluable talent for the great Six Sigma leader.

There is a dose of stamina and endurance required, plus the underlying wisdom to know when advocacy needs to be spirited and outspoken and when it should take a more subtle and influential form. It is perhaps this area of leadership that will most impact the smooth flow and the final success of the program.

Key Capabilities

An individual who has the ability to be a powerful advocate possesses these qualities:

- Assertiveness: A core behavior that allows the leader to champion a cause, confront a situation, and exercise command and control.
- Influencing Skills: The companion skill that enables the leader to win the collective buy-in of the organization to an important change event.
- Tenacity: The persistence and resourcefulness to unravel all barriers, overcome all resistance, and keep the team on track.

Target Conversation Centered on Insight Five

These questions will enable an assessment of the capabilities of a potential Six Sigma leader to act in a full advocacy role:

- What is it about your own management style that would qualify you as an effective Six Sigma Leader?
- Do people think of you as commander-type person who gives directions, sets expectations, takes a stand, that sort of thing?

- What is your best technique for persuading people to go along with your ideas and suggestions?
- Give me an example of a particularly difficult situation that you had to manage through when running a project like this.
- Let's say there is an intense division of opinion within your team regarding some issue. What is your approach for resolving this?
- At the conclusion of the project when people provide their reviews and feedback, how would you hope you would be described in terms of your greatest value and contribution to the effort?

Positive Evidence

Individuals who demonstrate these qualities rank higher on the potential to be a strong Six Sigma advocate and leader of change:

- Exhibits full range of capabilities; can take many roles according to the situation
- Expresses self as a firm, unafraid leader
- Shows close attention to individual attitudes and motivations
- Places priority on building open communications and collaboration
- Is innovative, adaptive, flexible - and builds that spirit in the team
- Will take an advocacy position with superiors and associates, and face any other obstacle to a positive outcome
- Wants to be viewed as having impacted people and results

Counter Evidence

These qualities, however, often indicate weakness in an individual's ability to be a strong advocate of change:

- Great difficulty in responding - has not "rehearsed" self into the job and can't relate previous experience into this setting
- Any sign of a one-size-fits-all approach for most situations
- Always the Commander or always the Influencer; can't adapt according to the situation
- Restrained about taking a high visibility advocacy role
- Prefers to avoid conflict or confrontation
- Thinks of this role as technical and process-centered; shows limited evidence of a personal leadership impact

Summary of Six Sigma Leadership Insights

This chapter has explored a model for effective Six Sigma leadership made up of five components, or insights:

1. Passion for Delivering Customer Value
2. Focus on Execution
3. Fact-Based Decision Making
4. Emphasis on Performance Metrics
5. Visible Advocacy for Breakthrough Improvements

These are the qualities that Motorola University consultants have repeatedly seen in leaders in organizations where Six Sigma has produced breakthrough results.

There are several ways that these leadership insights can prove to be useful as an organization launches a Six Sigma campaign.

First, they can enable executive leadership to identify its own "mindset". Effective Six Sigma leadership begins with some serious introspection within the senior leadership team itself. The entire team should explore the target questions presented in this chapter as a group. Together, leadership should examine what responses they hope to hear from the Six Sigma team leaders. This is a valuable discovery - to learn just how aligned executive consensus is with all that goes into a Six Sigma effort.

Second, they provide a basis for ongoing coaching of team leaders. Reality suggests that it is rare for any one individual to match perfectly with the "ideal" model. So there is the need for ongoing oversight and coaching. Reviewing these components of leadership on a regular basis will often help to pinpoint development priorities for advancing the capabilities of an individual team leader. In many cases, this model will also serve to highlight problem areas that may be hindering a team from reaching maximum effectiveness.

Finally, they facilitate initial evaluation and selection of individuals for team leadership. The target questions are ideally suited for use in an interview process that can be used to determine which individuals will be assigned to lead the various teams that will make up the Six Sigma campaign. Leadership can gain valuable insights into the suitability of candidates for this role.

Chapter

5

Six Sigma Management System Case Study

The two prior chapters in this part of the Six Sigma Black Belt Handbook explained the background and principles behind the Six Sigma Management System, described the leadership modes and activities of the Six Sigma Management System, and provided insights about Leadership in a Six Sigma organization. This chapter will present a case example to illustrate how an organization can use the Six Sigma Management System tools to begin to transform itself into a true Six Sigma organization. The chapter will describe leadership's align, mobilize, accelerate, and govern activities and discuss leadership's involvement in a team's DMAIC project.

Case Background

The organization, a printing company based in the United States, was facing increasing quality complaints from customers, shrinking market for its printed products, and increasing competition that was driving down selling prices. The organization's CEO had been talking with Motorola University consultants about Motorola's experience with Six Sigma. They suggested that the CEO attend a Six Sigma Foundations training session. The Foundations training was all it took to get the CEO's buy-in to the potential benefits of Six Sigma. The CEO had the rest of the executive leadership team take the Six Sigma Foundations training. Then, the leadership team decided to engage Motorola University to guide them through a Jumpstart into the Align Mode of a Six Sigma campaign.

Align Mode Activities

Baseline Audit

Leadership made the decision to enter the Align Mode with a baseline audit that they could use to help make the case for accelerated change to the rest of the organization. Assisted by Motorola University consultants, they completed a Six Sigma Rapid Assessment. Here is the result:

Category	Best Practices	Ratings
Six Sigma Management System	Clear leadership alignment to the "vital few" metrics; visible and active selection and review of the critical improvement projects; trained and committed resources supporting the projects	2 - Fair "Vital few" metrics are financials - no balance. Improvement project work fragmented.
New Product/ Service Development Process	A disciplined, customer-driven process exists for identifying and prioritizing new product/service requirements. All product/service development projects follow a stage/gate review process to ensure resource optimization and timely delivery. Long-term product/service development strategies are synchronized through a roadmap development process.	2 - Fair Have process for new product and service development, but not synchronized through roadmap and not very customer focused.
Supplier Management	Procurement and supplier management processes are synchronized across the organization. Supplier performance management is proactive and data driven. Suppliers are actively engaged early in the product design and development process.	2 - Fair Have good supplier management but processes not mapped. Suppliers not involved in product design and development.
Management of Operational Processes	Operational processes are routinely reviewed and plans are in place to drive the required rate of	1 - Poor Do not have regular process reviews - have

	improvement. Process characterizations, control plans, quality checks, and problem prevention measures are acted upon and updated. Pertinent methods of statistical quality and process control effectively and efficiently used.	ops reviews where separate departments are reviewed individually. Do more quality checking than problem prevention. Not using SPC or process control effectively.
Management of Customer-Facing Processes	Customer service, sales, marketing and field service processes are routinely reviewed and plans are in place to drive the required rate of improvement. Process characterizations, control plans, and quality checks are maintained and acted upon. Pertinent methods of statistical quality and process control effectively and efficiently used.	1 - Poor Same comments as for management of operational processes.Customer services and customer sales and acquisition not viewed as processes.
Data Driven analysis for prevention, problem solving and decision making	Organization uses data driven problem-solving methods across the spectrum, from executive decision-making to front line root cause analysis. Data collection processes are systematic and efficient. Teams demonstrate the ability to apply appropriate tools to the problem that is being solved.	1 - Poor Data collection not being done effectively. Leadership tends to make decisions based on qualitative information rather than quantitative data.

Figure 5-1 Rapid Assessment Results

From the Rapid Assessment, leadership could see they had a long way to go to become a Six Sigma organization practicing the Six Sigma Management System.

They were convinced that it was imperative that they begin aligning on their vision, mission, core values, and strategic objectives.

Refreshed Vision, Mission, Core Values

The leadership took a look at the vision and mission statements they had developed as a group at an executive retreat a few years earlier, and reviewed their credo that essentially had never been updated since the company was founded over a hundred years ago. What they found was that none of these reflected current market and competitive conditions. So, with guidance from the Motorola University consultant, they started from scratch and came up with the following vision statement, mission statement, and list of core values.

Vision Statement

"We aim to be the most innovative, responsive provider of solutions for improving the flow and management of business information.We will forge ever-deeper relationships with current and new customers.

We will out-gun our competitors with revolutionary products and service. We will build an environment that fosters creativity and imagination and focuses on continuous process improvement."

Mission Statement

We are a producer of print and electronic documents that enable our customers to communicate within their organizations, with their suppliers and customers, with their shareholders, and with regulatory agencies. Our products and services include:

- Custom and stock printed business forms and labels
- Electronic form design and programming services
- Document and office supplies warehousing and inventory management with just-in-time delivery
- Automatic identification software, printing systems, and data collection devices
- Short run color business documents
- Product packaging labels
- Printed promotional direct mail pieces
- Forms handling equipment

Our main markets are:

- Manufacturing
- Financial
- Health care
- Wholesale
- Distribution

Our core processes are sales and customer service (customer relationship management), marketing, information technólogy development (electronic information management products and services sold to customers), document management and warehousing, order management, order fulfillment, and delivery. Support processes are purchasing and inventory management, engineering and maintenance, and human resources management.

As the printed form becomes more of a commodity in the eyes of our customers and markets, and customers increasingly see the advantages of on-demand production of printed forms and documents, and of our digital document services, we are striving to surpass our competitors in this arena, both in terms of products and services.

Core Values

- Quality everywhere
- Focus intensely on surpassing customer expectations
- Exhibit trust, respect, and integrity in all relationships
- Foster change, continuous improvement, innovation, and creativity
- Work as a team with open, honest communication throughout the organization
- Value and reward each employee's contribution to the organization's every success
- Build a safe and productive environment
- Develop a diverse workforce with the greatest talent
- Invest in future technological bases for success
- Preserve and protect the environment, and replenish natural Resources
- Support all communities where we do business

Leadership was quite pleased with its renewal efforts and felt that their new statements of vision, mission, and core values could help lead and guide the entire organization through its Six Sigma journey.

Voice of the Customer

When it came time to align on the Voice of the Customer, the Vice President of Marketing tried to make the case that the organization had its finger on the pulse of its customers. The CEO, however, was determined that this had to be a fresh start. The organization was facing so many challenges that it was imperative to its future that they get this Six Sigma effort "right". The CEO commissioned each member of the executive leadership team and their direct reports to get out in the field and have face-to-face conversations with their customers, to see how and where their products were being used - and not used. The CEO stressed the importance of partnering with customers to uncover their real needs - those that were critical to the survival of their organizations.

These are typical of the statements leadership heard customers make:

- "Our customers are starting to demand we use RFID on shipping cases and pallets."
- "We're shipping more and more as web commerce grows."
- "We need communications solutions that will support our Six Sigma processes."
- "We will continue to evolve our paper documents to digital format."
- "We want to be able to view proofs via the Internet rather than having to wait for you to produce and send the proof to us."
- "In this economy, we need to lower our printing costs any way we can."
- "Our telemarketing results have been severely affected by the National Do Not Call Registry. We're considering going back to the mail, or moving to e-mail, to deliver our pitches."
- "We need quick order turnaround with fast delivery, shorter print runs, and real time access to our order status information."

Strategies

Armed with this wealth of customer information, leadership set out to review its strategies. The Motorola University consultant explained the concepts of the Balanced Scorecard and Strategy Mapping to them. The executives thought they had been taking a balanced approached to strategic planning, but actually discovered that their strategic objectives were all stated in terms of cost reduction and revenue and profit generation.

Leadership selected their key strategies and organized them into a strategy map. This is an excerpt from that map:

Customer Develop mutually beneficial partnerships with large customers.	**Business Process** Make manufacturing processes more time efficient with fewer errors.
Learning and Growth Expand Six Sigma training to sales consultants so they can understand customers' Six Sigma related needs.Develop sales consultants so they can understand and sell more high value printed products to meet customers' direct mail related needs.	**Financial** Increase profits through increased sales of high value products to large customers.Reduce costs and pass savings on to customers in form of reduced prices for commodity products.

Figure 5-2 Excerpt from Strategy Map

The Dashboard and Stretch Goals

The executives struggled a bit with setting stretch goals. They did not want to set the performance bar too high, or discourage the organization by setting unreachable goals. But the CEO was adamant about motivating the organization to undertake the challenge of re-inventing themselves through Six Sigma.

The executive team also struggled with setting goals that were measurable. For example, at first they said they were comfortable with a business process stretch goal of "minimize inefficiency, turnaround time, waste, defects, and human error". When the Motorola University consultant began to challenge the group as to how they would actually measure progress toward that sweeping goal on the Dashboard, the team began to realize the importance of concrete measures.

Leadership eventually developed the following stretch goals related to its strategic objectives.

Customer	**Business Process**
• Make it easier to do business with us by expanding the ways customers can provide us their print order specifications • Accept electronic order entry and all standard electronic graphics formats for the printed image directly from the customer by end of current FYReduce order cycle time on traditional printed products from 1 week to 2 days by the end of the current FY	• Improve order entry lead time from 4 days to 4 hours • Reduce the number of order errors due to order entry errors by 500% • Reduce press downtime by 20% • Reduce paper waste by 25%
Learning and Growth	**Financial**
Have sales consultants prepared with knowledge and skills to sell high value direct mail products to large customers by end of Q2	• Reduce order entry costs by 40% • Increase revenue from sales of high value direct mail products by 50% starting in Q4

Figure 5-3 Stretch Goals

Performance Driver Analysis

Once leadership had set their stretch goals, they went about trying to determine what they could do to cause organizational performance to change, in order to move the metrics on the Dashboard. They noted that they had set a number of goals related to the order entry process - the process where a sales order is officially entered into the organization's accounting system and translated into specifications for materials procurement and production. Given the stretch goals they had set for this process, there was a clear need to improve this process.

Leadership was aware that there was a lot of manual copying and rewriting of information as well as manual look-ups in the order entry process. These activities created room for error, added time to the process, and cost the organization money. Leadership believed that, if it could streamline the order entry process, then defects, costs, and lead time would all be reduced.

The Six Sigma Alignment Roll Up

Leadership formalized its renewed vision, mission, and core values, its strategic objectives, and the stretch goals into the Six Sigma Alignment Roll Up. The Roll Up also showed the one Six Sigma project that leadership has given priority to. Other priority projects should also be shown on the Roll Up.

This is the main vehicle leadership would use to communicate the message of the Six Sigma campaign to the rest of the organization.

Strategic	Initiatives and Six Sigma Projects	Operational Dashboard
Vision We aim to be the most innovative, responsive provider of solutionsfor improving the flow and management of business information.We will forge ever-deeper relationships with current and new customers.We will out-gun our competitors with revolutionary products and service.We will build an environment that fosters creativity and imaginationand focuses on continuous process improvement.	**Customer** Develop mutually beneficial partnerships with large customers **Business Process** Make manufacturing processes more efficient with fewer errors	**Customer** Expand ways customers can provide us their print order specifications Accept electronic order entry and all standard electronic graphics formats for the printed image directly from the customer by end of current FY

<table>
<tr>
<td>Mission

We are a producer of print and electronic documents that enable our customers to communicate within their organizations, with their suppliers and customers, with their shareholders, and with regulatory agencies.

Core Values
• Quality everywhere
• Focus intensely on surpassing customer expectations
• Exhibit trust, respect, and integrity in all relationships

• Foster change, continuous improvement, innovation, and creativity
• Work as a team with open, honest communication throughout the organization
• Value and reward each employee's contribution to the organization's every success
• Build a safe and productive environment
• Develop a diverse workforce with the greatest talent
• Invest in future technological bases for success
• Preserve and protect the environment, and replenish natural resources
• Support all communities where we do business</td>
<td>Financial
Increased prof

its; lower prices for commodity products through reduced costs

Learning and Growth
Expand Six Sigma training to sales consultants
Develop consultants abilities to sell more high

value printed products

Six Sigma Project
Streamline Order Entry process</td>
<td>Reduce order cycle time on traditional printed products from 1 week to 2 days by the end of the current FY

Business Process
Improve order entry lead time from 4 days to 4 hours

Reduce the number of order errors due to order entry errors by 500%

Reduce press downtime by 20%.

Reduce paper waste by 25%

Financial
Reduce order entry costs by 40%Increase revenue from sales of high value direct mail products by 50% starting in Q4Learning and GrowthHave all sales consultants prepared to present our organization's commitment to excellence through Six Sigma by start of Q3</td>
</tr>
</table>

Figure 5-4 Six Sigma Alignment Roll-Up

Mobilize Mode Activities

The Change Leaders

The Motorola University consultant briefed leadership on the typical make-up of a Six Sigma organization, and the importance of training to develop the organization's Six Sigma skills and expertise. Leadership then began the task of building its cadre of change leaders, and realized the challenge. The organization had no trained Master Black Belts, Black Belts, or Green Belts. No one in the organization had enough experience with Six Sigma to become the executive Champion. So the executives decided to bring in talent and experience from outside the organization.

First, the organization hired its executive Six Sigma Champion from an ink formulation company that was one of their suppliers. Their new Champion had formerly been that organization's Senior Vice President of Business Improvement and Chief Technology Officer. Leadership felt this individual had background in their industry, without being a direct competitor, and could become productive very quickly.

Motorola University consultants assisted leadership in hiring five Black Belt consultants and provided resources where the organization could tap into Master Black Belt technical expertise on an as-needed basis. The leadership agreed that eventually, the organization would develop Black Belts and at least one Master Black Belt within their organization, but they did not want to delay the campaign kick-off waiting to get Black Belts and an MBlack Belt trained.

Leadership did agree, however, that they wanted their own people who had expertise in their processes actually doing the Six Sigma DMAIC project work. The executive Champion and Black Belt consultants identified 15 Green Belt candidates, brought them on board with the campaign, and arranged for them to start training.

Planning the Six Sigma Campaign

Leadership had developed a hopper of projects. They prioritized those projects and made project selections, to make the best use of available resources. In qualifying candidate projects, leadership considered:

- the availability of the data required to study the problem
- the expected benefits versus the human and physical resources required to complete the project
- anticipated length of the project

Leadership also evaluated the potential synergy and soft impact of project ideas. They wanted to avoid the trap of isolated and even conflicting improvement efforts that the organization had fallen into in the past.

They put together a cohesive campaign of projects. Some of these were related to error and lead-time reduction - projects that would have direct impact on the organization's customers. Others were related to generating new, high value products. Still others were focused on selling more high value products to targeted large customers.

Selecting and Empowering Project Teams

Leadership next identified the owners of the affected processes - like an owner of the Order Entry process - to act as DMAIC team Champions. This task was not as straightforward as it sounded. The order entry process, for example, touched many different arms of the organization - sales, manufacturing, information technology, and accounting. Within manufacturing, order entry took place in several different geographic locations throughout the county. Leadership determined that the Vice President of Manufacturing really had the central responsibility for order processing, and established that executive as the process owner and Champion.

The team Champions then selected team leaders from the five Black Belt consultants. With the Black Belt consultants, they identified team members from among the 15 Green Belt candidates and known process experts. Leadership met with all of the project teams to emphasize the critical nature of the Six Sigma campaign, explain the rationale for the campaign, and motivate the teams. Leadership took care to make sure team members knew that being part of a Six Sigma team indicated recognition that the individual was a superior performer and an organizational leader. Leadership was also careful to set clear boundaries for the teams, but also to empower them with the authority to investigate and act.

Training

The executives recognized training as a key Mobilize Mode activity. Leadership ensured that everyone was getting the training they needed. Green Belts went to training, focused on the Six Sigma improvement projects to which they had been assigned. All top-level executives and department heads also got champion training. Champions got additional training on coaching techniques and team dynamics.

Aligning the Rest of the Organization

Leadership used its Six Sigma Rapid Assessment to communicate the case for accelerated change to the rest of the organization. They told Six Sigma success stories to convince the organization of the potential of Six Sigma. The Six Sigma Alignment Roll Up provided the basis for communicating and promoting the Six Sigma campaign that leadership had planned.

The executives looked for signs of resistance to change, barriers that needed to be broken down. They looked for reactions like, "Not another quality program", "We'll just wait this one out - it will be gone in six months just like all the other three letter acronym programs", and "This will all change when management changes". Executives worked feverishly at conveying their own enthusiasm and commitment, as well as the pressing need to change. Leadership recruited Champions from every functional department to reinforce those messages on a daily basis.

Accelerate and Govern Mode Activities and DMAIC Project Work

With the campaign kicked off, leadership was poised to enter the Govern and Accelerate modes of the Six Sigma Management System - to guide the activities of the DMAIC teams and manage for accelerated results. The work of team Champions to actively and visibly support the individual teams became a cornerstone in the success of the Six Sigma Campaign.

Order Entry Process Improvement Team

One of the first teams to be formed was the Order Entry Process Improvement Team. The Champion, the Vice President of Manufacturing, met with the Black Belt consultant who was the team leader, and together, they identified the people they wanted on the team. They selected one newly trained Green Belt who was an order entry supervisor, one printing plant expert (internal customer), one Information Technology analyst who knew the Order Entry (OE) computer system, and five OE process experts. (There were five different jobs in the Order Entry process, so they picked an expert in each job.) These people were all assigned to the project full-time.

The Champion, the Black Belt, and the Green Belt met to prepare for the first team meeting. They defined the team's assignment in the form of a draft Team Charter. The draft Charter they presented to the rest of the team at the first meeting is shown below.

The Champion's involvement in this draft Charter was critical to ensuring the team's work started off in alignment with leadership's strategic objectives, and supported achievement of the organization's stretch goals. The Champion also needed to ensure that the project would get done in six months or less, and the plan did call for it to be completed in 24 weeks, or roughly six months. As project work progressed, the Champion would attempt to accelerate the schedule, if possible, in order to obtain accelerated results.

Business Case	**Opportunity Statement**
Reducing the order entry time and errors on print orders will • allow customers to do more just-in-time purchasing, thus increasing the loyalty of our largest and most profitable customers • reduce the cost of the order entry process and allow the company to remain a low price leader· increase customer satisfaction with product quality • reduce cost of rework due to poor quality	A year ago, all printing plant order entry operations were centralized from 18 plants into three regional OE Centers. The order entry process itself was never streamlined and it currently takes four days to move an order through the order entry process. The purpose of this project is to streamline that process and reduce order entry time and eliminate order entry errors that cause errors in manufacturing printed products for customers.

Goal Statement • Reduce order entry lead time to four hours • Reduce order entry errors by 500 percent	**Project Scope** Start - order arrives at Order Entry Center End - manufacturing instructions transmitted to printing plant
Project Plan • Define - 3 weeks • Measure - 5 weeks • Analyze - 6 weeks • Improve - 7 weeks • Control - 3 weeks	**Team Members** Champion - V. P. ArnoldTeam Leader -Richard Gross (Black Belt) Member - Sam Oakley (Supervisor OE and Green Belt) Member - Doris Deter (Screener) Member - Charlie Hansmith (OE Clerk) Member - Lisal Pantner (Pricer/Planner) Member - George Beach (Purchasing) Member - Juan Ramos (IT) Member - Chester Bristol (Scheduler) Member - Marie LaForge (Printing Plant)

Figure 5-5 Project Team Charter

Define Phase - First Team Meeting

Except for the belts, the team members had received only Six Sigma orientation training, so they knew basic Six Sigma measurement concepts and the structure and purpose of DMAIC methodology, but were not trained in statistical tools. They were, at first, very unsure of what they could contribute to the project. The Champion recognized that the team was in the forming stage, and had to direct them to view their roles as process experts and rely on the trained belts to teach everyone else as the project progressed. The Champion indeed intended to utilize the Action Learning methodology to ensure everyone on the team got follow-up training and coaching as needed.

The Champion also explained the strategic objectives of retaining and adding new large customers and customers' expectations of quality, short order cycles, and low price. The team thought that the goals were daunting, but agreed to begin the Define Phase with the understanding that they could revise the goals after they had gathered more information. The team discussed the OE process; almost everyone was intimately familiar with it. They agreed that the scope was correct and that they could get through the project in the time allotted.

The Champion told the team that their agreed-to draft Charter would be published to the rest of the organization, to make their work highly visible. The Champion also informed them that their results would be publicized to the rest of the organization. Part of leadership's "campaign management" strategy was to publish project schedules that tied completions to strategic objectives, track progress, and to publicize success and on-schedule performance. All Champions would also publicize their teams' progress and results to the rest of the organization to keep the organization informed and to encourage and promote future change.

The Champion reiterated that their selection for the team was recognition of each member as a superior performer and organizational leader, and that their work was critical to meeting customer requirements. The team decided that they all wanted to understand the customer requirements better. So they began by analyzing the Voice of the Customer.

Define Phase - Voice of the Customer

The team asked permission to talk directly to customers. The Champion approved the idea since the DMAIC methodology calls for project teams to understand the Voice of the Customer, and obtained the names of contacts to call from the Vice President of Sales. Four team members were selected to conduct the interviews using a list of questions that the entire team had prepared and the Champion had approved. They talked to 20 large contract customers selected from each of the company's five main market segments. These customers each placed multiple print orders monthly, generating a large amount of revenue with a high profit margin.

The customers said that the best quality at the best price was their fundamental requirement. For the most part, customer companies employed professional print buyers who were expected to get the lowest prices overall. These buyers also demanded error-free printed products. The customer buyers would not switch suppliers over a single order price, but overall they expected to get the best prices for what they considered a commodity printed product. Customer buyers also stated that being able to order printed materials on a just-in-time basis was becoming more and more important because they did not want to maintain any printed materials inventory at their facilities. A number of customers said that it took too long to manufacture their printing orders. The team recognized that short order-to-delivery manufacturing cycles were a critical customer requirement.

Define Phase - Voice of the Business

The team also talked to the people in each internal department that had to use outputs from the OE process. They used this as their Voice of the Business data, along with the restatement of the top-level corporate strategy to reduce errors and increase efficiency in manufacturing.

Define Phase - Process Maps

The team developed the following SIPOC (Supplier-Input-Process-Output-Customer) Diagram very quickly.

Suppliers	Inputs	Process	Outputs	Customers
Customer	Price Quote	Screening	Final price	Billing
Sales Rep.	Proof	Spec Entry	Instructions for:	Pre-press
	Art or Copy	Price Check	• Pre-press	Press
	Order Specs	Job Planning	• Press	Finishing
		Inventory Check	• Finishing	Shipping
		Manuf. Schedule	• Shipping	Inventory Control
			• Inventory pulls	Printing Plants
			Equipment loading schedule	End Customer

Figure 5-6 SIPOC Map

They then took several days to gather the data for and draw an as-is Functional Deployment Map.

Every order moved through six activities in the process:

1. Incoming Order Screening
2. Spec Entry into the OE System

3. Price Checking
4. Job Planning, which generated instructions for all phases of the manufacturing process, product packaging instructions, labeling and shipping instructions, and a list of materials required for the job
5. Inventory Check - Doing a current materials inventory check and generating inventory pull-orders to stock handlers
6. Manufacturing Schedule - Assigning a plant and group of equipment and scheduling the order to run in available time slots.

Each of these activities was done as a separate job in the process, except one job included both the Price Checking and the Job Planning activities.

The Order itself was a group of documents in a Job Folder that moved from station to station. Different types of information were generated and entered into the OE computer system at each station. By the time the OE process was complete, all order information, pricing, manufacturing instructions, materials requirements, and plant equipment assignments and schedules were in the OE system. OE was done with the order when all the required manufacturing information was transmitted to the assigned plant by the scheduler.

Critical Measures

While they were doing that mapping, the team was looking for good metrics to measure the achievement of their goals. The Order Entry Center did measure, or track, the internal movement of orders inside the order entry process. The team learned that data were available on the average number of orders arriving at each OE Center daily, and the numbers of orders completed each day by each station. With the help of their Black Belt leader, the team decided that they could calculate and use lead time for an order from the time it arrived at an OE Center until the time the order manufacturing instructions were transmitted to the assigned plant. They also determined that the historical records were available to calculate the number of order errors caused by order entry that were reported and/or rejected by customers. They also found that they could access records of order entry errors caught and reported by the plants. These error rates were tracked by all three OE Centers and had been quite stable over the past six months, so the team felt comfortable using these records.

Weekly Status Meetings

The team had agreed to a three-week timeframe to complete the Define Phase of their project. At the end of each week, the Champion held a weekly project status meeting as part of Governance. The Champion made sure the team was on track.

It was toward the end of the second week when the team was discussing critical measures. Since the team was discussing using lead time as a critical measure, the Champion arranged for the all team members (aside from the Black Belt) to get a couple of hours of training on how to calculate lead time and what affects it.

During these weekly status meetings, the Champion noted that the team was developing as a cohesive group, and they required less direction. The Champion anticipated that the team would soon become fully formed, and move into the storming stage where they would require more coaching.

Define Phase Tollgate Review Meeting

Near the end of the third week, team members were satisfied that they had completed a satisfactory definition of the purpose of their project. So, a meeting with the Champion was held to review the team's deliverables. This is a summary of that meeting with questions from the Champion and answers from the team:

Champion: "What is the Big Y that will be influenced by this project?"
Answer: There are two Big Ys impacted - strengthening partnerships with large customers and making manufacturing more efficient with fewer errors.
Champion: "What Voice of Customer data were used to establish Critical Customer Requirements? How were the data validated?"
Answer: Corporate VOC statements were validated by direct interviews with key large customers.
Champion: "What Voice of Business data were used to establish Critical Business Requirements? How were the data validated?
Answer: The Corporate Strategic Goal for manufacturing was validated by talking to all the internal manufacturing customers of the OE process.
Champion: "What are the boundaries of the process to be improved?"
Answer: Receipt of order documents at OE Center to transmission of manufacturing instruction to assigned plant.

Champion: "What is the specific problem being addressed?"
Answer: The Order Entry process takes too long and makes too many errors.
Champion: "Has this problem been tackled before? What was learned from that attempt?"
Answer: Order Entry was centralized from 18 plants to 3 OE Centers several years ago. It was believed that specialists could develop high degrees of skill in the technical jobs. It still takes too long, and there are errors being made.
Champion: "How do the little y's directly or indirectly influence the Big Y?"
Answer: Shortening lead time for OE will shorten the total delivery printed product delivery cycle to large customers. That is one of their important satisfiers and will help retain their business and acquire new customers.
Champion: "What are the goals, in measurable terms, of the project? Are they achievable in the time frame established?"
Answer: Goal one is to shorten the OE lead time to from four business days to four hours. We believe that can be done. Goal two is to reduce errors by five fold. We need to measure the types of errors and where they occur before we can promise to reduce errors by five times.
Champion: "Did we choose the right members for this team? Do you understand your roles and responsibilities?"
Answer: We have the process and technology expertise to understand and do our tasks. We are learning tools and skills as we go.
Champion: "Were team guidelines established? How are violations of the guidelines handled?"
Answer: We have written guidelines about meeting conduct and responsibility to complete assignments.
Champion: "Who are the stakeholders that will be affected by this project? What level of communication or involvement is necessary for each stakeholder group?"
Answer: We have begun to talk to people from each department that gets manufacturing instructions from OE. Of course, they want to be kept informed and have a say in any changes to the documents that control their operational methods.
Champion: "What concerns may the stakeholders have? How will the team prevent these concerns from becoming obstacles?"
Answer: The 18 plants are still resentful that their old order entry departments were disbanded. They don't like the central authority dictating changes. We will need to communicate a lot about the positive reasons and benefits of any changes we make.
Champion: "What quick wins have been identified? What is the plan for implementing quick wins? What are the plans for ensuring that the quick wins work? What effect will the quick wins have on the goal?"

Answer: We want to immediately eliminate the function of the screener. We can see no value-added work being done by the screening step that is not duplicated in the next two steps in the process. We can test this approach for a week in one OE Center and compare lead times with and without a screener. We need to measure these as-is process activities anyway. If it doesn't reduce lead time significantly, we can easily put the screening step back in to the process. The current Screener is cross-trained and can do a different OE job during the week. Screening looks like a choke point. Elimination of the screening step should substantially decrease order lead time, because there is one screener who feeds work to four order entry clerks.

Figure 5-7 Summary of Define Phase Tollgate Review

Champion's Approval and Support

The Champion accepted the team's report on their Define Phase work and approved them to proceed to the Measure Phase. The Champion also approved a trial of the Quick Win plan proposed by the team. OE Center #1 was selected to drop the screening step from their process. The Champion communicated to the manager of that OE Center, and won agreement to the trial period of two weeks.

The Champion met later that week with the leadership team where the status and direction of multiple Six Sigma projects was reviewed. The leadership team, as a group, reviewed the Charters for all projects to be certain that their strategic objectives were being supported and that there were no conflicts or overlaps among the projects.

Measure Phase - First Team Meeting

The two key items on the agenda for the team's first Measure meeting were to review the critical metrics they had chosen to measure and develop a measurement plan. The team had chosen Lead Time and Errors in Order Entry (order errors) as the two critical metrics. They assigned each metric to a sub-team to develop a measurement plan.

Order Error Measurement Plan

The sub-team assigned to the order errors metric first had to develop an operational definition (OD) of an error. The OD they developed was any reported product defect or production problem that they could trace in the order record to an error in the OE process. The team planned to calculate the yield of error free orders, the DPMO, and the sigma level to establish a baseline error measurement.

To gather as-is measurements of errors, they planned to review the 500 orders produced by each OE Center. They would sample 100 orders from each of the prior five months from each OE Center. Monthly samples would be drawn equally from the first, middle, and last weeks of the each month (first five days, middle five days, and last five days) because the team knew that there were regular cycles in the incoming order demand rate. High demand could pressure people to work quickly, and that could increase errors. In each sample they planned to check to see if any errors had been detected and reported by the customer or by any internal department. They planned to measure error rates separately for new versus repeat orders, and individually for new orders of five different product types, ranging from very simple to very complex.

Since this was a big task, they asked their Champion to get them help in gathering and sorting data. The Champion agreed that the plan was worth doing and provided OE expert people in each OE Center to be trained and led by one team member.

Lead Time Measurement Plan

The goal of the sub-team assigned to the lead time metric was to uncover variations in lead time to the printing plants. The sub-team determined that they needed to measure the incoming demand rate of orders to each OE Center. That information was available as historical records going back two years. In addition, the OE Centers kept records of the number of days each order was in the OE process before it was transmitted to a plant. This sub-team coordinated with the "errors" sub-team, and they agreed to track the lead times for the same samples of orders. So the lead time sub-team would be able to determine actual lead times of new versus repeat orders, and lead times by product complexity. They could also measure lead time for orders with and without errors. The Champion agreed that the extra expert help that was gathering error data could gather lead time data.

Measuring Error/Lead Time Source

Finally, the whole team agreed that for error measurements, they could track the error back to the station that made the error and, thereby, define the error source more accurately. However, for lead times, the sub-team needed to get some additional measure of where delays were occurring and causing long lead times. This team was helped by the Black Belt consultant, who showed them how to estimate lead time without having to measure the actual movement of orders through OE. He explained that the team would have to observe and measure the completion rate (number of orders completed in a standard unit of time) of each station in the process. Once they had the completion rate for each workstation, they could use the demand rate for any day and determine work-in-process queue sizes for each station, and then the lead time for each station. The team determined that they would gather the completion and demand rate data themselves by observing actual OE job performers. They developed a sampling plan that included observing people from all three OE Centers. The Black Belt showed them how to create observation worksheets and had them practice data gathering on each other before doing actual observations.

Governance and Change Facilitation

At the weekly status review meeting, the Champion approved the measurement and data gathering plans. At that point, the Champion wanted to know about the plan to explain to the OE Center personnel the purpose of the visits and observations by team members. In formulating the team's response, the team members went through some storming. Some team members did not think it was necessary to do much in the way of communicating their activities to the OE Center personnel. The other team members disagreed. The Champion stepped into the role of coach, and got the team to compromise. The team finally chose one team member to meet with station operators from each OE Center who would be observed, prior to beginning the data collection.

Before data collection started, the Champion also held meetings with the managers and top-level supervisors of each OE Center to explain the purpose and air any concerns. The concerns aired in those meetings were documented and shared with leadership. The most frequently voiced concern was that the Centers would be downsized as a result of this project. Leadership developed a videotape telling everyone that they believed that, given shorter order cycles and higher quality, sales and marketing could acquire new strategic customers and make use of any increase in capacity that resulted from increased per person productivity. They also surveyed the OE Centers to gather information about their improvement ideas to make OE work less stressful or easier to perform.

Collect the Data

The team proceeded to execute their measurement plans. They worked a great deal on their operational definitions to make sure that the data were valid. The data gathering took several weeks and the sub-teams reported status, successes, and barriers at the weekly status meetings. Meetings were kept quite short and mostly were done by telephone conference call. The Champion was present for all meetings and reported status to the leadership team.

Team Meeting - Review As-Is Process Measurements

The team met to consolidate their results. The sub-teams presented interim results.

The order errors sub-team presented these measurements.

OE Errors on Orders The As-Is Order Entry Process		
Samples of Orders	**DPMO**	**Sigma**
Repeat Orders	55	~ 5.3
Breakout of New Orders by Complexity		
Product A (New and Very Simple)	3.5	6.0
Product B (New and Simple)	108	5.2
Product C (New and Average)	4000	~ 4.2
Product D (New and Complex)	22,750	3.5

Product E (New and Very Complex)	96,801	2.8
Breakout of New Orders by Time of Month		
First Week (First 5 Days)	22,750	3.5
Middle Week (Middle 5 Days)	80	~ 5.3
Last Week (Last 5 Days)	308,536	2.0
Breakout of Repeat Orders by Time of Month		
First Week (First 5 Days)	75	~ 5.3
Middle Week (Middle 5 Days)	75	~ 5.3
Last Week (Last 5 Days)	175	~ 5.1
Breakout of Orders by OE Center		
OE Center 1	1,355	~ 4.5
OE Center 2	1,355	~ 4.5
OE Center 3	1,355	~ 4.5

Figure 5-8 Order Entry Errors

The team concluded that OE errors that caused "rejects" by the external customer or internal customers:

- did not vary by Order Entry Center
- did vary by time of the month (first week, middle week, and last week) where the last week experienced the highest error rate (and the busiest in terms of order volume), the first week had the next highest error rate (and next highest WIP volume), and the middle week had the lowest error rate (and the lowest order entry volume).
- occurred at low rates for repeat orders where a great deal of the order entry data had already been key entered into the OE computer system and, for the most part, only changes had to be entered.
- occurred at very different rates for different product types. When processing simple products with few specs, the OE process achieved a six sigma level. As product complexity, and order entry requirements, became more complex, the error rate of the OE process rose dramatically.

Next, the lead time sub-team presented their results. They had calculated the measured time in days and hours that it took from the time the order was stamped as received in the OE Center by the mail/fax clerk to the time the system recorded that the order was transmitted to a printing plant. The results were straightforward.

- The mean lead time for all orders sampled from all Centers was exactly 4 days. The range was 2 to 10 days. Even using the existing upper specification limit (USL) standard (4 days), the as-is performance level for on-time order entry was less than 1.5 sigma.
- The three OE Centers did not vary significantly in lead times, all being about 4 days.
- Repeat orders averaged 3.5 days in lead time.
- New orders averaged 4.1 days in lead time.
- Very simple new orders averaged 3.2 days in lead time.
- Very complex new orders averaged 5.5 days in lead time.
- Lead times for other order types fell in between lead times for very simple and very complex orders.
- Week of the month showed the most dramatic differences in lead times:

Mean Lead Times Shown By Week of the Month that Order Arrives at OE Center			
Product Type	**First Week**	**Middle Week**	**Last Week**
Repeat	3.7 days	3.0 days	4.2 days
New, Very Simple	3.6 days	2.8 days	4.0 days
New, Very Complex	5.0 days	4.0 days	7.5 days

Figure 5-9 Lead Time

As the errors sub-team had noted, this sub-team also noted that total numbers of orders that arrived at the OE Centers during the first, middle, and last weeks of each month varied a great deal. In the as-is OE process, heavier volumes of incoming orders were associated with longer order lead times. More complex products seemed to suffer more. The team explained that the OE Centers did not sort or expedite any order type officially, but OE personnel were known to "cherry pick" easier orders in time of heavy volume.

Finally, the team reviewed the results that had been gathered through observation of processing actual orders in the OE Centers. The team had observed and measured the average completion rates of for each OE process step. They noted that these completion rates were highly variable and that the variation needed to be understood and controlled. They used historical data to estimate the demand rate on the process - that is, the number of orders that arrived at the OE Centers each day.

They presented three scenarios showing the processing flow of orders in a week and times they found.

Expected Flow of Orders through OE Process in an Average Week (90 orders/day)					
Day 1	**Step # and Name**	**# of Stations**	**Completion Rate (mean)**	**Total Capacity**	**WIP - at end of day**
Incoming Orders = 90	#1 - Order Spec.Key Entry	4 people	20 min./order 3 orders / hr	12 orders/hr 90 orders/day	00 orders
Day 2	**Step # and Name**	**# of Stations**	**Completion Rate (mean)**	**Total Capacity**	**WIP - at end of day**
Orders from Step #1 = 90	#2 - Price and Plan	6 people	30 min/order 2 orders / hr	12 orders/hr 90 orders/day	00 orders
Day 3	**Step # and Name**	**# of Stations**	**Completion Rate (mean)**	**Total Capacity**	**WIP - at end of day**
Orders from Step #2 = 90	#3 - Check Inventory	2 people	05 min / order 45 orders / day	12 orders/hr 90 orders/day	00 orders
Day 4	**Step # and Name**	**# of Stations**	**Completion Rate (mean)**	**Total Capacity**	**WIP - at end of day**
Orders from Step #3 = 90	#4 - Order Scheduling	2 people	05 min / order 45 orders / day	12 orders/hr 90 orders/day	00 orders

(WIP = Work In Process)

Figure 5-10 Completion Rates

From this chart the team could see that each OE Center was staffed to handle a maximum capacity of 90 incoming orders per 7.5-hour workday. OE Centers operated on just one shift, so orders waited overnight (24 hours) between each of the four steps. The team noted that where the screener job was still in place, there was one screener per OE Center and that step added an extra day to the process on average.

In the final week of each month, order volume would jump to into the range of 100 to 120 orders per day. The team charted this scenario.

Expected Backlog of Orders through OE Process in Heavy Week (120 orders/day)					
	Step # and Name	**# of Stations**	**Completion Rate (mean)**	**Total Capacity**	**WIP - at end of day**
Day 1 Incoming Orders = 120	#1 - Order Spec.Key Entry	4 people	20 min./order 3 orders / hr	12 orders/hr 90 orders/day	30 orders delayed extra day
Day 2 Incoming Orders = 120	#1 - Order Spec.Key Entry	4 people	20 min./order 3 orders / hr	12 orders/hr 90 orders/day	60 orders delayed extra day
Day 3 Incoming Orders = 120	#1 - Order Spec.Key Entry	4 people	20 min./order 3 orders / hr	12 orders/hr 90 orders/day	90 orders delayed extra day
Day 4 Incoming Orders = 120	#1 - Order Spec.Key Entry	4 people	20 min./order 3 orders / hr	12 orders/hr 90 orders/day	120 orders delayed extra day
Day 5 Incoming Orders = 120	#1 - Order Spec.Key Entry	4 people	20 min./order 3 orders / hr	12 orders/hr 90 orders/day	150 orders delayed extra day

Figure 5-11 Backlog Analysis

The team noted that this backlog and the increasing lead time for orders in process was true for every job step. Whenever the volume of incoming orders exceeded 90 orders, a backlog queue would accumulate and move through the OE process like a lump. The OE Centers did not have FIFO processing. So the team could conclude that, by day 5 of a very busy week, 150 orders out of 600, or 25 percent of orders, would have had 5 days added to their lead times just to get to step #1 in the process. Then, depending on the incoming volume the following week, additional days of delay would be added in queue waiting to get to other steps. They found orders that were the last through a lump had been delayed by as much as 14 days.

Measure Phase Tollgate Review Meeting

The team members were satisfied that they had met the requirements of the DMAIC Measure Phase. So, a meeting with the Champion was held to review the team's deliverables. The Champion reviewed the team's data, and then used the Tollgate Review questions to make sure they had covered all the bases. This is a summary of that part of the meeting with questions from the Champion and answers from the team:

Champion: "Has the charter been updated? If so, how?"

Answer: Yes. The definition of the opportunity is more precise - that is, we know better where and when the performance gaps in errors and lead time are occurring.

Champion: "Has the scope changed?"

Answer: No. The start and end steps of the process are the same.

Champion: "What x data were collected?"Answer: The x data include the completion rates measured for each job step and the hold times between each step. The x factors of a heavy workload, order complexity, and amount of new information to be captured from a new versus repeat order. Champion: "What are the operational definitions?"

Answer: There were operational definitions for each little-y and each x variable studied. These were written down and applied in practice rounds by everyone who collected data.

Champion: "How much data were collected? Are the data collected representative of the population?"

Answer: A total of 1500 orders were studied to get error data. These were randomly sampled from several orthogonal sub-populations, including OE Centers, weeks of the month, and repeat versus new types of orders. Observations were conducted in parallel in three OE Centers. Teams observed each job performed 20 times in each OE Center, using 6 different job performers.

Champion: "What was done to assure the reliability and validity of the measurement process? If the data collection were repeated, would the team get similar results?" Answer: Yes. As far as reliability is concerned, the different teams observing job performers in the three Centers measured very consistent performance. Also, the historical measures on errors from the three different OE Centers were in substantial agreement. We think the operational definitions of the measures used are simple and direct, and have face validity. The teams practiced with the operational definitions and were able to get the same results with practice data observations.Champion: "Has the data collection provided consistent information throughout the data collection period?"

Answer: Yes. The period of actual observation was three weeks, where OE personnel were working under "normal" demand loads.

Champion: "Do the data collected provide the information needed?"

Answer: The data appear to show strong relationships to support our y = f (x) theories of the factors that were the chosen driving process performance metrics. Now we need to look for the root causes of the x factors.

Champion: "What is the baseline value for the data?"

Answer: The USL set for lead time in our project goals is 4 hours. The as-is process yield is less than 50 percent. The baseline sigma for OE errors causing "rejects" is very different for different products. If fact, the OE process is made up of at least 5 different processes that are running concurrently through the OE process steps. This complexity is a contributor to variation all by itself.

Figure 5-12 Summary of Measure Phase Tollgate Review

Champion's Approval, Support, and Encouragement

The Champion accepted the team's report on their Measure Phase work and approved them to proceed to the Analyze Phase to identify root causes. The Champion also complimented the team on their Quick Win, noting it had effectively trimmed one day from order lead time in the OE Center where it was piloted. The Champion approved the extension of the Quick Win trial to the other OE Centers, and planned to publicize the team's accelerated results by doing so. The team had been at work for a number of weeks now, and the Champion urged them to move quickly so as not to lose momentum. The team agreed to move as rapidly as they could in the Analyze Phase.

With that decision, the Champion noted that the team had reached its norming stage. The team evidenced a sense of group identity and cohesiveness. Members were comfortable sharing ideas and feelings, and giving and receiving feedback. They had a shared commitment to their assignment and their own goals. The Champion could assume more of a participative role.

After the Tollgate Review meeting, the Champion communicated to the manager of OE Center #1 to keep in place the process change eliminating the screening step. The Champion then communicated to the managers of the other two OE Centers to put the change in place, explaining why. The Champion also emphasized the importance of placing all screeners in equal or better jobs.

The Champion met later that week with the leadership team where the status and direction of multiple Six Sigma projects was reviewed. The leadership team, as a group, reviewed the Measure Phase plans and results for all projects to be certain the teams' measurements were relevant to the organization's stretch goals. The leadership team also reviewed all the Quick Wins teams had implemented, and published these to the rest of the organization to keep the organization informed and to encourage and promote future change.

Analyze Phase - First Team Meeting

In the Analyze Phase, the team had to identify the root causes of the measured OE process performance gaps. What could be causing the large amount of unwanted variation in order entry lead times and in error rates? How could they change or eliminate those underlying factors to reduce variation and improve process performance?

In Measure, the team had pulled the OE process apart into steps and looked at the contribution of each step to backlogs. In their first Analyze meeting, the team's agenda was to generate root cause ideas for each of the following x variables that they had identified as sources of unwanted variation. Prior to the meeting, each of three sub-teams had been assigned to prepare their ideas on one area.

1. What are the root causes of incoming variation in order volume between the last week of each month and the first and middle weeks? How can we change these root causes to eliminate unwanted variation?
2. What are the root causes of variation in completion rates for each step in the OE process and variation in the overall process lead times? How can we change these root causes to eliminate unwanted variation?
3. What are the root causes of variation in order errors in the OE process? How can we change these root causes to eliminate unwanted variation?

Incoming Order Volume Variation

The #1 sub-team reported that they had brainstormed possible root causes and developed two ideas, including: a) customers were restocking inventories on a monthly cycle, and b) salespeople were incented to get orders in at the end of each month. Sales quotas were measured monthly and quota periods closed on the last workday of each month. The sales organization constantly ran contests with prizes and bonuses, and the books always closed at the end of a month. The #1 sub-team was asked to validate these possible causes and report back to the whole team about ideas on how to change these drivers.

Variation in Completion Rates and Errors

The #2 and #3 sub-teams had banded together because they agreed that the underlying causes for both their process variables were the same and should be dealt with in a coordinated manner.

First, they identified "process complexity" as a root cause of both errors and long completion rates. Each step in order entry processed a mixture of six different order types:

1. Repeat Orders

and New Orders that were

1. Very Simple
2. Simple
3. Average
4. Complex
5. Very Complex

This made the nature of the job for the person working at each stage more complex. There had to be a change in "thinking mode" and a physical set-up time delay to get different resources each time the product type changed. In addition, the complexity of different order types made it impossible to use standard work procedures with every order. That kind of complexity, the Black Belt said, was a known cause of increased error rates in any manual job. The group believed that assigning OE people to work on only one type of product at a time would reduce errors and speed up completion rates.

This group reported that they had a second hypothesis about the causes of variable completion rates and errors for each stage in the OE process. While observing the steps in the process, team members had begun to question the value of many activities performed at each OE step. Their hypothesis was that at least half of the actual processing time spent at each step was wasted doing non value-added (NVA) work. For each product type, the group had taken a set of order entry forms, computer entry screens, and process outputs, changed them up to eliminate NVA items, and then reviewed them with each internal department in the company that was a customer of the OE process. They also determined that the paying customer, who never saw any of these internal documents, had no direct interest in these OE activities or outputs. The paying customer was interested only in getting an error free product in a short cycle time. The two sub-teams felt that they had come up with a validated, streamlined OE process for each order type that should be tested.

Finally, this group proposed to eliminate the holds between the four stages of the process (Spec Entry, Price Check and Job Planning, Inventory Check, and Manufacturing. Schedule) by having one person do all four stages of order processing without interruption. In the four days that an order was in the OE process, only 70 minutes of combined total live processing was done at all steps. Eliminating sequential step processing and running parallel non-stop order processing would eliminate all these holds. They noted that all OE personnel were already cross-trained to fill in each other's jobs to cover for vacations and absences and to handle backlogs at any stage.

Weekly Status Meeting

The team presented their list of ideas to the Champion at the weekly status meeting. The Champion told the group that they had generated a good list of creative ideas, and that their proposed changes sounded logical and could work. The Champion then asked the team to figure out how they could run experiments to test their theories and show if the theories worked and how much gain was possible. The next weekly meeting was set to present and review plans for experiments. Some Action Learning was required, since only the Black Belt team leader had been trained in designing experiments. The Champion delegated to the Green Belt the assignment to arrange for himself and the rest of the team to start training immediately, and requested that the Black Belt coach them as needed during the week.

By this time, the team had reached the performing state. The team was making most of its decisions independently. The team had attained a high degree of autonomy in pursuing their DMAIC tasks and was able to function without direct participation of the Champion. The Champion was able to delegate more assignments and responsibilities to the team.

After the status meeting, the Champion also reported to the rest of the organization the progress the team was making. The team members felt they were making a substantial contribution to the organization, and were excited about acquiring some new Six Sigma skills.

Experiments on Non-Value-Added Work and Complexity

The team successfully completed their training, and the Black Belt led them in designing the experiments to test their improvement ideas.

At this design review meeting, the team proposed their plans for the experiments. First, the team reported that the Black Belt had showed them how to simulate mathematically the effect of combining the separate stages of the process into one continuous job activity. The team then presented these results, concluding that in normal mid-month weeks where volume averaged less than 90 orders per day, the OE Center running at their current completion rates could make a breakthrough gain, reducing order processing lead time from four days to four hours. Everyone, including the Champion, agreed that without further testing, that change should be considered for implementation in the Improve Phase. The tough issue for this change was going to be how to implement it.

Next, the team addressed how they could test their combined theories about reducing all the individual stage completion rates and also reducing order entry error rates. The experiment was proposed to test both the effect of eliminating NVA work and removing complexity from the process at the same time. The team had looked running two experiments to test the "NVA" and "complexity" theories separately, but the time and cost of separate tests looked greater than the benefit. The team felt that the combined gain of eliminating these two root causes would be greater than the sum of the individual gains, and could safely be combined in one test.

The team proposed to run the experiment with all process stages combined into one continuous job. The test would include a test group for each order type and a control group that ran the orders for the different types of products in a mixed sequence. Each of the six single-order-type test groups and the one mixed-order control group would be further divided into two sub-groups. One of these sub-groups would run the original

Design of NVA and Complexity Experiment	Run Current Process Tasks for all 4 stages of OE	Run Streamlined Process Tasks for all 4 stages of OE
Repeat Orders	2 people each run 20 Repeat orders	2 different people each run same 20 repeat orders
Very Simple Product	2 people each run 20 Very Simple orders	2 different people each run same 20 Very Simple orders
Simple Product	2 people each run 20 Simple orders	2 different people each run same 20 Simple orders
Average Product	2 people each run 20 Average orders	2 different people each run same 20 Average orders
Complex Product	2 people each run 20 Complex orders	2 different people each run same 20 Complex orders
Very Complex Product	2 people each run 20 Very Complex orders	2 different people each run same 20 Very Complex orders
All Product Types Mixed	6 people each run 20 mixed orders randomly selected from other groups	6 different people each run same 20 orders randomly selected from other groups

Figure 5-13 Plan for Designed Experiment

process tasks, and one would run the streamlined process tasks that had all NVA work removed. This made 14 total groups. Current OE personnel would be randomly assigned to each group to run test orders. They would process 20 orders using the current method and the streamlined method that the sub-group had prepared.

The team laid out their null and alternative hypotheses. The first null hypothesis (the hypothesis to be disproved) was that the streamlined subgroups' method would not produce faster completion rates and fewer mistakes than the subgroups using the current process. The first alternative hypothesis (the hypothesis to be proved) was that the streamlined subgroups' method would produce significantly faster completion rates, on the order of twice as fast, and would make significantly fewer mistakes. The second null hypothesis was that the groups running single product types would not show faster completion rates or lower error rates than the mixed product control groups. The second alternative hypothesis was that the groups running single product types would show significantly faster completion rates and significantly lower error rates than the other single product groups.

The Black Belt assured the Champion that the experimental design was sound. The Black Belt and Green Belt would run an Analysis of Variance (ANOVA) test to test the statistical significance of the lead time results of the experiment and a Chi Square test to test the differences in the probability of getting errors on orders run by the various groups.

The experiment's design called for training personnel in the new method before starting test runs. The Champion agreed to have OE personnel run the orders. The IT analyst had gotten the commitment from the IT department to build a working, mock up interface of the OE computer system that would match the streamlined changes proposed by the team. The printing plants agreed to print the new paper forms that were needed for the test. The team's implementation plan and schedule for the experiment were approved by the Champion.

Validation of Hypotheses about Root Causes

Incoming Order Volume

Sub-team #1 reported that a survey of customers showed that it was not customer restocking patterns that were driving the end-of-month bulge in incoming orders. In meeting with top-level company sales executives, the team Champion and sub-team #1 learned that the quota closing periods and special incentives were used to encourage sales to get last minute orders into Order Entry. They negotiated a solution plan where the sales force was divided into four groups. Each group was assigned a different week in which their quota period closed. The Vice President of Sales approved the implementation of this change.

Completion Rate Results of NVA and Complexity Experiment (minutes per order)	**Run Current Process Tasks for all 4 stages of OE**	**Run Streamlined Process Tasks for all 4 stages of OE**	**Grand Means Single-Type versus Mixed-Type**
Repeat Orders	Avg. = 21 minutes	Avg. = 20 minutes	GM = 44.0 minutes
Very Simple Product	Avg. = 40 minutes	Avg. = 25 minutes	GM = 44.0 minutes
Simple Product	Avg. = 46 minutes	Avg. = 28 minutes	
Average Product	Avg. = 65 minutes	Avg. = 32 minutes	
Complex Product	Avg. = 79 minutes	Avg. = 38 minutes	
Very Complex Product	Avg. = 90 minutes	Avg. = 44 minutes	
All Product Types Mixed	Avg. = 70 minutes	Avg. = 35 minutes	GM = 52.5 minutes
Grand Means Current versus Streamlined Process Tasks	GM = 61.22 minutes	GM = 32.4 minutes	

Figure 5-14 ANOVA Test Data

The ANOVA results showed that the streamlined process groups were significantly faster in completion rates by a margin of 32.4 minutes (Streamlined Process) to 61.22 minutes (Current Process). Also the overall average completion rates for all single-type product-processing groups (44.0 minutes) were measurably faster than the mixed-type processing groups (52.5 minutes). The team concluded that they had validated two root causes of lead time variation in order processing.

Error Rate Results of NVA and Complexity Experiment (number of orders with rejectable error)	Run Current Process Tasks for all 4 stages of OE	Run Streamlined Process Tasks for all 4 stages of OE
Repeat Orders	0 rejectable error*	0 rejectable error*
Very Simple Product	0 rejectable error*	0 rejectable error*
Simple Product	1 rejectable error*	0 rejectable error*
Average Product	1 rejectable error*	0 rejectable error*
Complex Product	1 rejectable error*	0 rejectable error*
Very Complex Product	2 rejectable errors*	1 rejectable error*
All Product Types Mixed	12 rejectable errors**	4 rejectable errors**

*out of 40 orders **out of 120 orders

Figure 5-15 Chi-Square Test Data

The results of the Chi Square tests showed that there was a significantly smaller chance of an order error occurring in the streamlined process groups versus the current process groups. The team observed less errors with the single-type product groups versus the mixed-type product groups. The team concluded that they had validated two root causes of order error variation in the OE process.

Analyze Phase Tollgate Review Meeting

The team members were satisfied that they had met the requirements of the DMAIC Analyze Phase. So, a meeting with the Champion was held to review the team's deliverables. The Champion reviewed all the results of the team's experiments, and then used the Tollgate Review questions to make sure they had covered all the bases. This is a summary of that part of the meeting with questions from the Champion and answers from the team:

Champion: "Have then been any revisions to the charter? Has the scope changed?"

Answer: We believe that the goal of 4 hours for order processing is now attainable if the changes we approve are implemented. The goal of a 5 times error reduction in OE can't really be predicted at this point because the team has reduced the number of opportunities for error differently for different product types. However, we can say that there will be a significant reduction in the number of orders with errors, at least 4 fold.

Champion: "What was the approach to analyzing the data? Why were these tools chosen? What worked well/did not work well about these tools?"

Answer: Both the experiment and the math simulation were chosen because they were the least cost methods of testing our hypotheses. We want to develop a better measure of order entry process quality (fewer errors).

Champion: "What are the root causes of the problems? How were these conclusions drawn?"

Answer: The root causes of process performance problems identified were unnecessary process complexity (mixing order types in the same processing stream) and a great deal of non value-added work that had evolved into the system over 25 years. The root cause of the input problem, variability in volume of orders, appears to be sales quota schedules. The variation should decrease greatly in the next few months if the hypothesis is correct.

Champion: "How did the team analyze the data to identify the factors that account for variation in the process?"

Answer: We used a factorial experiment and analyzed our data with ANOVA and Chi Square tests.

Champion: "Now that you know how to fix root causes, what is the strategic opportunity represented by making the fixes? What is the impact on customer satisfaction, retention, and loyalty?"

Answer: Addressing these OE problems will allow customers to depend on shorter order cycle times and control their own JIT inventories more closely. Reduced errors will prevent customer dissatisfaction and improve retention of key large customers. The company should experience a reduced cost of order entry per product due to a reduction in WIP orders and increased productivity per OE worker.

Figure 5-16 Summary of Analyze Phase Tollgate Review

Champion's Approval, Support, and Encouragement

The Champion accepted the team's report on their Analyze Phase work and approved them to proceed to the Improve Phase. The Champion reported to leadership, as well as to the rest of the organization, the great progress the team had made. The Champion also started working closely with the key stakeholders in the OE Centers to begin to prepare them for change. The OE Center personnel who had participated in the team's experiment, particularly those groups who focused on single products and used the streamlined process, were enthusiastic about the coming changes. They said they felt it had made their work much easier and less stressful, and they felt good about being more productive. The Champion enlisted these individuals to champion the improvements the team was poised to plan making.

Improve Phase - First Team Meeting

The team conducted its first meeting of the Improve Phase. The agenda was to plan the implementation phase, and begin to finalize the solution they would recommend to the leadership team in a formal presentation. The Black Belt outlined the way they would want to structure the presentation of their recommended solution.

- Review their Team Charter
- Review the flowchart of the as-is process and its measured performance gaps versus customer and business requirements
- Present the root causes of the gaps
- Describe how the team arrived at their final recommended solution
- Present their solution, with should-be process maps and strategic benefits
- Summarize the implementation plan
- Summarize the risk management strategy
- Summarize the process control strategy
- Summarize the culture change strategy
- Questions and discussion

The Black Belt pointed out that all these points had to be considered when they formulated their final recommendations. Then the team was turned loose to begin solidifying what their solution would look like.

Solution Options

Implementing the improvements that the team had identified involved changes in job definitions, changes in staffing levels, changes in the OE computer system, changes in the physical workflow, and changes in the OE Center office layout. As they thought about the complexity of the implementation, the team decided that they had to run a pilot at only one of the three OE Centers, versus starting with a full rollout to all three Centers. If they ran into unplanned problems they would not shut down the company's ability to manufacture orders.

Team members threw out a surprising number of ways to implement their fundamental changes. For example, they discussed options on whether people should become specialists at one type of product or rotate between product types. For competing solution options, they weighed the positive and negatives.

Solution Options	Positives	Negatives
Specialist	• Higher skill, accuracy, and speed • Simpler training requirements	• Loss of knowledge of other products • Less variety in job
Rotating	• Flexible assignments to handle variations in volumes of different types of products	• Increased error rates, especially at the beginning of rotations

Figure 5-17 Weighing Solution Options

Ultimately, they chose a cell model, where each of six cells was dedicated to one product type. They discussed how to sort and distribute the order types to the cells and settled on a pull system, where any member of each cell who was not busy would go pull orders for the entire cell. Cells would be responsible for managing and improving their own process after implementation. People would rotate through cells, but no more than half the people in a cell could be newly introduced to the cell at the same time.

They used a mathematical model that simulated the flow through the process, given the completion rates they had obtained in their experimental runs. This allowed them to estimate the staffing levels for each cell that would be required to process projected incoming volume for each type of product. They also assessed options for changing the OE computer system to fit their streamlined process design. The method they used to evaluate these options was a cost/benefit analysis, developed in partnership with the IT department.

With great care and total agreement, the team settled on an integrated solution that they all could support.

Champion Status Review

Before the team went any farther with the implementation plan, they wanted to be assured that the Champion accepted and would support the solution they had finally settled on. The presented their final solution idea, and the Champion approved the plan. Then it was time to figure out how to "sell" their solution to the people who would run the process and the people who would be impacted by the changes.

Selling the Solution

The team held another meeting, and started with a stakeholder analysis. First, they identified all the people who were stakeholders. Then they listed possible concerns of these groups that needed to be addressed up front, and positive changes and benefits these stakeholder groups could expect.

Stakeholder Group	Anticipated Concerns	Potential Benefits
Press Department Finishing Department Shipping Department	Changed, less detailed manufacturing instrutions means more training for operators	Easier to use instructions with focus on only important specifications simplifies jobs
OE Center Job Incumbents	Uncomfortable with new team-based process responsibilityWill I lose my job?Will I be able to do this new job?	Will be more productive and valuable to the companyThe job will be less complex and therefore less stressfulNew tools will make risk of error much less

OE Center Management	Will my operation be disrupted? For how long? What are the risks of major problems?	Higher per employee productivityLower error rates in OEIncreased job satisfaction and lower turnover
Printing Plant Management	Will my operation be disrupted? For how long? What are the risks of major problems?	Shorter OE cycle times mean fewer disruptive and costly emergency order expedites in the plant Fewer OE errors mean fewer re-runs of rejected orders that cut into plant profitability

Figure 5-18 Stakeholder Analysis

With a lengthy stakeholder analysis in hand, the team assigned people to plan communications and open communication channels with all stakeholder groups. Key representatives from these stakeholder groups were engaged as advisors in preparing the details of the implementation plan that impacted their organizations.

Champion Support

While the team was working on the communications plan, the **Champion** assisted in obtaining the support of key stakeholders. The **Champion** also carried the message of the need for change and buy-in to the implementation plan up through the leadership group.

Cost/Benefit Analysis

Using the description of their solution, the team sought the help of the accounting department to help estimate the costs of both implementing the solution and continuing to operate it. The main cost elements were the training of OE personnel and plant customer department personnel, and the implementation of the changed OE computer system. The team also the estimated financial gain in terms of the reduced Cost of Poor Quality (COPQ) and the reduced cost of entering an order that would result from the changes. Using these numbers, they prepared a summary cost/benefit analysis for presentation to the leadership group.

Potential Problem Analysis

The Black Belt now told the team that their last major task before finalizing their recommendation and presenting it to the leadership was to identify and plan for potential problems. Employing the *Action Learning* methodology, the Black Belt demonstrated and coached the team on several tools.

They began with a Risk Assessment Matrix. Here the team brainstormed to identify all possible risks. Then they rated and rank-ordered these risks.

Risk Description	Severity of Impact Low 1 - High 5	Probability of Occurrence Low 1 - High 5	Weighted Risk Value and Rank
Computer system major bugs	5	1	Value = 5 Rank = Third
Incoming demand will not be leveled by changes in quota periods	3	2	Value = 6 Rank = Second
Inadvertently eliminated some critical order entry information from the streamlined system that will cause problems	3	4	Value = 12 Rank = First
New types of errors will be introduced by changes in the process	3	1	Value = 3 Rank = Fourth

Figure 5-19 Risk Assessment Matrix

For each risk, the team looked for ways to control or mitigate the impact using a risk control analysis matrix.

Risk	Control	In Place	Capable	Action Required
1. Critical OE information eliminated	Test system on real orders	No	Yes	Manual test of new. process on a sample of real orders
2. Incoming order volume does not level out	Track incoming volume closely	Yes	Yes	Adjust staffing levels
3. System bugs	Software testing	Yes	Yes	System testing with real orders by users during development.Run dual system for test period
4. New errors introduced	FMEAPoka-yoke	Yes	Yes	Do FMEA analysis on process for possible errors and poka-yoke to error proof

Figure 5-20 Risk Control Analysis Matrix

Improve Phase Tollgate Review Meeting

The team felt it was ready to make its presentation to leadership. The Champion conducted a Tollgate Review meeting with the team before they made their presentation to the entire leadership team. The team practiced their presentation. The Champion asked many questions.

Champion: "What criteria were used to evaluate the potential solutions? How does the preferred solution address the root causes of the process performance problem?"

Answer: The team generated several ways to implement the needed changes. The final solution was chosen as least costly and most acceptable to the people who will operate it. The team-operated single-product cells will own and operate an order entry process for each type of product. The streamlined process will reduce errors, completion rates, and total OE lead time to the promised levels.

Champion: "Did you conduct a cost/benefit analysis? What assumptions were made? Did a financial subject matter expert validate the cost/benefit analysis?"

Answer: Yes. The implementation plan cost/benefit analysis and the long term operating *pro forma* were developed in partnership with the accounting department.

Champion: "How will the team answer the stakeholders' "What's In It For Me" question? What can be done to mobilize their support? How is this reflected in the communication plan for implementation?"

Answer: The concerns and potential benefits for each stakeholder group were identified, and a targeted communications plan was developed for each group.

Champion: "What training is required to ensure the people affected will be able to support the new process design with minimal frustration and maximum preparedness?"

Answer: Training objectives are being developed for each group, and the implementation plan includes time for both classroom and hands on training for all OE personnel and users of OE documents and output information in other departments. Champion: "What are the risks associated with the rollout? How are they being controlled?" Answer: We will pilot the new solution in only one OE Center for a six-week period. Then the other two OE Centers will cut over. A list of risks was generated and control mechanisms and contingency plans were put in place.

Figure 5-21 Summary of Analyze Phase Tollgate Review

Governance and Acceleration

The Champion approved the team's presentation to leadership. The Champion then arranged a time for the team to make their presentation to leadership. The team took care of all of the other logistics, and presented their recommendations to leadership at the appointed time. Leadership approved the implementation of the pilot as soon as the team could possibly start. Everyone was anxious to see the results.

Control Phase - First Team Meeting

After the team received approval from the leadership team to implement their solution, they met to plan the Control Phase of the project. They had been working together for four months - the project was actually ahead of schedule thanks in part to the acceleration activities of the Champion. The team shared a big sense of accomplishment at this meeting. The Champion, who attended the meeting, went around and acknowledged each individual's special contributions and proposed a team party for members and families that would be paid for by the company.

Their Black Belt leader reminded everyone that real success would not happen until their pilot was successful and their solution could be fully implemented across all three OE Centers. They had to get the implementations up and turned over to the owners to run and maintain. Then, all lessons learned had to be documented and shared so beneficial improvements could be understood and possibly duplicated in other processes through the company. The Black Belt emphasized that the purpose of this last process improvement phase was to guarantee successful process performance that would endure after the team was finished and disbanded.

The team broke into sub-teams to deal with the different elements of the implementation. There was a group focused on the OE computer system development, a group focused on training, communications, and change management, and a group that focused on the final development of tools, poka-yoke, and documenting the streamlined processes. The team leader kept the overall schedule and coordinated all the activities. The Champion made sure that the team got the cooperation they needed from other departments and kept the leadership team appraised of the progress.

Process Control Plan and Results of Pilot

The pilot implementation took six weeks. While planning for the pilot, the team worked on a process control plan that they could use for the pilot, and that the owners of the streamlined processes could also use going forward. They treated each product-type as a separate process in their control plans. Each cell regularly monitored a sample of orders going through their process. The project established the protocols for the cells to follow in gathering and charting x and y data for process control.

Lead Time Control Charting

During the pilot, the team measured and control charted the lead time for each cell. Applying *Action Learning* to process workers, the team also taught the cell members to chart these results themselves.

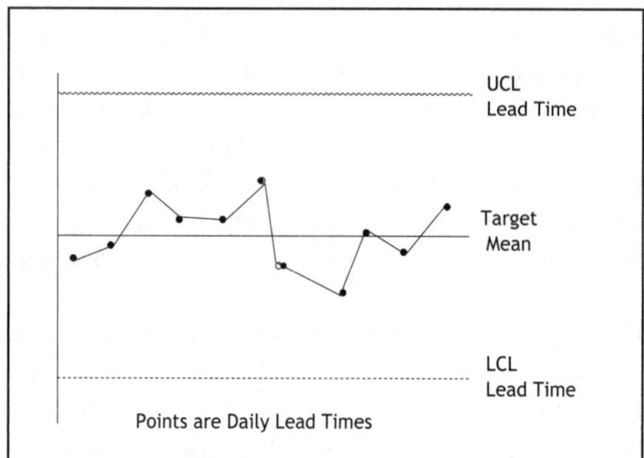

Figure 5-22 OE Lead Time Control Chart - Cell #1

Along with the lead time (a y-variable), the team measured and tracked two related leading x-variables: completion rates for each cell and incoming demand rate of orders to the cell. The incoming demand stabilized over the pilot period of six weeks, so that the end-of-month bulge in incoming orders was reduced from 120 per day to 90 per day. That was within the control limits of the new process as long as the completion rates also stayed within the control limits.

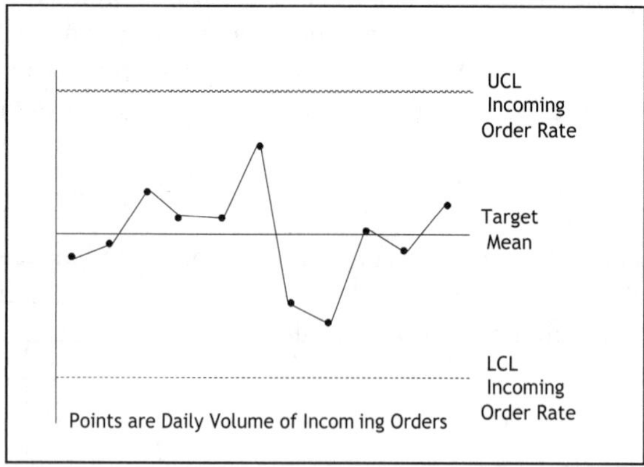

Figure 5-23 OE Demand Control Chart - Cell #1

The completion rates for the cells became remarkably stable and the team was able to measure that completion rates and lead times were controlled at six sigma levels over the six-week pilot.

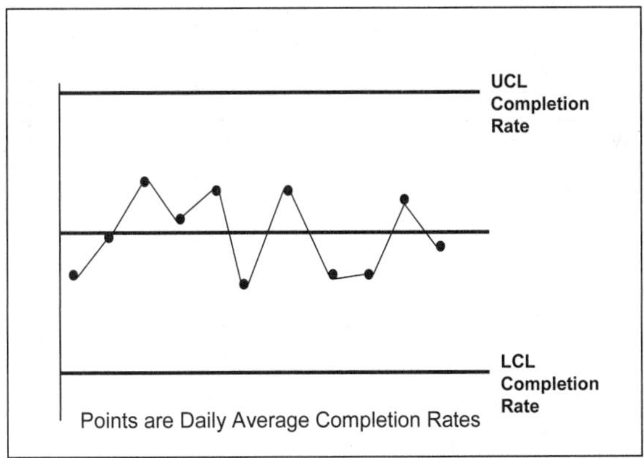

Figure 5-24 Order Completion Rate Control Chart - Cell #1

Controlling Errors in the Streamlined Process

During the pilot, the team tracked order errors for each cell. They counted errors on orders that caused the order to be rejected (reworked or sent back) by printing plant departments or caused actual defects in products that were caught by the customer or someone internally. They also tracked and Pareto charted the sources of all errors made in each cell. After six weeks of tracking, the team showed that all product types except Complex and Very Complex had attained six sigma levels of quality.

Pilot OE Center #1 Errors on OrdersThe Streamlined Order Entry Processes		
Processes Measured	**DPMO**	**Sigma**
Cell #1 Process (Repeat Order)	2.0	6.0
New Orders		
Cell #2 Process (Very Simple Product)	3.0	6.0
Cell #3 Process (Simple Product)	3.4	6.0
Cell #4 Process (Average Product)	3.4	6.0
Cell #5 Process (Complex Product)	150	5.1
Cell #6 Process (Very Complex Product)	200	5.0

Breakout of New Orders by Time of Month		
First Week (First 5 Days)	3.4	6.0
Middle Week (Middle 5 Days)	3.2	6.0
Last Week (Last 5 Days)	3.4	6.0
Breakout of Repeat Orders by Time of Month		
First Week (First 5 Days)	1	6.0
Middle Week (Middle 5 Days)	2	6.0
Last Week (Last 5 Days)	2	6.0

Figure 5-25 Sigma Levels by Product Type

The team identified and classified the errors that did occur, and set the new cell teams to work at fixing the root causes of these types of errors. All cell teams were trained to track their own error rates on control charts and watch for trends and patterns.

Implementation Plan

As the pilot approached the end, the team updated their process flow charts, the process procedures, and the training materials that were being used. The new OE computer system proved bug free. No other substantial problems materialized. People operating the cells were actually excited about the new way their jobs worked and the success they could see.

As soon as the pilot operation became stable in its third week, the team presented a company-wide implementation plan to the Champion. They wanted to roll out the improvements to the other two OE Centers one at a time. The implementation plan called for bringing key people from one of the other OE Centers to get hands-on experience in the last week of the pilot. The selected OE Center would send one team leader from each cell, and a representative from each management function. While that was happening, the people who remained behind would be trained and participate in the installation and setup of the new system in their OE Center. The remaining OE Center and the pilot OE Center would pick up the extra order entry load during that cut over week. After four weeks of running the second OE Center, the last OE Center would be cut over using the same implementation model. The Champion approved this plan.

Team Evaluation

The team's job was almost done. The Black Belt brought them together to do a "lessons learned" exercise.

The team did a self-evaluation. The team reviewed what they had done well, in addition to problems that they had experienced at each of the five phases of the DMAIC method. The covered many issues, including:

- Were meetings run with to the point, effective agendas?
- Did sub-teams always perform their assignments?
- Did all team members make useful contributions? Did everyone stay involved all the way through the process?
- Did the team manage disruptive behaviors?
- Was the DMAIC methodology applied throughout the project?
- Did the team make data-based decisions? What were those key decisions?
- Is the process fully understood?
- How well did the team meet its goals as stated in their Charter?

After this group evaluation, the leader asked everyone to do a peer evaluation of other team members and of the Black Belt himself. This was a structured checklist evaluation that covered contributions to the solution and interpersonal and team relations. Later that week, the Black Belt ran a one-on-one feedback session for each team member.

Recognition and Reward

The leadership team invited the entire project team to an awards banquet for this team and two other teams that had been operating in other areas of the business. All individuals on the teams received recognition awards, and a few team members received special awards for special efforts.

Prior to the final Tollgate Review meeting, the Champion met with each member of the team and gave each his new assignments. The OE Supervisor was promoted to a new job as Manager of Six Sigma Operations for Manufacturing and scheduled to go to Black Belt training. The Screener, whose job had been eliminated, was promoted to be the OE Supervisor. The other team members who were from the original OE Center were given temporary jobs as cell trainers for all three OE Centers.

They were all promoted to Senior OE Specialist job levels. The IT analyst was promoted by her boss, the IT Systems Manager, and put in charge of the OE computer system. And finally, the printing plant representative was promoted to a senior manufacturing technician job.

Control Phase Tollgate Review Meeting

The Champion met with the team for their final Tollgate Review. At the beginning of the meeting, the Champion expressed satisfaction with the results so far, and was pleased that the team had finished the entire project in slightly under the six months they had initially projected. The Champion had this wrap-up conversation with the team.

Champion: "How will the implementation plan for the other OE Centers be monitored to ensure its success? Who is accountable?"

Answer: The OE Center managers have assumed responsibility for the implementation of the solution in their Centers.

Champion: "I am promoting Sam from OE Center Supervisor to the job of Six Sigma Operations Management for all Manufacturing including Order Entry. After Sam is finished monitoring this implementation, he is going to Black Belt school."

Champion: "How will the responsibility for continued review be transferred from this team to these OE Center managers and the cell teams? What is the review process?"

Answer: Management and cell leaders will observe and be trained on the process control and review process used in the pilot OE Center. OE Center supervisors will be taught to review status with the cells on a daily basis and report weekly status to the Center manager.

Champion: "What is being measured? What evidence does the team have that would indicate the process is "in control"? How well and consistently is the process performing?"

Answer: The pilot processes are operating at six sigma for lead time, and at six sigma for order errors except for complex and very complex processes. The out-of-control plan in already in action for those processes.

Champion: "How has the process been standardized? Have the process changes been documented?"

Answer: All six streamlined processes are now standardized in terms of work steps, tools, and resources. The team did poka-yoke error proofing and taught the cell members to use it too. Updated process documentation is ready.

Champion: "How has the training been improved from the lessons learned in the pilot? How is the effectiveness of the training being measured?"

Answer: Training is blended with performance on the job and the process results are the final measure of the power of the training. There are tests to measure that people do understand the new process as well as their roles in continuous process improvement.

Champion: "What are the current attitudes of the people in the other two OE Centers?"

Answer: People remain concerned that the increased productivity will eventually lead to a loss of some jobs.

Champion: "The sales force is already promoting our new shortened order cycle capabilities to targeted large customers. The promotion announcement will hit the OE Centers and Plants next week. The sales force is going through Six Sigma Foundations training and we have Six Sigma projects to improve sales performance. We are letting all the Plants and OE Centers know that we are expecting increases in order volume in the next month."

Champion: "What other barriers do you see to successful change in the OE Centers?"

Answer: After several weeks of the pilot, some of cell members have asked to be transferred to jobs that are not team based. The people say they are not comfortable with team process and the disagreement that it generates. Twenty percent of the people in cells have been counseled about this and agreed to stick it out for at least a quarter. We think this problem bears watching, but will become less common as the teams mature and people adjust to the change.

Champion: "What gains or benefits have been realized from the pilot implementation? How can the improvements be replicated elsewhere in the organization?"

Answer: The immediate productivity and quality improvements are a result of eliminating unnecessary work that crept into our processes over many years. The strategy of eliminating all work that is non value-added could be applied to all of our business processes. Secondly, we let our OE process become terribly complex. It was never one process. It was several processes lumped into a series of steps. The tasks to be performed for each product type at each step were quite different. We suspect that other so-called single processes in our company suffer from the same complexity problem.

Champion: "As a team, what did you learn from this DMAIC problem-solving experience?" Answer: We learned that we could do things as a team that we could not accomplish as individuals. After we learned to trust each other we became much more efficient and effective. For a while it seems like we might just blow apart. But keeping focused on our goals and making progress help give us confidence. It is hard to believe that our job as a team is ending. It has been fun, and a challenge individually as well as to the team.

Figure 5-26 Summary of Analyze Phase Tollgate Review

Summary

The organization continued to apply the *Six Sigma Management Process* to its organization in its drive for continuous improvement. Leadership eventually realized breakthrough results throughout the organization through its align, mobilize, accelerate, and govern activities and the rigorous work of the DMAIC continuous process improvement teams.

By the end of 18 months, the organization had developed 10 Black Belts and 25 Green Belts to sustain their initiative. The leadership had defined six core business processes where Six Sigma projects were focused:

- Critical (Large Contract) Customer Partnership Process
- Critical (Large Contract) New Customer Acquisition Process
- Integrated New Market/Product Development Process
- Emerging Manufacturing Technology Identification and Application Process
- Streamlined Pressroom (Printing) Process
- Key Supplier (Paper and Ink) Partnership Process

Six Sigma process performance improvement teams had completed 10 projects and another 10 projects were underway. In their annual report, the organization attributed the following results to the campaign and the *Six Sigma Management System*.

- $25 million increase in revenue from new customers and sales of higher value products
- $30 million in cost savings from reduced COPQ and per order processing and manufacturing costs
- 25% reduction in turnover of large customers
- $15 million in EBIT directly attributable to Six Sigma project improvements

The entire organization came to understand and see the power of the *Six Sigma Management System*.

Part Two:

Six Sigma and Lean

This part of the *Six Sigma Black Belt Handbook* includes just one chapter. It focuses on how two powerful improvement methodologies, Lean and Six Sigma, can complement one another and strengthen an organization's performance improvement efforts. The chapter begins with a quick overview of quality improvement methodologies used for the last twenty years. They all express a common set of themes. The chapter then goes on to review Lean as a business strategy and describes how that may differ from traditional thinking. Next, it describes some of the foundation elements of any successful Lean initiative, both in traditional manufacturing operations as well as in transactional settings. Applying Lean in transactional environments is somewhat different than its application in manufacturing. So a few case studies/examples highlight typical opportunities and actual results.

Next, this chapter provides an outline on how to implement a Lean enterprise or manufacturing initiative. Several different rollout methods are shared. The DMAIC model is used as a framework for describing typical implementation steps.

Finally, this chapter closes with a number of examples of Lean projects and a suggestion on where Lean is likely to evolve over the next several years.

Chapter 6

Six Sigma and Lean

It's interesting how much things change, yet as time goes stay the same. People interested in performance and quality improvement have been having the same arguments for a long time. Much of the discussion revolves around, "Which improvement tool is best?" or "This is the latest improvement tool, so it must be the greatest!"

It might be surprising to know that Lean has been around for at least 100 years. It wasn't called Lean in the beginning. In the early 1900s, industrial engineers developed methods of "mass production" geared to low-cost manufacturing of things like cars and appliances, which previously had been out of reach to the average consumer. The Ford Rouge plant was a model in Toyota's early benchmarking. A wave of new concepts, including Lean, were (re)introduced in the 1980s when high interest rates, increased demand for product quality, a global recession, and global competition converged to require renewed emphasis on operating fundamentals that had been under-utilized for decades.

Many different approaches have been taken to improve business performance over the last several decades. Like anything in life, each approach has its own set of strengths and weaknesses. Each one also offers a slightly different perspective on how to improve performance. Some of these different perspectives may be beneficial to organizations trying to implement broad based improvement efforts. So before digging more deeply into Lean, consider the major improvement initiatives over the last 20 years:

Figure 6-1 Major Improvement Initiatives

A Brief History of Performance Improvement Initiatives

Quality Circles - Involve your people. They know!

Most companies gained benefit from Quality Circles, however, the literal translations by a team from Lockheed were somewhat misguided. They did not understand how Circles fit within the larger process, though the basic idea of involvement is on target

Total Quality Management (TQM term coined by Armand Fiegenbaum)

Dr. W. Edwards Deming, Dr. Joseph Juran, and Phil Crosby, brought forth the idea that high quality is actually cheaper, not more expensive. For the most part, organizations derived some good from TQM. Crosby defined quality as "meets requirements." Japan's Quality Award is called the Deming Prize. Juran wrote "The Quality Handbook" and really promoted the use of team activities for improvement. Most organizations did this programmatically and hoped that once the program was implemented, they could move on to other things.

Cost of Quality - Quality is Free! (Crosby)

There are some great concepts within this. Identify your cost of conformance and non-conformance. Costs are typically broken into categories: appraisal, prevention, internal failure, and external failure. The quality costs increase for each successive classification. Years ago, ASQC published a study that stated a $1 investment in prevention avoided $16 in cost for failure in the field. The cost of poor quality is ultimately measured by lost customer bids, declining market share, and declining profit. People went a little crazy with this and actually installed a second accounting system to track it.

Customer Quality Management - TQM with More Customer Focus

This did help people to see that customer requirements exist at several levels (specifications, expectations, and delighters). It also helped people realize that multiple customer groups existed and multiple internal departments have some conflicting requirements versus other constituencies (e.g., purchasing, engineering and operations).

Statistical Process Control, or SPC (W. Edward Deming)

Deming gave the world a process centric view of the world - the whole concept of reducing "variability." These ideas were well known by statisticians, but like a new science to the management world. Good things came from it, but it was never totally embraced by general management. Perhaps it was too technical, and not sexy enough.

Reengineering - Truly Cross-Functional Process Improvement

Hammer got the fame, but many people were practicing this prior to his book - definitely a sexy title. Hammer took the idea of processes and applied them to the whole business. Deming talked about processes, but Hammer gave a clearer, easier to see picture of how these concepts applied to overall business practices. This really sounded cool and was embraced by many organizations. Unfortunately, much of the focus was on getting smaller (reducing cost, not business growth), and management's support systems to sustain change (communication, accountability, measurement, etc.) were largely ignored.

Six Sigma - Started at Motorola in the 1980s

Six Sigma evolved into a hot, sexy, new tool when GE worked with Motorola to implement Six Sigma. Jack Welch's strong advocacy definitely popularized and advanced Six Sigma. It is a great tool for looking at complex process interactions and really takes SPC to the next level of impact - this is SPC with a Goal. Six Sigma is where "Green Belts" and "Black Belts" were spawned. Today, many self-appointed training and certification groups exist, so the phrase "Buyer beware" certainly applies.

Theory of Constraints (Eliyahu Goldratt)

At the risk of over simplification, Theory of Constraints is largely used to find and eliminate the bottlenecks to increase capacity. This is a cool thing to do if one can use the new capacity. The basic argument is that improvements in non-bottleneck processes do not increase throughput and, therefore, are generally not good projects to work. However, improvements to non-bottleneck areas may reduce costs, so they are not all bad. A good book, "The Goal," was a great story but was not widely embraced by the management world. This improvement approach was largely viewed as manufacturing approach, although the Jonah's are trying to expand it now into a larger process perspective. Opportunities exist to apply this concept in engineering and the sales ordering process, as well as other areas.

Lean Manufacturing, Lean Thinking (Womack and Jones)

Womack and Jones are now trying to foster the concept of the "Lean enterprise" rather than just the production floor, but it could be difficult to regroup though after the horse is out of the barn. Lean is based on the Toyota Production System visual approach, use of *Kaizen* teams, continuous movement (flow), elimination of waste as defined by Ohno and Shingeo. Womack and Jones took Richard Schonberger's work from the 1980s and expanded on the use of production cells and Lean flows.

Other Tools/Techniques

A myriad of other improvement techniques have certainly bounced around the periphery:

- Quality Function Deployment (QFD)
- *Kaizen* as a stand alone
- Time Management (Stalk & Covey)
- Hoshin Planning (i.e., manage the exceptions a little more closely),
- Postponement Strategies
- Managing the White Space
- Visual Controls
- 5S
- And many others

Status of Performance Improvement Initiatives

So what does it all mean? All of these approaches focus on a few basic concepts. The basic intent of most of the above initiatives is to do work:

1. better
2. faster
3. cheaper, and
4. to create more meaning in our work.

Perhaps a fifth element could be added today regarding environmental consciousness - to do no harm to the environment (Natural Capitalism; Hawken, A. Lovins, & L. Lovins - Little Brown & Company, 1999). Most companies/organizations say they want to do these five things. However, only a few really focus on the fourth and fifth elements.

People trash these initiatives once they are replaced by the next new thing, but think about it. Other than for nostalgic purposes, would you rather buy a car made in 1975 or 2005 for everyday driving? Would most cars manufactured in the 1970s drive further than 100,000 miles? Probably not! The expression at the time was, "built in obsolescence." And the amazing thing is people accepted it.

Companies have taken the above initiatives and truly remade their organizations. There is no comparison between today's car and yesterday's in terms of quality, durability, reliability, etc. The same is true in many other industries. The frustrating part of this story is all of the competitors did the same thing, and some of them probably did it better! So companies then begin a search for the next new thing in order to get a jump on the competition. So the cycle continues.

People may poke fun at the various improvement programs that have come and gone over the last 20 years. But the facts are: companies have gotten much better, competition has gotten tougher, the marketplace is more global, customers are very demanding, and the game continues to increase in difficulty.

A Motorola executive once said years ago that if they could go back and only focus on one thing it would be "time." It makes sense, although it can't be that myopic. People can take any one of the above tools and implement them to such an extent that they are doing counter-productive things and contributing more problems than solutions. So every tool needs a sense of balance. The other idea is to stop compartmentalizing the basic concepts.

As Deming, Juran, Russ Ackoff, and countless others have indicated, everything is part of a larger process. It all has to work together - whether it is the supply chain or delivering a meal in a restaurant. It is not possible optimize every subcomponent individually and have the process work. The pieces of a process all need to work together. All of the improvement tools described in this chapter offer different perspectives. Each has its own respective strengths and weaknesses.

If there was only one way to do business improvement, everyone would do it that way. Fortunately, there is not just one way. There are trade-offs for every path. There is no single answer, but there are a lot of good ideas.

Six Sigma and Lean

Some people ask which is better - Lean or Six Sigma? That is not the right question. Is a hammer better than a screwdriver? It depends on the application. The same is true with business improvement methodologies. They all have their benefits. If someone uses a screwdriver as a hammer, it will never be an effective usage of the tool. The nail may still go into the board, but there were better ways to do it. Both Lean and Six Sigma have their own strengths and weaknesses. Organizations should draw on the strengths of each.

Lean methods tend to focus on speed and eliminating waste. Many of the "traditional Lean tools" are simpler to use than hard core Six Sigma analysis, so it is easier to get a group of people up to speed on how to use them. One can make an argument that it is best to use Lean techniques first to eliminate waste from the process and then apply Six Sigma concepts to reduce variation and stabilize a process. A Six Sigma team should take advantage of "Lean tools" and use them as part of their improvement approach.

Six Sigma's Evolution and Approach

The Six Sigma approach was born out of W. Edward Deming's work with Japanese companies from 1950 to 1970. Deming had a background in statistics. His work strongly influenced strategies of analyzing the variations of the processes. Measurement of variation was an indicator of how well a process performed. It was represented, in terms of standard deviations, by the Greek Sigma (s) symbol. The Control limits for a process could be calculated and people could measure if the process was "in control". SPC charts allow users to understand when assignable or special cause variation has been introduced into a process, and if a process has the potential to produce product within specifications. If a process is "in control," however, it does not necessarily mean it is producing "quality" products or services. When done properly, SPC can be an extremely useful tool, but like any tool it is only as good as the people who use it.

In the 1980s, Bill Smith, an engineer at Motorola, began working with Deming's concept of variation and took it one step further. He encouraged the organization to reduce variation as a way to improve performance. He made statistical process controls and design of experiments practical quality tools. The Six Sigma approach became the key focal point of Motorola's quality effort. Bob Galvin, Motorola's CEO, introduced Six Sigma as a way of doing business in everything that they did. As a result, Motorola reduced in-process defect levels by a factor of 200, and reduced manufacturing costs by $1.4 billion.

At Motorola, the Six Sigma process was expanded into the *Six Sigma Management System* after their work with General Electric in the late 1990s. This was described in more detail in Part One of this book.

Six Sigma is an outstanding process for solving difficult problems and finding answers that are not easy to see. As a management system, Six Sigma brings a rigorous measurement analysis that is missing in most process improvement efforts. Measurement is also a weakness of traditional Lean methods. In a typical Lean initiative, "management" is asked to accept on faith that things are getting better. People involved on the floor can see improvement, but if you are not close to the action, it is difficult to know what is happening or to make appropriate adjustments. A few key metrics can provide cover for Lean improvement activities and drive new behaviors inside the business. This concept is further explored in the financial/measurement chapter.

Lean as a Business Management Strategy

A big question is, "Does the organization wish to do *Lean* as a strategy?" A leadership group should be able to explain how Lean will competitively differentiate an organization. While Lean applies to the overall enterprise, its roots are in operations. The ideal Lean production system has almost no work-in-process inventory and is able to produce single-unit quantities as cheaply as large quantities. "Mass customization," the term used to describe this type of system, has become a reality in industries where product complexity dictated large lot sizes and large inventories only a few years ago. One key to mass customization is short changeover times which allow for small lot sizes and immediate response to the next customer-configured order.

Basically, the strategy part of Lean looks at balancing multiple value streams (i.e., typically, a family of products or services) and integrating the work done in operations and in the rest of the organization (be it a factory, a hospital, a software development company) with the customer in mind. The concept is simple. "Lean" describes any process developed to a goal of near 100% value-added with very few waste steps or interruptions to the workflow. That includes physical things like products and less tangible information like orders, request for information, quotes, etc.

Ultimately, Lean seeks to produce or provide exactly what the customer wants, when the customer needs it, at a profitable price, with zero waste, and in a safe manner. When industries have excess capacity - and this is true for most industries - the paying customer reigns supreme. Internal customers, or secondary customers, must balance their wants against the paying customer's requirements.

Lean is typically driven by a need for quicker customer response times, the proliferation of product and service offerings, a need for faster cycle times, and a need to eliminate waste in all its forms. Some people interpret these factors as "reduce costs"; others interpret them as find new ways to grow the business. Whatever the reason, it is essential to communicate why the initiative is important to the people who work in the organization.

Lean requires new ways of thinking to align the flow between the enterprise and its customers. This is the challenge for Master Black Belts, Black Belts, and Green Belts to influence leaders to understand new ways of thinking:
The Lean approach challenges everything and accepts nothing as unchangeable. It strives to continuously eliminate waste from all processes - a fundamental principle totally in alignment with the goals of the *Six*

From	**To**
Batch operations	Units of one
Wait time is invisible	Eliminate wait time
Traditional accounting	Lean accounting
Labor and machine utilization	Only working on what the customer needs today
Production efficiencies	Visual controls
Lots of inventory	Minimal inventory

Sigma Management System. These methods are especially effective in overcoming cultural barriers where the impossible is often merely the untried.

Lean, like any other major business strategy, is best if driven from the top, linked into the organization's performance measurement systems, and used as a competitive differentiator. That is the ideal. Sometimes reality differs. In those instances, the Champions driving this approach should look for pilot areas in the organization to test the concept and see if a business case for Lean can be built over time. One cannot just flip a switch and get the whole organization doing this anyway, so starting small and building from there can be a valuable approach. A high level roadmap for doing Lean is laid out in the "Implementing a Lean Initiative" section of this chapter.

The execution is a challenge. It impacts "traditional" ways of looking at performance. The complexity of production processes, whether on the shop floor or in an office, often become a barrier to improvement. This complexity is viewed by traditional business cultures as unchangeable, "this is how it has to be done." Lean seeks to break out of this way of looking at a business. Traditionally, machine up time (or utilization) is a good thing. Lean recognizes that making parts that are not needed is not a good thing. So the related metrics need to be changed. Lean directly attacks traditional batch modes of operation whether manufacturing large batches or doing transactions in batches (e.g., accumulating several invoices before processing).

Key Elements of Lean

Standard Work

Lean is the latest key attribute of the Toyota Production System that people point to as "the secret!" If one were to say Lean is about one thing, it would probably be the idea of "Standard Work." Certainly, Standard Work is a key foundation of Lean. It describes a best practice for how work gets done, makes work repeatable, and decreases variation during the process. It is amazing how much variation exists in the ways people go about doing their work. This holds true on a factory floor, and the variation is at least ten times that for administrative processes, including processes like software design. The concept of Standard Work is further explored in the *Kaizen* and Lean Team chapters.

Value Stream Mapping

Most Lean initiatives start out with a Value Stream Mapping (VSM) exercise. (See the Lean Teams chapter of this book for more information on this tool.) A VSM typically focuses on one product or service family. The reason for this is to find enough critical mass to be able to devote a set of assets and, hopefully, a group of people to do a stable amount of work.

Cells

In a Lean factory, people often talk about "cells." A cell is simply a dedicated piece of the enterprise that typically makes products for a value stream (product family). For Lean to be successful there needs to be a relatively steady state of work that can be run through a cell or work group.

Visual Management

There is a desire to make "work" more visual so that it is easy to know if things are going well, or if problems exist. If management and associates can easily and quickly see when problems arise, they can take action much more quickly.

Simplification and Elimination of Waste

To further understand Lean, consider Toyota's metaphor of lowering the water level in a stream to expose the rocks at the bottom. The water level represents the work-in-process inventory that "flows" (sort of) through a manufacturing plant. It flows lazily when the water level is high, rising to the banks. But the water will pick up speed when the dam is removed and the water level drops to the riverbed.

However, that speed is constrained by the rocks on the bottom that appear as the water level drops. The rocks represent non-productive elements in the production system, like inappropriate equipment, mistakes in production, and excessive production floor space that creates long travel distances, etc. It is easy to imagine how removing the rocks can lead to faster, more efficient flow in both the stream and the plant.

To a degree, the rocks also illustrate the concept of "value-added" versus "waste." Clearly the rocks add no value to enhancing the water flow.

Similarly, in many production plants, it is easy to identify non-productive equipment and floor space that add no value to the product flow and clearly impede the production process.

Most of the tools used to start a Lean initiative are outlined in the Team chapters of this book under *Kaizen* and Lean Teams. Other books describe ways to calculate kanbans and safety stocks. Dennis Butt dedicates a website [1] to Taiichi Ohno, the key architect of the Toyota Production System, where he outlines a number of the calculation formulas so they will not be touched upon here.

Lean in the Office

Lean in an assembly manufacturing plant tends to focus on one-piece flow. In a process or job shop, it tends to focus on eliminating wait time. The idea of eliminating wait time and defining "standard work" also applies to the office and administrative environments. Most overhead departments and activities do not have effective metrics. Standard work does not exist for most tasks. Most overhead departments would score poorly in a 5S assessment (Sort, Set-in-order, Shine, Standardize, and Sustain). Many administrative or transaction type systems are typically designed to handle the most complex transactions. These problems can cause excessive rework, delays in processing, and confusion. A few examples:

- An accounting close typically takes a long time to do because a flood of transactions take place at the end of the period: journal entries, special analysis, allocations, report preparation, etc. The excessive transactions cause accounting departments to be somewhat chaotic places at the end of the period. Adopting Lean in this world is different from a factory, but the goal is still stable amounts of work, flexible operations (within defined parameters), and pull from the customers of the process.
- Imagine a purchase order going through a process. Ninety-nine percent of its processing life is going to be "wait" time. It may also have rework problems as people try to get the information right, and in terms of workload balance, some purchase orders are more difficult to do than others. This is not so different from what goes on in the factory. Many of these problems can be individually addressed

[1] "http://www.toyotaproductionsystem.net/Personal%20Web%20Page.htm

using Kaizen and Lean Teams.

- Multiple re-inputs of information into Excel spreadsheets, Access databases, requirements generators, etc. Or the different languages used inside a business for the same physical product. Purchasing has a different numbering scheme than engineering, which has a different numbering scheme than accounting. And someone is supposed to keep a matrix up-to-date that maps these relationships - now there's a value adding activity from a customer perspective!

Real Life Cases

The Credit Department in a Financial Services business had a responsibility to review credit for new and existing accounts - an important responsibility for certain. They perceived their role as protecting the company's assets - period. They did not see their role in generating new business or in customer service. So every time a salesperson would try to open a new account, it seemed as though Credit was inventing the credit process. They would sit on information, they would ask for one document from a customer, and then come back some time later and ask for more information. When existing customers came up for review, it was as though they all were new high risk customers and they were marched through an inflexible, non-customer responsive process. With a new, high risk customer, everything the Credit Department did was appropriate. Perhaps it could have been more customer friendly, but they had to do the steps. However, with a lower risk (especially existing) customer, many of the steps were not relevant, thus distracting from the organization's competitive capabilities.

A very similar thing happened in a company that does custom work. Engineering reviewed and spent design time on every order, as though it were custom, even though 80% of most jobs were essentially for standard products. When Lean is applied to the office and administrative processes, typically new value streams get defined. Multiple (flexible) processes replace rigid structures.

Implementing a Lean Initiative

How does an organization use Lean concepts to improve business performance? The answers to that question are not as simple as buying the latest "magic system" and plugging it in. Lean is not an object. It is a business culture, top to bottom.

The Basic Lean Organization for implementation of Lean may include:

- **Executive Leadership** to set key business goals, allocate resources, and manage four Continuous Improvement functions:
  - Measurement of progress on Key Performance Indicators and process indicators
  - Communication/education about Lean, Quality, and Productivity
  - Systematic search for opportunities
  - Improvement actions via individuals and teams (below)

- **Improvement Action Projects** with comprehensive project plans:
  - *Kaizen* teams for local departmental, immediate opportunities
  - Lean/Critical Process Teams for large or cross-functional opportunities
  - Six Sigma Teams for deep process analysis and variance elimination

- **Departments** to adopt Lean methods in routine operations:
  - Mid-Management
  - Front-Line Supervision
  - Work Teams

The *overall* size of the organization will determine how much *Lean* organization (above) is necessary to launch and sustain a Lean business culture. Many of the components in the Lean organization structure are similar to those found in the *Six Sigma Management System.* So Lean and Six Sigma can go hand-in-hand together.

Types of Lean Initiatives

There are many ways to roll out a Lean initiative. Three alternatives include:

1. **5S** - Use the 5S tool (Sort, Set-in-order, Shine, Standardize, and Sustain) as the foundation for a Lean initiative. This is an inexpensive way to get a Lean initiative started.
   Benefits of this approach:
   - inexpensive, involves people
   - not particularly threatening
   - allows the organization to move up the learning curve more slowly

   Drawbacks:
   - requires much discipline from the management group
   - takes a while to show a payback

2. **Bootstrap Method** - This is typically done by launching a series of *Kaizen* teams. Each *Kaizen* team implements an improvement.

   Benefits of this approach:
   - provides some early gains to the business
   - builds credibility
   - people can see the changes
   - provides short-term measurable improvements

   Drawbacks:
   - can end up with disconnected islands of improvement
   - improvements in one area may result in negative impacts on another
   - sometimes loses focus on external customer.

3. **Major Initiative** - This is typically done with some type of an executive kick-off. The leadership team may have initially attended some type of an improvement workshop to get started.

   Benefits of this approach:
   - a strategic direction gets set
   - the leadership team is supportive of improvement efforts
   - improvement teams can hit the ground quickly and get started
   - resources are dedicated to improvement
   - the Lean transformation can happen more quickly

   Drawbacks:
   - may start with an internal view of why Lean is necessary rather than an external customer perspective
   - sometimes management groups lose focus after the launch

- usually more expensive front-end investment in consultants, training, etc.

Stages of Lean Implementation

The major phases of the DMAIC model are conceptually very similar to the steps in the rollout of a Lean Initiative. The comparison will be shown below to highlight the similarities between Lean and Six Sigma.

Assessment

Lean should start with some type of an assessment. The assessment needs to be scoped so that it does not take too long before the organization begins implementing changes. Implementation is the ultimate test of ideas. Don't waste a lot of money doing a six-month study or endless "as-is" value stream or process maps. The assessment starts by looking at the business goals. If the goals are not relevant, everything that follows will be weak. If the goals are based on the "voice of the customer" and they take into account key operational changes that need to happen to better meet customer needs, this can be a powerful exercise.

Project Selection

The most important outcome of the assessment is where to get started. One way to prioritize from a strategic level is to use an Impact Analysis Matrix. On a spreadsheet, line up the possible projects down the vertical side, and list the company goals across the top. The impact of each activity is weighted as high, medium, or low at each intersection, and activities with the highest total impact are potential projects.

Simple Impact Analysis Matrix

Business Goals	Cash Flow	Market Share	New Products	Etc.	Total Points
Potential Projects					
Project One	2	2	3		7
Project Two	1	0	0		1
Project Three	2	3	3		8
etc.					
Total	5	5	6		16

Scale:
3 = Very Important
2 = Important
1 = Nice to Do
0 = Not important

On this simple scale Project three is most important

Figure 6-2 Simple Impact Analysis Matrix

Define

The projects selected should include a definition of the benefits expected, who should sponsor the project, and a timeline. This information would go into a Lean team's charter. Determining how success will be measured provides clarity. Different results will be accomplished depending on whether the initial target is a complete rollout of 5S or quick start *Kaizen* teams. The more metrics link to customers and real world business issues, the greater the likelihood of the organization sustaining the gains and providing appropriate resources to improvement. One of the key problems with activity-driven rollouts like 5S is that the business results take a while to see.

The roadmap does not need to be overly prescriptive. Experience will move an organization along the learning curve. There should be some type of criteria for judging the initial success, a person or group responsible for assessing the results of the pilot (beyond just the sponsor), and a person or group responsible for continuing the rollout. This can typically be done in a few weeks.

Measure

Establishing baselines of the current level of performance is a key part of Measure. Measurement is generally a weak area of Lean the way most people practice it - looking for instant results and never getting down to the process details.

In a typical Lean implementation, most of the initial focus concentrates on eliminating waste and operational measures switch to more use of visual controls. This is good for the people in the area who can see what is happening, but lost on people outside of the area. Some elimination of waste (reduction of material defects or customer returned goods, reduction of lost orders, etc.) does yield a measurable return on the financials. But making more time available (which is a key benefit of most Lean initiatives) does not yield immediate benefits. Freed capacity only yields a benefit to the business when the organization actually sells more. For example, if previously the organization could make eight an hour and now it make ten an hour, the P&L will only benefit when the extra two are sold. If they go into inventory, or if the equipment sits idle, no P&L benefit exists.

Lean initiatives may also negatively hit the financials in the short term due to inventory reductions. This is mostly an accounting anomaly, and it only happens one time. The benefits of reducing inventories and building processes flexible enough to meet customer demand can provide much benefit, but timing needs to be a consideration.

Analyze

This is where the real work gets done. The rollout pathway selected determines the amount of effort required. For the 5S pathway, Analyze is pretty much bound by a work cell, a department, or a work area. This type of analysis can happen fairly quickly, but it only looks at a small area of an operation. If the pathway is an overall organization rollout, it becomes more complex.

Jamie Flinchbaugh [2] describes The Four Rules followed by the Toyota Motor Company:

1. Structure every activity
2. Clearly connect every customer/supplier
3. Specify and simplify every flow
4. Improve through experimentation at the lowest level possible towards the ideal state

A major rollout in an organization typically targets one or two major value streams for improvement at the start. A Value Stream [3] is typically a family of products or services with a common set of equipment to produce, people to make, and people to sell. A Value Stream is typically like a mini-profit center in the organization. In some ways, it could be viewed as a free-standing business.

Organizations should identify the value streams with the highest potential increase in shareholder value (streams that will have most impact on revenue and costs) and that involve recurring activities and repetitive cycles. They may or may not elect to start with the most important VSM with their Lean efforts.

A more holistic approach might consider all of the specific actions required to bring a product (whether a good, a service, or a combination of the two) through the three critical management tasks for any business:

---

[2] "BEYOND LEAN Building Sustainable Business and People Success through New Ways of Thinking" By Jamie Flinchbaugh, Lean Learning Center

[3] Lean Thinking, Womack and Jones, Simon & Schuster, 1996

1. Problem-solving tasks running from concept through detailed design and engineering to production launch,
2. Information management tasks running from order taking through detailed scheduling to delivery, and
3. Physical transformation tasks proceeding from raw materials to a finished product in the hands of the customer.

Whatever approach is selected, it's important to avoid becoming trapped in the "as-is" analysis cycle. Find something meaningful to do and make it happen. Use that as a learning opportunity and build from there.

Improve

So many improvement initiatives and projects stall at this stage. It is critical to avoid over analysis and too much "as is" activity. Minimize burnout and save time for implementation. Improvement ideas that are not implemented do not count. Great ideas are a dime a dozen. Great implementation is a precious gem.

Improve in the Lean model is no different from any other. A Lean pilot project has the following key attributes:

- The process selected is meaningful to the business (it does not need to be most important, but it needs to be important)
- One team and one Champion should largely be able to drive the effort, pulling in a few part time additional resources on a limited, as needed basis
- The project should not require any major new equipment or information system expenditures
- The pilot project should be up in running in less than 60 days time, preferably less than 30 days
- Strong lines of communication with executive leadership, with the associates impacted by the changes and interaction with key suppliers are critical

Control

Avoid becoming enraptured with any one Lean tool. 5S is great, *Kaizen* teams are wonderful, but the use of additional tools can further increase the organization's degree of success with Lean. The leadership team should have established some type of criteria at the very start of the endeavor. Hopefully, the criteria get used for guidance along the way.

Now is the time for the leadership team to assess the results of the pilot project and to determine the next steps. One of the areas requiring special attention is sustaining the gains. It is simply too easy to revert back to old ways of operating. What actions are being taken to avoid this problem? If special attention is not paid to sustaining the gains, they will not get sustained.

It is also appropriate at this point to give some thought to how to move forward:

- What is the ongoing role of the current team?
- Will some members of the initial Lean team participate on a second round?
- Where is the next logical place to implement the new processes?
- Will there be just one area or more than one, for the next stage of a rollout?
- What feedback if any has been received from customers on the revised processes?
- What is the updated timeline for moving forward and who is accountable?

Continuous Improvement and Team Methods

Two features about the Basic Lean Organization need special emphasis:

- Holistic Continual Improvement (CI) Process
- Team-Based Lean Implementation Methods

First, to make Lean sustainable in the long run, it must become a formal process of CI embedded in the business culture, actively managed to align the organization with the ever-changing key business goals. There are four sub-processes in the formal process:

1. Measurement
2. Communication/Education
3. Search for Opportunities
4. Improvement Actions

When all four sub-processes are operational, then CI becomes a closed-loop, self-reinforcing process that guides the organization in the most productive direction for current business conditions. Holistic CI is a characteristic of industry leaders that leaves their competitors wondering, "How do they do that?"

Second, most everything about Lean relies on team-based business methods. A major reason for this is found in research findings from a few years ago, but not well-publicized:

- Technologies are easily copied by the competitors.
- People-based business differentiators are difficult to copy but provide sustainable competitive advantages.

If there is such power in team-based processes for continuous improvement in products, services, business processes, and working relationships, why aren't they more universally applied? Perhaps the answer is in a common behavior pattern. Technical backgrounds often cause individual to over-emphasize whatever has already been mastered (technical processes, IT, etc.) and under-emphasize whatever has not been mastered (people processes). Team based process are described in the Team chapters of this book.

Team-based business methods are not "feel-good team-building" exercises. Rather, they are the nuts-and-bolts of practical business team development. Industry leaders make sure that people processes and teamwork are key elements in their continuous improvement processes.

Examples of Lean Projects

Remember the visual nature of Lean processes. If queues of materials, products, paperwork, files, etc. in any process exist - in a plant or office - then the process is not Lean and almost certainly has major opportunities for improvement, including:

- Reduced inventory (working capital investment)
- Reduced floor space (fixed assets and period expense)
- Quicker response times and shorter lead times
- Decreased defects, rework, scrap
- Increased overall productivity

It is important to note that these improvement techniques have been applied not only to manufacturing and supply chain processes, but also to engineering design and overhead processes (such as finance) with significant results.

Invoicing

A VSM was created for the Invoicing Process. Customers would not pay when the invoice was wrong (terms, price, amount, etc.) or because the invoice went to the wrong place, etc. One team focused on invoice accuracy since it supports delivering the right product to the right place. They involved the inside sales group along with the accounting department. Days of sales outstanding (DSO) were reduced 17% and the number of customer orders on hold by 50%. They reduced (simplified) the number of terms by 60% and are working on streamlining the price lists. The teams are still working to further reduce too many error opportunities.

Payroll

A VSM was done for generating an employee paycheck. The team was surprised to learn just how many different payrolls existed. The organization was running 22 payrolls a month - some weekly, some monthly, some bi-weekly, and some bi-monthly. They reduced the number of payrolls run by 64%. Much of the paper printed, sorted, stuffed and mailed was eliminated.

Healthcare

Using the Toyota approach, a hospital traced problematic infections in some patients to their source, prompting two intensive-care units to change the way they insert intravenous lines. The result was a 90% drop in the number of infections after just 90 days of using the new procedures. By reducing infections, the new procedures have saved almost $500,000 a year in intensive-care-unit costs. [4]

In a factory, the Toyota approach emphasizes the smooth flow of people, gear, and finished goods. In hospitals, it emphasizes rapid flow of patients and staff.

At a clinic, a dozen staffers were trying to cut a typical 61-minute office visit as well as staff overtime by 50%. They produced a 25-foot wall map charting a pneumonia patient's typical office visit. They concluded that 17 steps were valuable and 51 steps were not valuable. In the latter category, for instance, patients walk to a separate laboratory to get blood drawn. By the end of the day, the team had concluded that six assistants no longer should be assigned to individual doctors; instead, they should be pooled. The team also proposed that assistants do blood work-ups in examining rooms. No one was laid off. People were reassigned to other pressing responsibilities.

Manufacturing

Many manufacturing examples demonstrate the benefits of Lean and flow. Typically inventory levels are drastically reduced, customer response times are dramatically improved, and it becomes much easier to see work being done because clutter has been eliminated and large amounts of non-value adding time are eliminated:

- Changeover times on large 4,000 ton injection molding presses reduced from 6 hours to 15 minutes
- Inventory levels cut by 70% or more
- On time and complete deliveries from increased from 70% to 99%
- Customer order response time (to shipment) reduced from 16 days to 3 hours
- Non-value added time reduced by 80%
- Reduced time to respond to a quote from up to 5 days to under 30 minutes

[4] Bernard Wysocki, Jr.; The Wall Street Journal, April 9, 2004, Page 1

Inventories

In the early years of the Lean renaissance movement, inventories became the primary target. Low inventories are indicative of Lean operations - in a manufacturing plant, low inventory is visually obvious.

The steps involved in reducing inventories are usually clear-cut. In general, the reasons for large inventories can be traced to the perceived or potential cost of each replenishment transaction. Therefore, the steps to reducing inventory focus on reducing equipment changeover times, order handling costs, and other replenishment transaction costs.

The examples seem almost incredible until you stop to analyze how business processes in the 1970s and 1980s had become so bloated with wasteful steps that removing them was like opening the floodgates. Until you have experienced it first-hand, such quantum improvements can be difficult to grasp.

Lean Tomorrow

Much of the Lean work being done today is driven by the power customers have over suppliers - "If you do not meet my needs, I will switch to a different supplier." If the economic climate changes, so will this balance of power. The basic Lean principles will still apply, but people will talk about them differently.

Sometimes we rush to the future, only to find that we must return to the past. The wonderful thing about Toyota is they have largely ignored economic cycles. They are far and away the most profitable automotive manufacturer in the world. They have had a steady rate of improvement for more than 50 years. But even at Toyota, they periodically have to go back and revisit the basics. "The Wall Street Journal" ran an article by staff reporters Norihiko Shirouzu and Sebastian Moffett titled, "Bumpy Road as Toyota Closes in on GM, Quality Concerns Also Grow." Toyota realized that they were not giving their American workforce the skills they need for the future. There is a tendency for American managers to take on too much responsibility. Essentially, Toyota is initiating a "back to basics" approach for improving quality and reducing costs.

Many of the tactics Toyota will pursue are described in this book. As they get back to basics and improve even further, it simply puts more pressure on the rest of the industry to change - do it cheaper, better, and faster, do only work that is meaningful, and don't harm the environment.

Lean is just a word. Doctors and healthcare providers in hospitals shudder when they think there is something they can learn from the Toyota Production System. Initially, they do not understand the core concepts are about performance improvement. The core concepts apply to any industry. The key is to have a management and leadership system that drives performance improvement.

As time goes by, most initiatives will evolve to:

- Truly move up and down the supply chain
- Analytical areas like engineering, administrative paper, and office processes
- Simplify accounting systems to provide more useful, timely information
- Build accountability and responsibility into operating teams responsible for
  product families - most likely some degree of decentralization
- Integrate Lean into management processes, folding projected savings into budgets

Companies need to drive their Lean performance improvement activities based on their strategic situation. A seasonal business will adopt performance improvement strategies that differ from a company that produces one of a kind products or one that has a steady year round demand. And this makes total sense. The whole concept of customer "pull" is a beautiful concept. Each organization needs to adopt this concept in a way that challenges current ways of thinking (paradigms) and in a way that makes sense for their business.

There are no guarantees of lifetime employment in a global competitive environment. The only guarantee is organizations that do not adapt to the changing competitive environment will be left behind. Unfortunately, there is no "magic pill" or one way to do effective organization performance improvement. The most exciting part may simply be seeing perspectives change. People literally grow as they make change happen. Of course, that is what makes it so rewarding.

Summary of Six Sigma and Lean

Six Sigma and Lean are really part of the evolution that quality, productivity and process improvement initiatives have taken over the last 20 years. Each approach has its own unique perspective regarding performance improvement. But the goals are the same: better, faster, cheaper, more meaning in work, and don't harm the environment.

A Lean manufacturing or enterprise initiative can be enhanced by Six Sigma's powerful focus on metrics and gathering baseline data. Six Sigma initiatives can be improved by Lean's use of simple analytical tools, a "do it now" philosophy and the strong emphasis on letting customers pull production, rather than traditional batch manufacturing practices.

The steps to apply Lean concepts to the office and administrative functions are not significantly different from the operational side of a business. Hospitals have adopted Toyota Production System (TPS) techniques, although there are cultural difficulties with using the TPS name in this environment.

A variety of approaches exist to implement lean initiatives, including 5S, Kaizen and Lean teams. Typically, organizations first eliminate the obvious waste (low hanging fruit) from a process before using statistical methods. Statistical methods may jump to the front of the line if an organization is experiencing quality problems that have a customer impact. The Six Sigma Management System incorporates both of these approaches. Toyota's emphasis on "standard work" does not radically differ from Six Sigma's goal of reducing variation.

At the end of the day, it does not matter what you call it. Organizations are still looking for simplicity, the ability to better meet the requirements of process customers and the elimination of waste in all its forms.

Part Three:

Process Improvement Teams and Tools

When an organization adopts the *Six Sigma Management System* as their business process improvement model, it becomes an umbrella business strategy that should include a variety of team-based initiatives. The purpose of this part of *The Six Sigma Black Belt Handbook* is to address various team approaches that have been proven to be successful as part of a Six Sigma system, to present management's varying roles and responsibilities with each team approach, and to illustrate the common tools used by teams to implement improvements and types of measurements that drive performance.

The following chapters will:

- Provide an overview of the different team-based approaches, including *Kaizen*, Lean, and Six Sigma (Chapter 7)
- Explain the Champion's roles and responsibilities with each type of team (Chapter 8)
- Describe *Kaizen* teams (Chapter 9)
- Describe Lean teams (Chapter 10)

An entire chapter has been devoted to the role of management and Champions because when teams fail, more often than not the root cause sits with management and the processes they are using for managing teams.

Chapter

7

Introduction to Process Improvement Teams

The Power of Teams

There are multiple things occurring whenever an organization tries to accomplish significant, meaningful improvement. Without a doubt, actions of the leadership team set the pace and guide the direction of change. When Art Byrne became CEO of Wiremold in 1991, the first thing he said in his presentation to employees was "Productivity = Wealth." This is a simple yet powerful statement. Productivity is a key economic factor in terms of fighting inflation, staying competitive on a global basis, and promoting business growth. This is what the Toyota Motor Company has done so well year after year. In order to be competitive, all organizations need to work with an ongoing sense of renewal.

While the transition is taking place to "something new," work still needs to get done, customers still need to get served, and day-to-day problems still need to get addressed. Implementation of the transformation needs to happen somewhere.

One of the most meaningful ways to differentiate an organization is to have the people inside gain a better understanding of the Voice of the Customer and to take a more active role in transforming the organization to meet those needs. This is equally true for publicly held companies, private enterprises, not-for-profits, and governmental organizations. Cross-functional teams are a key method for doing this. They provide an opportunity for people to take ownership of finding new ways to do work. The names may vary - Lean, Blitz, or Breakthrough teams - but these are largely cross-functional, multi-discipline teams that look at business processes.

Types of Process Improvement Teams

There are three types of teams described in this Part of the book. *Kaizen* teams, Lean (or Process Improvement) teams, and Six Sigma teams. All three focus on performance and process improvement. They each bring a different perspective to the world of improvement.

Physically Observable	Horizontal flow across the process Alignment toward requirements	Deep Analysis Stabilize & Eliminate Variation
***Kaizen* Teams**	**Lean Teams**	**Six Sigma Teams**

Simple/Visual >>>>>>>>>>>>>>>>>>>>>>>>>>>>>*Complex/Hard to See*

Figure 7-1 Performance and Process Improvement Continuum

- *Kaizen* teams tend to address the physical processes that lend themselves to the use of visual analytical tools. If the team can observe what is being done to find improvement opportunities, this is a good team approach.
- Lean/Process Improvement teams tend to focus on cross-functional projects, with requirements that are not clearly understood or agreed upon between the different functional players (departments), or by the people working the process. They take a horizontal focus across the process to understand requirements and eliminate waste. Waste is anything done that does not contribute to meeting requirements.
- Six Sigma teams take a vertical approach to process analysis and go deep inside the process where the root cause of the problem is not easily understood. They are more analytical than a Lean team. Six Sigma teams use sophisticated tools to discover the "root cause" of problems, eliminate variation, stabilize processes, and sometimes even design a new process.

While some projects are definitely most appropriate for a specific approach, many lend themselves to a blend of the tools. So it is best for an organization that is using the *Six Sigma Management System* to be familiar with all three approaches because they each have their strengths and weaknesses.

Deploying Teams with Varied Timing and Purpose

1. *Kaizen* Teams
   a. Short duration (one week or less, full-time)
   b. Visual project analysis

2. Lean Teams
   a. Medium duration (30 to 90 days, part-time)
   b. Cross functional in nature
   c. Time is a major improvement factor
   d. Requirements are usually not in alignment

3. Six Sigma Teams
   a. Longer duration (60 to 120 days, part-time)
   b. Seek to reduce variation, eliminate errors
   c. New product/service development (proactive design)
   d. Process stabilization and elimination of "root causes"

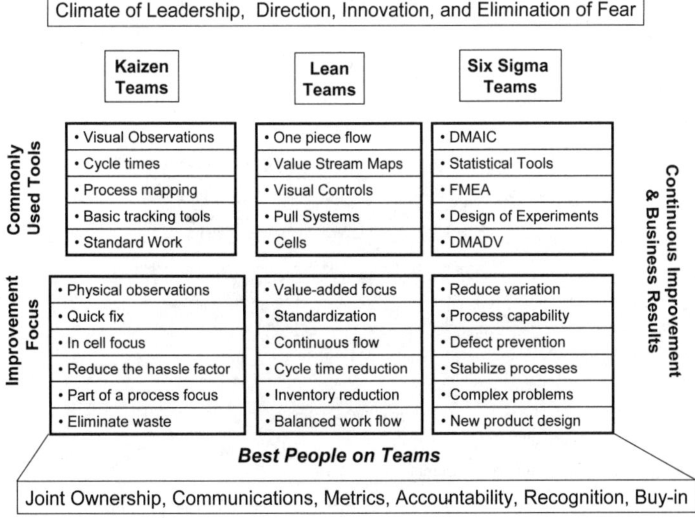

Figure 7-2 Types of Teams

Timing is largely related to the scope of the project, allocated resources, and project complexity. Consider three typical projects:

1. *Kaizen* team - reduce changeover time on a machine that stamps car doors, going from a two door vehicle to a four door vehicle. Scope: last good two-door product to first good four-door product. Time: one week, full-time team. Complexity: waste in the process that is happening right in the work area. Can usually physically observe waste in the process and use simple tools to capture metrics.
2. Lean team - reduce time in the customer order entry to delivery process. Scope: order received to product delivered at customer site. Time: part-time team, one day per week. Complexity: cross-functional, much information processing, requirements not always clear. Takes time to surface relevant data and information, and people may have differing views about requirements.
3. Six Sigma team - reduce the variation in customer order response time. Scope: from the time order is received to time product is delivered. Complexity: cross-functional, much information processing, variation in order mix, variation in order volumes. Tends to be a different perspective or way of looking at the same problem the Lean team above would address.

The bottom line is that teams must organize their work around the most effective methodology for the task at hand.

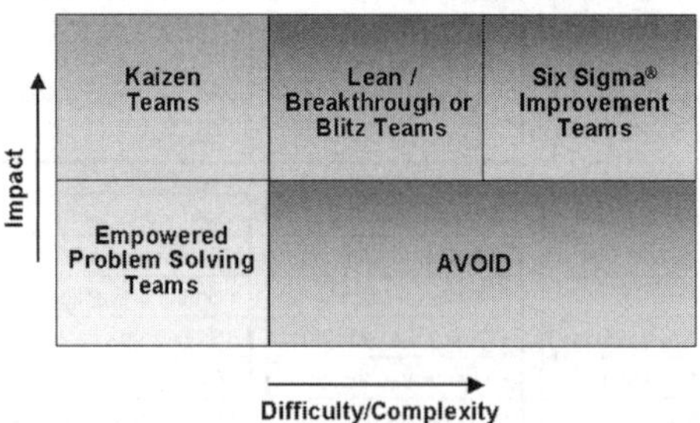

Figure 7-3 Different Teams for Different Tasks

Summary of Different Types of Process Improvement Teams

Different teams and team problem-solving tools sets are appropriate, depending on the nature of the problem or opportunity. This chapter has provided an overview of three team approaches: *Kaizen*, Lean, and Six Sigma.

Chapters 9 and 10 will discuss *Kaizen* and Lean teams in more detail. Those chapters will discuss the work done by *Kaizen* and Lean teams in the framework of the Six Sigma DMAIC problem-solving methodology for consistency. In practice, most *Kaizen* teams are moving so quickly to get their work done that normally they would not think in terms of the DMAIC phases. But at the end of their assignment, they still will have defined problems and opportunities, measured baselines and benefits, analyzed a process, improved a process, and put in place the groundwork for controlling and maintaining the gain over time. *Kaizen* usually would not call that work DMAIC - different terminology is used, but the context is the same.

Chapter

8

Leadership Roles in Deploying Teams

Management has a few "critical to success" factors common to all three types of teams. The most important is leadership! At the start of any new performance improvement initiative, management typically selects improvement teams. Motorola uses a Jumpstart Workshop where the leadership team reviews improvement opportunities across the enterprise. Depending on the decisions made regarding strategic objectives and listening to the Voice of the Customer, certain improvement opportunities have a higher priority than others.

Process Improvement Teams are the front line change agents in the adoption of the *Six Sigma Management System*. Initially, it is leadership's role to mobilize, accelerate, and govern teams. As time goes by and more people learn about Six Sigma and other problem-solving tools and skills, the process may become less formal and just become "the way we do things around here." Ultimately, improvement opportunities should less frequently require the launch of a formal freestanding team. If a problem arises, the appropriate analytical method/tool gets used.

Leadership Keys to Mobilizing Successful Teams

Solid Team Charter

The first factor critical to success is a meaningful Team Charter. (Team Charters were highlighted in Chapter 3 as an activity within the Mobilize mode.) All three types of teams should have a clear Team Charter. The charter should outline the team's reason for existence (purpose), the scope of their project (where it starts and where it ends), a set of goals or expectations for improvement at the completion of the project, a rough timeline for work, and a list of the people on the team, including the project Champion. The charter is outlined in more detail in each of the team sections.

Right Champion and People on the Team

The second factor critical to success is people. In Jim Collin's book *Good to Great*, he talks about how important it is to "have the right people on the bus, in the right seats." The same thing is true for any type of project team. The team's Champion (or Sponsor) should be a person with enough clout in the organization to largely approve team member recommendations. If Champions cannot solely approve team recommendations, they must be in a position to exercise major influence over their peer groups. Otherwise, the team is reporting to an "idea" salesperson and that is usually not a good thing. The backgrounds of people that populate these teams are discussed under each team type.

Time to Do the Work

The third factor critical to success is time. Too often organizations convince themselves that everything is equally important. And the most important of the important is "whatever I talked to you about last." When people are put on teams, there is a reluctance to face the fact that they may not have time to do everything else that was previously on their list. If good people are put on the team, they are busy people. It is unfair to say, "Ok, you are now on this team, and oh by the way, I still want you to do everything else that was on your list." If that happens, good people will still usually do it, but the organization is not treating them fairly and risks burning them out over time.

Kaizen team membership is usually a short period of time - three to five days with full- time participation. The other two types of teams usually require 20% of a person's time, at a minimum, to be effective. While team members may not get 1:1 relief, they should get some relief - move some work to another person, extend due dates for other assignments. It really is a sign of respect for the person and the process.

Danger Signs

Management needs to avoid launching too many teams and inappropriate teams. How many resources are available in terms of people on the teams, people supporting teams, and people doing work to serve customers? If a leadership team, in its excitement to get started, launches more teams than the organization can absorb, or if it launches teams that are not working on key priority projects, the entire Six Sigma improvement campaign is at risk. The initial projects should all be directed toward the Voice of the Customer and key strategic objectives. The initial teams should all have a visible, measurable payback to the organization. For example, a team launched to increase revenues or to decrease costs is much more appropriate at the outset than a team launched to improve a support system, like new employee orientation. It is not to say that the latter project is not important. But think about this from a Voice of the Customer perspective. If teams get launched that do not have visible line of sight impact on the customer, it becomes difficult to ascertain over the short term (three to six months) if the project results are meaningful to the business.

The Process Improvement Team Champion's Roles and Responsibilities

The DMAIC model that teams apply is really iterative. It is sort of like a continuous loop of concentric circles that lay next to one another. Once you have gone through the loop one time, the team goes back through it again, at a deeper level. In the Align mode, leadership does its own version of DMAIC just to define each candidate project and launch the team. The leadership teams do a good deal of project definition in Jumpstart Workshops.

Just like the team members are expected to spend time working their project, the team's Champion must also spend time supporting the project. With a *Kaizen* team, this is pretty instantaneous. While the team is working the area, the Champion should periodically drop in to see what is happening. In the case of longer term projects, "dropping in" looks like one-on-one meetings with the project lead, and meeting with team members on a weekly or bi-weekly basis. Strategic breakthrough projects require even more involvement from the Champion. The post-project launch questions provide guidance. The Champion's role and responsibilities are further outlined in the following pages.

A key thing to consider is that the Champion really has three different responsibility channels that often operate in parallel:

1. Project Governance: guiding, getting started, making certain solutions align with business and customer objectives, reviewing projects regularly, ensuring the methodology is followed appropriately, and providing reasonable protection for team members to do their task.
2. Change Management: providing resources, helping to address conflict on an as- needed basis, preparing the organization for change, providing assistance as needed, encouraging team performance, and gaining support from peers.
3. Advocacy: supporting the team, making certain communications are active and relevant, aligning other people in the organization that have the clout to support or resist proposed changes, removing roadblocks, identifying and rallying key players, and challenging the status quo.

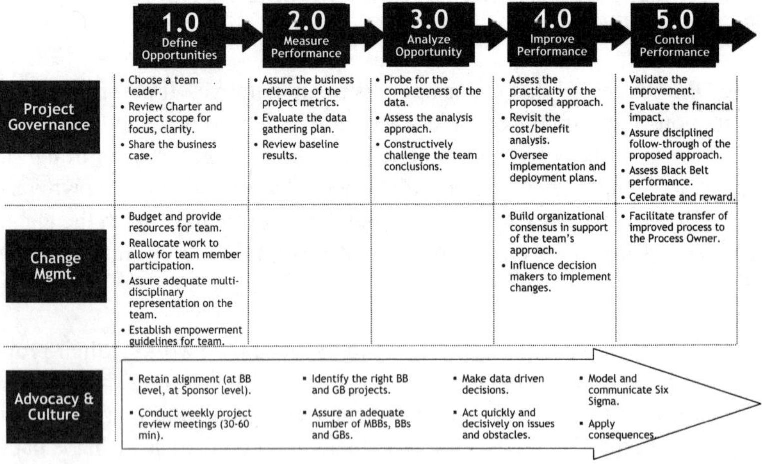

Figure 8-1 Champion's Role in DMAIC

Motorola uses the DMAIC Phase chart shown here for Six Sigma teams to highlight the Champion's responsibilities. This chapter will highlight

Champion responsibilities at each phase. Note that the phases and Champion responsibilities largely apply to all three types of teams. Obviously, the depth of answering or addressing each question/task will differ. A one-week or less *Kaizen* team project will not require the same level of critical thinking that a four to six month Six Sigma team would need in terms of guidance. The team's Champion should make certain the following is done. Note that some of these activities are done even before a team gets created.

Champion's Roles and Responsibilities in the Define Phase

An actively involved Champion working with a Black Belt Team Leader is likely to do all of the following to get a team launched in the Define Phase:

- Draft a project description that includes a Team Charter.
- Prepare a simple "Briefing Package" for all team members with key information: linkage to strategic objectives and Dashboard, the customers' requirements (there may be more than one customer), capacity statistics, historical performance, area layout, staffing data, error rates, etc.
- Compile information that may be useful to the team. However, the Champion must be careful not to overwhelm them with paperwork and must guard against generating improvement idea - that is more appropriately left to the team members.

The team Champion is responsible for establishing a clear project scope or boundaries. How far upstream or downstream does the team go? What is included in the team's scope and what is outside? Many teams have wasted considerable time getting started when they were not presented with a clear project scope. The scope of the project should be significantly different for each type of team. For example:

- *Kaizen* teams - reduce changeover time from last good piece of outgoing product to first good piece of the new product being made. This scope would include everything that happens in between the time it takes to change over the line or machinery, and it would exclude anything outside of that time frame.
- Lean Teams - decrease time and complexity of the engineering

change-order process from the time a request is put in for a change, until the implemented change is documented and stored.

- Six Sigma Teams - decrease the number of times a new change is needed to an engineering change order within twelve months of a revision. Six Sigma teams are similar to Lean teams; they use many of the same tools. But Six Sigma teams usually have a different perspective that focuses on variation reduction, process stabilization and perhaps even defining a new process.

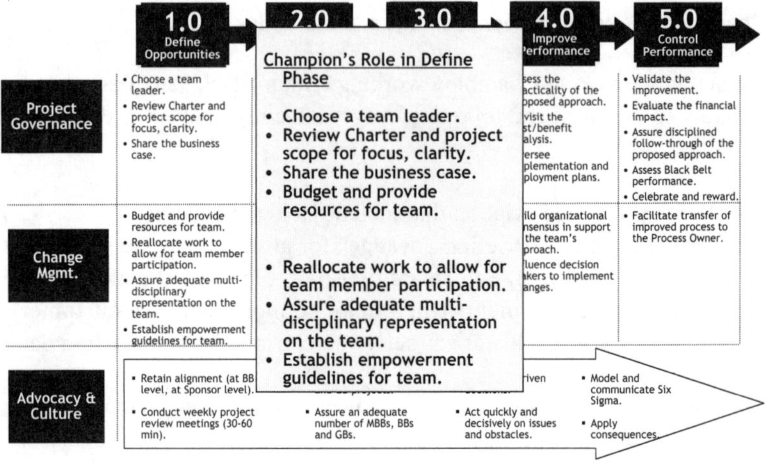

Figure 8-2 Champion's Role in the Define Phase

There are also several questions to address, from a management perspective for each phase, that apply to all three types of teams. While the Champion and the leadership team may give initial thought to these questions before launching a team, given the iterative nature of DMAIC, the same questions can be used as an indicator of how effectively the team completed this phase of the project.

Define Phase Questions - Management Perspective

<u>Pre-project Launch</u>

1. Who is the customer and how will this project benefit the customer?
2. What is the impact on the business?
3. What is your problem or opportunity statement - what are you trying to fix or avoid?

4. How will this project help the business?
5. Has the project focus been sufficiently narrowed to complete in an appropriate time frame (different for each team type)?
6. Are there sufficient resources in place to ensure project success?
7. Taking into account the opportunity statement, goals and scope outlined above, what is the team's charter?

Post-project Launch

8. Describe what you learned from the as-is process mapping.
9. What are the Critical to Customer Requirements?
   a. How did you determine them?
   b. How are you measuring them?
   c. How good is the measurement system?
   d. Have you verified them with the customer?
   e. Do the current specifications reflect them?
10. What are the Critical to Business Requirements?
11. What are the "Quick Wins?"
12. Has the team put together a communications plan to let relevant people in the organization know what is happening and to regularly solicit (appropriate) inputs?
13. What are the next steps?

Champion's Roles and Responsibilities in the Measure Phase

Clear metrics are a key part of success. The phrase "what gets measured is what gets done" was coined for a reason. For better or worse, measures provide focus. A key part of metrics is the goal. In the Measure phase of DMAIC, the team gathers hard data about the current actual performance of the target as-is process. Metrics that teams establish in Measure provide both a baseline and evidence that a problem/opportunity exists.

Based on new measurement data, the goal statement from Define is usually refined to become more specific. The right measures will make a significant difference. The team Champion should watch the metrics selected, be sure that the metrics measure significant process gaps, and assure that the metrics tie to strategic objectives.

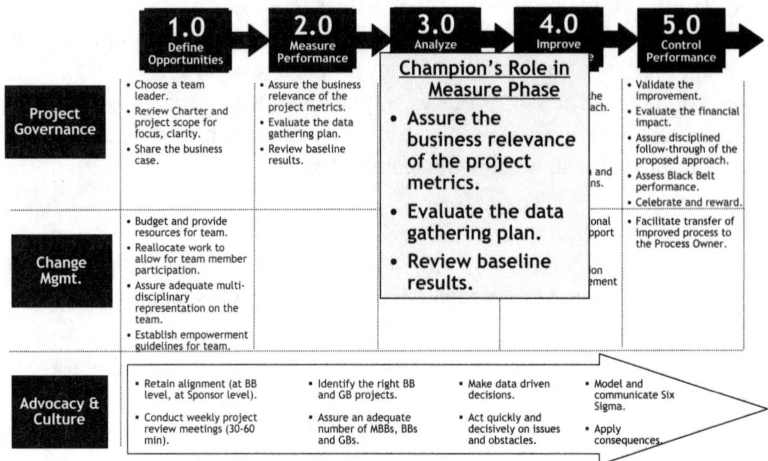

Figure 8-3 Champion's Role in the Measure Phase

The team Champion should be tightly linked into all updates of performance improvement goals. It is often best to set some type of a stretch goal, or BHAG (Big, Hairy, Audacious Gooals). For example, improve cycle times by 50%, decrease error rates or set-up times by 90%, increase customer face time (sales) by 70% are all BHAGs.

Measure Phase Questions - Management Perspective

Goal Questions to Revisit (Answers May Change)

1. What are the key goals for this project?
2. What is the impact on the business? What strategic objectives will be impacted and how?

Questions about Metrics and Measurements

3. What Critical to Quality (CTQs) metrics has the team identified and how do they know the customer is impacted by them?
4. What Critical to Process (CTPs) metrics are important to the business?
5. Which CTQs/CTPs does this project focus on? Why?
6. How do you know that the data collected are representative of the process?

7. Is the team rigorously following the appropriate methodology?
8. Do they have adequate resources to complete the project?

Questions Relevant to Six Sigma Teams

9. What have they learned about the source of the variation from the initial data gathered? Is the process in control?
10. What is the current process sigma?

Questions for All Teams

11. Did the team re-scope the project as a result of the measurement phase?
12. What are the next steps?

Champion's Roles and Responsibilities in the Analyze Phase

The Analyze Phase is about finding root causes of problems. The Champion does not want to do the analysis for the team. Analysis can be the most difficult DMAIC phase. It is the Champion's role to gently push (or occasionally shove) the team to dig deep into their process to find causes. This encourages higher levels of team performance. Champion involvement is also key to keep the team on track and focused on the charter. The Champion is responsible for making certain a team does not jump to a solution in the Analyze phase. Evidence must be gathered to support the team's root cause hypothesis - what they believe to be true. Teams pressured for fast actions sometimes like to bypass this step and implement their first solution idea for a fix.

Implementing premature, unanalyzed solutions is one of the primary reasons that so many performance improvement efforts have a poor record. A solution gets implemented, but it only takes care of a symptom, not the root cause, and within a short period of time the problem resurfaces. For example, one organization was looking to increase sales "face time" with customers. The team decided to take the administrative work the salesperson was doing and give it to a customer service assistant to handle. Not a bad idea on the surface, but unfortunately, the process was rife with exceptions and redundancies. The salesperson was filling out over 15 different information reports. When a pilot implementation project ended up failing, the team went back and addressed root cause issues. They reduced

15 forms to eight, simplified data entry by only entering information once (common elements were automatically populated on other forms), and they gave the customer service assistant training in the new process. The resulting change was a 50% increase in face time and a 30% increase in revenues per salesperson.

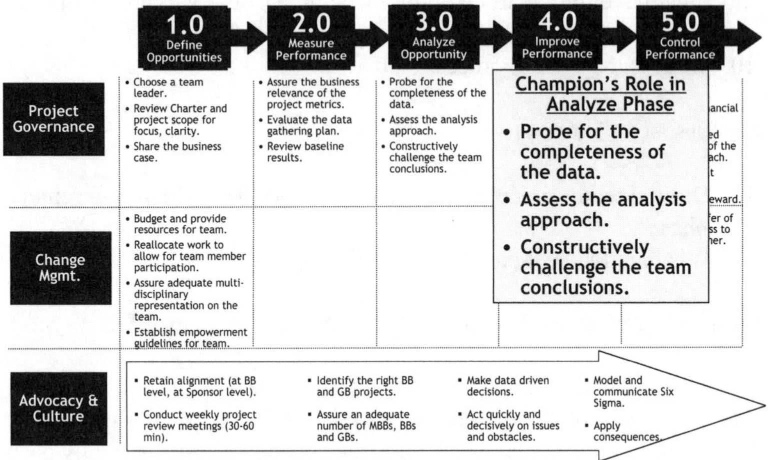

Figure 8-4 Champion's Role in the Analyze Phase

Analyze Phase Questions - Management Perspective

Questions about Causes and Effects

1. What is the statement of the problem in terms of cause and effect?
2. What are the vital few factors causing variation in the outputs?
3. What analysis exists to verify the root causes?
4. Are you being open minded and creative in your team thinking about causes?

Questions Mostly Relevant to Six Sigma Teams

5. How much of the problem is explained with the vital Xs? How much unexplained variation exists?
6. Are the vital few Xs statistically significant? Are the effects of practical significance?

Questions for All teams

7. Are any of the current learning's transferable across the business? Is an action plan for spreading the best practice appropriate?
8. Do you have adequate resources to complete the project?
9. What are the next steps?

Champion's Roles and Responsibilities in the Improve Phase

A surprising number of teams falter when they get to this phase. This is a critical point in the process. The Champion needs to make certain the team transitions from analysis to "doing it." Too often teams stall saying, "We need more information." or anyone of a dozen other perceived crises that can delay getting something implemented. This is the whole purpose of having a team do the analytical work. Motorola is big on the use of pilot to test concepts and fine-tune before a full rollout. The learning from a pilot is valuable, and adjustments can be made accordingly. The trade-off is that the implementation process takes more time. There is no one right way to do this.

The other thing teams tend to do relative to implementing improvements is the exact opposite - to run with their first idea. Even if they had the patience to wait until the analysis was completed, there is a tendency to try to get something implemented quickly. Consider the "Rule of 3" - a team's third solution tends to be more of a breakthrough change. The first solution tends to address the primary problem that the organization is experiencing. If the team has the energy to come up with a second workable solution, it tends to be an extension of the first idea. If they have the mettle to press through and create a third workable solution, it tends to be more of a breakthrough idea, a different way of working. This holds true for all three types of teams - *Kaizen*, Lean, and Six Sigma.

A short example: A Lean team was working on an Enterprise Resource Planning (ERP) integration project. One of the problems was how to get equipment operation information to people in the field (customers, maintenance, and modification shops). Their first solution was to publish a paper or digital manual. The second was a toll-free hotline that anyone could call. The third was using a web-based solution that was menu and keyword driven. They were able to make direct links to OEMs for relevant parts and direct connections into their organization. All three solutions would have worked. The third solution provided differentiation for their services that the customers valued, thus increasing their market share.

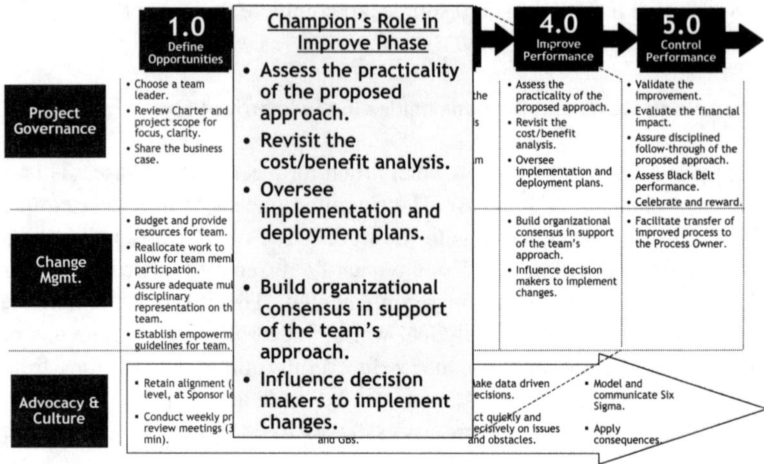

Figure 8-5 Champion's Role in the Improve Phase

At this point, communication with the rest of the organization is critical. The team and the Champion should have been communicating along the way to gain buy-in and support for their ideas to be implemented. Most teams make a presentation to management shortly after they start their work.

Improve Phase Questions - Management Perspective

<u>Questions for All Teams</u>

1. What alternative solutions did the team consider for solving the problem?
2. How was the current solution selected?
3. What are the expected benefits from the selected solution? What are costs?
4. How will you communicate the solution to the organization?
5. Will the team make use of pilot project teams or do they have a complete fix?

6. Do you have adequate resources to complete the project?
7. What are the next steps?

Champion's Roles and Responsibilities in the Control Phase

The "easy" part of the project is completed. A team is in place to monitor the changes, to do the analysis, and to make certain something is happening during the course of the project. It is after the team disbands when the wheels can fall off the cart. Organizations that do not plan on how to control the gains ahead of time risk losing the gains shortly after the improvement team is disbanded. Again, this concept applies to all three types of teams. From project day one, the Champion and the team should be giving consideration to how the gains (the improvements) get sustained over time.

A number of studies have been published over the last ten years indicating rates of failure ranging from 30% to 70%. Is this simply due to resistance to change? Probably not. Using the Six Sigma Governance process, leadership will monitor and detect processes that drift back. But, more importantly, there are proactive actions the Champion should ensure the team takes before its job is done. Think about how hard it is to learn something new to a high skill level. If you have ever played tennis and did not initially learn to watch the ball all the way into the racquet, you know how hard it is to change that behavior. You know you should do it, and if your opponent hits a soft shot you probably can, but if the ball comes whistling back across the net, it is all too easy to revert to your old ways. The same is true with organizations. People are usually willing to try the new way. But what happens during the first crunch period or crisis? We move to automatic, or someone says, "I know we are supposed to follow this new procedure, but if we just make this one exception and do it the old way, I can get it out much faster." One exception follows another and we simply drift back to our old ways, then "Poof!" - the gains are gone.

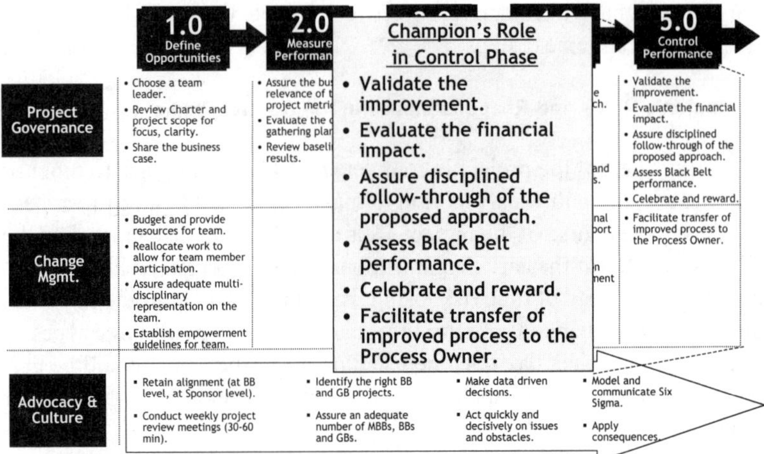

Figure 8-6 Champion's Role in the Control Phase

The Champion needs to plan ahead of time and to work with the team during the course of their project to identify ways to sustain the changes. Simply posting the results and developing a few measures to maintain the gains over time can go a long way toward sustaining the gains. It is also necessary to keep a spotlight on the changes until the new behaviors are more routine. New habits are hard to learn. It's not that people don't want to think, but things can happen faster if we know what needs to be done and we do it automatically. Dr. W. Edwards Deming once said, "It takes 16 repetitions to burn in a new habit." If something is only done once a month or once a week, this can take a long time.

To change behaviors, one literally needs to forge new neural pathways in the brain and once they're initially formed, they need constant reinforcement. New scanning methods show that this deep section of the brain lights up when we develop and express sequential motor acts. It also lights up in response to rewards. With the new ability for researchers to see the brain's electrical activity while learning is in progress, they can actually see patterns of activity change permanently after learning takes place. Learning a habit is different from other kinds of learning. We often are not aware of developing a habit, and we develop it slowly over time. "The process doesn't seem to go in reverse, or else we don't have access to the means to reverse it," Professor Graybiel [1] said. "Unlike an association between an object and a word ("Oh, so that's a blue jay!"), learning a

[1] http://web.mit.edu/newsoffice/tt/1999/sep22/habit.html

habit is very slow. It takes many repetitions, often reinforced with positive feedback, before an action or series of actions become a habit." Strong positive or negative motivators help develop or break habits. Positive feedback is more effective than negative. "The brain has an absolutely fabulous system for getting reward signals," said Professor Graybiel. The system is so sensitive that researchers have seen nerve cells fire in response to a single word, evoking a craving long after an old habit has been kicked. [2]

The degree to which communications have been maintained throughout the project can go along way toward preparing people in the organization for change. A periodic walk-through by a senior manager to check on "Did the change stick?" is valuable and communicates the importance of the project. Clear lines of accountability will also help to sustain changes once the spotlight of the team event has passed.

Control Phase Questions - Management Perspective

Questions for All Teams

1. Did the team conduct a pilot of the new process? What was learned and changed as a result of the pilot?
2. What is the new process result (sigma)? Has the project achieved its goals? Has the team demonstrated the new results with data?
3. Is the learning transferable across the business? What is the action plan for spreading the best practice? What lessons did the team learn?
4. What are the variables being monitored to assure that process performance improvement continues?
5. What training has been conducted to assure the new process runs as expected? How are new habits or behaviours being learned?
6. What are the financial results of the project? How have they been calculated and documented? What changes is the customer seeing?
7. What are the next steps?

[2] TechTalk, MIT News Office, Deborah Halber, 9/22/99, http://web.mit.edu/newsoffice/tt/1999/sep22/habit.html

Summary of Leadership Roles in Deploying Teams

This chapter has focused on the leadership requirements during the implementation of Six Sigma, Kaizen, Lean, and most other team-based process projects. In particular, the roles and responsibilities of the team Champion in the deployment of teams were emphasized. The Champion has Governance responsibilities as the representative of the leadership team to the project team. The Champion has Change Agent responsibilities as the management level representative of the team to the organization at large that is affected by the team's work. The Champion also has an Advocacy responsibility to communicate and support the assignment, the goals, and the greater objectives of the team, and to get resources required for the team to do its job.

Chapter 9

Kaizen Teams

The essence of *Kaizen* is simple and straightforward: *Kaizen* is the Japanese word for "continuous improvement" (Masaaki Imai). This improvement tool was introduced into the U.S. in the mid 1980s. Pratt & Whitney, General Electric, and a number of other U.S. companies were early practitioners. In Japan, *Kaizen* is the foundation for all improvement endeavors: TQM, Quality Circles, Quality Function Deployment, etc. The Japanese *Kaizen* process emphasizes continuous, ongoing improvement. If you walk into a factory or service operation, it should *not* be operating the same way it was a year before. Things should change, becoming faster, better, and cheaper. *Kaizen* incorporates a process view in addition to the traditional American pursuits of innovation and results.

Common *Kaizen* Team Deployments

Kaizen teams are a key part of any Lean initiative. Lean or flow concepts were originally adopted by manufacturing companies. Today, they are spreading into other areas including administration, new product development, and accounting, and into other industries such as banking, health care, and governmental operations. *Kaizen* teams tend to use visual observations and simple data gathering tools to drive waste out of processes. Like any other type of improvement team, *Kaizen* teams have the greatest impact when they are linked to an organization's overall strategic direction. Organizations seeking growth and operational flexibility often target core business processes as strategic change vehicles to satisfy customers and stay ahead of competitors. *Kaizen* teams complement and improve the effectiveness of these changes.

Typical Operational Improvement Objectives	Typical Project Areas of Focus
Teams should strive to accomplish "stretch" goals: • Reduce space by 50% • Reduce inventory by 70% • Reduce overall (total) costsby 20% - 30% • Achieve a 20% productivity gain annually • Cut lead times from weeks to days to hours to minutes • Improve quality ten-fold • Improve safety by 50% • Improve turnaround time (document processing) by 60%	Initial candidates are visual processes with a lot of steps or movement: • Order release time to production floor • Changeover and set up times • Packing and shipping times and materials • Lot sizes, work-in-progress (ideally reduced to units of one) • Scrap, rework, and/or errors • Cell layout/material flow • Equipment reliabilityo Staffing requirements • Paper/information flow

Figure 9-1 Kaizen Projects in Support of Objectives

Kaizen Team Project Selection

A *Kaizen* team is best used when the project meets the following criteria:

- Project involves physical process operations with visual observations possible
- Analysis and implementation can be done quickly (three to five days in duration)
- Focus on waste elimination, faster cycle time, and flow
- Linked to meaningful business strategies
- Opportunity to develop people and gain buy-in to continuous improvement by the people impacted by the changes
- Goal is to reduce the "hassle factor" in getting work done
- Can get three to eight team members, full-time for duration

Kaizen Team Project Success Factors

- *Kaizen* projects should be visual in nature and improvements should primarily be identifiable using physical observation tools versus in-depth analytics. They tend to be narrow in scope. These projects should relate to customer or business needs, just like any other improvement project, and should have a scope large enough to allow several people to work together. Ideally, *Kaizen* projects permit teams to experiment with changes on the spot.

- Best practices would include a videotape of the as-is process prepared ahead of time to give the team a running start. A videotape record makes detailed analysis much easier to complete. It is wise to show the timer on the videotape for easier analysis by the team.
- If a team is observing something that only happens periodically, the timing of the team's observations needs special attention and management by the team's lead and Champion. Think about observing a production change over or the opening of mail in a department. Those are events that do not happen continuously throughout the day. They can be simulated, but it is best for a team to observe real work being done in real time. So the leader running the *Kaizen* event needs to make certain the team is ready for their observations when the event is happening. This is another reason videotapes are important, because they are a great backup. Sometimes a project Champion can put pressure on people who schedule work to schedule something convenient for the *Kaizen* team, but that is probably not the best way to make it happen. However, the *Kaizen* team must have something to observe and somewhere to test solutions, even if it is inconvenient for the organization.
- Teams should be fast tracked for implementation support of improvement ideas. They may elect to move equipment, reconfigure the work area, etc. during the course of their work.

Strenths and weaknesses of Kaizen teams can be summarized as follows:

Strengths	**Weaknesses**
Minimal team training needs Rapid deployment Simple data gathering tools Implement changes quickly Very visual More involvement of the people who do the work	Only works for physical, visible waste May disrupt other processes No time for in depth analysis May miss "root cause" of problem Later accountability for recommendations not initially implemented

Figure 9-2 Kaizen Team Strengths and Weaknesses

Kaizen Team Membership

Kaizen teams usually have five to eight members. They include employees from the work area, one or two employees that are not familiar with the process (often they ask insightful questions), and possibly a process customer or process supplier. Teams typically include the manager or supervisor with the most vested interest in the *Kaizen* team's results. This individual helps to remove barriers, obtain resources as needed, and share background information on the current process when requested. The supervisor should not dominate a team or try to overly influence recommendations. Do not overload the teams with professional staff. A couple of people on the *Kaizen* team should have strong analytical skills to lead the data gathering.

Kaizen Team Project Focus

The focus of *Kaizen* teams is "Do it!" The true emphasis is on getting things done now. The biggest difference between a *Kaizen* team and other types of teams is that *Kaizen* teams are very rapid implementation teams. Recommendations may be leftover for more complex changes, but teams are encouraged to implement to the greatest extent possible during the three to five days of their existence. As a result, the scope of a *Kaizen* project is narrow and the tools the team uses for analysis are simpler.

Kaizen Team's Work Plan

Kaizen teams typically follow a fairly standard roadmap for doing their work. The steps can be customized for the unique requirements of a team's project mission, improvement goals, and assignment. The traditional Six Sigma DMAIC project phases can apply to *Kaizen* teams.

Define Phase for the Kaizen Team

The purpose of this step is to develop a clear understanding of the improvement task, to understand "What is important?" from a business perspective. The team should meet with the project Champion (or

Sponsor) at the outset. The expectations on what is to be accomplished should be clear. Too often, teams will delve into solving a problem without a clear definition of what they are trying to change. In the Jones and Womack book *Lean Thinking*, they talk about "learning to see." That is a key part of what is happening in this step for people who do the work. In some instances, this is the first time that people who do the work have actually been able to take an objective view of the work they perform.

Deliverables in the Define Phase

The primary outcomes for the Kaizen Team's Define Phase include:

- Team Charter is understood and accepted
- The process has been physically walked through by the team
- Sub-teams have formed to take responsibility for the different analytical tools.

This chapter describes the actions required to accomplish these results along with other tools and techniques that are appropriate for *Kaizen* teams.

The very first thing the team should do after a brief training session (typically two to four hours) on *Kaizen* tools, is to get organized and oriented, and to get to know teammates. As with any project improvement team, the *Kaizen* team's project responsibilities begin with the project charter. To get organized and oriented quickly, the team should meet with the Champion to review the charter and learn his or her expectations and goals.

Kaizen Team Charter

The *Kaizen* Charter is not significantly different from the charters for the other two types of improvement teams. The same charter could be used for all three types of teams. The one shown here is a slight variation from the Six Sigma Team Charter. The team's Charter includes: the mission (purpose) for the team, business case, project goals, project scope, Champion's name, team members, and exclusions from the scope. Timing is usually not critical to a *Kaizen* team because they essentially complete their work in one week or less period of time.

The Team Charter should address why this team exists. What is its purpose? The mission statement becomes a starting point for measuring the effectiveness of everything the team does. If they know where they are headed, they can figure out what to do to get there. Typical *Kaizen* mission statements might read:

- Reduce the time and difficulty to change over the molding machine.
- Increase the amount of time sales people can spend with customers.

The *Kaizen* Business Case and Project Goals will put numbers on the mission purpose.

Kaizen Business Case and Opportunity Statement

The Business Case and Opportunity Statement can actually be combined into one category. This factually describes why the organization is taking on this project. It helps the team see why their work is important. Typical business case statements might read:

- We have been losing customer orders as a result of late deliveries. Currently 12% of all deliveries are late.
- Customers reaching the call center have reported a high degree of frustration with our inability to quickly understand their issues, causing a number of them to buy services from our competitors.

The business case begins to clarify the "pain" being experienced by the organization. If numbers are known about how just how painful it is, they should be stated.

Kaizen Project Goals

The goals simply state what success looks like. The goal statements clarify what is important. They also guide the team in selecting metrics.

Goal Statement	Statement Evaluation
Reduce cycle time	Does not state by how much or for whom.
Reduce changeover cycle times for the finish and coater machines by 50%	This begins to clarify what the team is to do. (With a Six Sigma project, it might be even more specific.)
Decrease number of customer call backs	Which customers? For any type of call?
Decrease the number of customer call backs regarding delivery times by 70%	Again, this begins to be more specific. It might be appropriate, but, depending on the situation, it may also need to state the product or service family or for a specific customer group.

Figure 9-3 Evaluation of Goal Statements

A clear goal lets everyone know when the objective has been accomplished.

Kaizen Project Scope

Project teams lose more time due to muddy or unclear scopes than for any other reason. If the scope is not clear, team members waste a lot of time arguing about the scope or working on the wrong things. The scope describes where the project starts and where it ends. This is usually pretty straightforward for a Kaizen team. It gets murkier for the other two types of teams. Typical scope statements:

Step	Change Over Project	Salesperson "Face Time" with Customers
Starting Step	Last good piece of the job currently running	Beginning of salesperson's work day
Ending Step	First good piece of the new job up and running	End of salesperson's work day

Figure 9-4 Kaizen Project Scope Statements

The right scope is an important project Champion responsibility. With *Kaizen* teams, it is particularly important to avoid a scope that is too large, since the team only has one week to get its job done. There is no one "rule of thumb" but experience helps. The scope should be broad enough to implement a meaningful change in a period of one week or less. It is amazing what people can actually do during this brief timeframe.

Kaizen Project Exclusions

Unfortunately resources are not unlimited. So on most projects, there are some things that cannot be changed at that point in time. Key items should be clarified for the team. This is largely an extension of the statement of project scope. Typical exclusions from *Kaizen* projects include: large capital expenditures, changing information systems, and adding people to the process. Key exclusions should be noted.

Process Maps

If a *Kaizen* team is working a process on a factory floor, they typically do not need a formal process map because physical maps of the observable process are more common. If the team is working on an administrative or service process, a formal process map can be very helpful. The map can draw the flow that is sometimes hard to see in administrative or service processes.

Kaizen Team's Work Schedule

Kaizen teams only have a week (or less) to complete their project. It is important that the time be well managed or the team will not get its work done. The team should develop a working schedule as part of their initial activity. This is a typical one-week plan:

Activities for the Week
1. Meet with team Champion to review the charter, learn his or her expectations and team goals 2. Attend a half day training session to learn about Kaizen analysis tools 3. Get to know your teammates 4. Observe the operation by physically walking through the flow of one part or document, led by the team's internal leader. 5. Review any videotapes made of the process before hand 6. Divide the project work among the team members to get started:

Scorekeeper, Map Maker(s), Process Time Analyst, Quality Auditors

7. For a large process, sub-teams can start simultaneously at different points along the process.
8. Create a wall display for documenting the work flow
   a. Process steps
   b. Process characteristics data (time, space, inventory, quality, value-added, etc.)
   c. Process maps
9. Determine what to measure
10. Post an Opportunities Log sheet to record metrics as they are identified
11. Map the process steps
12. Record the sequence of operations (process steps)
13. Take baseline measures pf as-is process
14. Analyze the process characteristics to identify improvement opportunities
15. Do a 5S (sort, set in order, shine, standardize, sustain) assessment
16. Brainstorm improvement opportunities
17. Prioritize improvement opportunities
18. Brainstorm potential solutions for the key opportunities
19. Prioritize the solutions
20. Experiment with solution in the work area to test for viability
21. Take new measures to capture the results of the experiments
22. Review the results of the experiments
23. Fine tune and adjust or try new solutions (repeat the above couple steps until a satisfactory solution is discovered)
24. Do a new 5S assessment
25. Document the results
26. Document any open action items
27. Prepare for the management/Champion presentation
28. Present to management
29. Turn over project documentation to appropriate party
30. Celebrate success!

Figure 9-5 Sample One-Week Plan of Activities

Activities in the *Kaizen* Team Define Phase

Kaizen team members should have a convenient place to work during the course of their project. It should have adequate wall spaces to mount the information displays used for team working sessions. Visual wall displays make it easier for the *Kaizen* team members to communicate within the team and to bring people in from outside the team for updates and questions.

Process Walk-Through

It is important for the team to get out on the floor/work area and observe the operation by physically walking through the flow of the part or document, typically led by the team leader. On the first pass, people should simply observe the process. They do not need to be taking time studies or drawing the activities during their first walk through.

Sub-Team Assignments

Team members have a tendency to want to do everything together. After the initial observation is completed, it is important to divide the project work among the team members to get started. For a large process, subteams can start simultaneously using different tools at varying points along the process. Typical roles include:

- Scorekeeper: Organizes the main project information display
- Map Maker(s): Creates a wall display for documenting the work flow
- Process Time Analyst(s): Records the sequence of operations (process steps)
- Inventory Analyst(s): Identifies and documents the inventory levels of materials or documents at each process step
- Requirements Analyst: Gets the process output requirements from the process customer(s) and posts a summarized version
- Pacer(s): Measures the number of feet, meters, paces, etc. traveled for each step in the process.

This work is often done best in teams of two or three. For example, when conducting time studies, two people might record the process steps while one person records time. On a team of eight or fewer people, team members will need to be responsible for more than one role.

Studying Process Video

A videotape prepared ahead of time can be a valuable tool to a *Kaizen* team. It is difficult in many instances to capture times, especially for something that is not continuous, like a set up. In most cases, a video recording helps a team:

- analyze their subject process more effectively because they can see the steps and times in detail, in slow motion if necessary, to give them time to fully understand the process, using process analysis and design check-lists, and
- avoid personal resistance to change since the video reinforces that the project is focused on the "process" not the "people."

Just a few guidelines will ensure the usefulness of videotape recordings of operating processes to be improved by a project team:

1. Use the in-frame time stamp display in hours, minutes, and seconds (HH:MM:SS). This is critical. Without the time stamp, team members will waste a lot of time in separate stopwatch recordings. Worst of all, they will have trouble establishing consistent time breakdowns, particularly while doing slow-motion studies when a real-time watch is useless. Double check that the in-frame display includes seconds, not just minutes and hours.

2. Work with the operators in advance to plan camera angles that will show as much of the work details as possible. Strike a balance between angles that are close enough to see details but wide enough to show the overall work area and how things are positioned relative to each other.

3. If helpful, create title cards that can be held up by operators to indicate the start of each new process segment, or other important information. This may be a by-product of creating the process outline.

4. Decide, in advance, if several different process cycles need to be recorded. For example, the process methods and time may vary considerably from one product to another in the same family. An average unit and two extremes might be a good initial range.

5. Meet with everyone in the work area in advance so they know the

purpose of the video - that it's focused on the process not people - and that they may be helping the project team to interpret it while looking for process improvement opportunities. Ask everyone to retain their normal work pace or methods for the video. Normal conditions are best.

Kaizen Team Building

Kaizen teams are only together for one week. Usually they do not experience the same degree of conflict as teams that work together for a longer period of time. The *Kaizen* experience can still be intense and exciting. Team members are typically assigned to this full-time for the duration of the *Kaizen* event, so this type of team will not experience the same types of problems other teams do. Still, it is worthwhile for Kaizen team members to agree to a few rules for working together. Team norms are simply an agreement on the way we wish to work together, as a team. They might include:

- Active listening - making an effort to listen without getting defensive or wishing the other person would be quiet, so you can talk. Concentration on what is being said by others.
- Play flat out - if you have something to say, say it. But do it in a way that you do not say something hurtful to other people in the room. "Don't trash your neighbor!"
- Manage by agreement - practice making and keeping clear agreements with other team members, a foundation for trust.
- Be open - the success of the Kaizen event partially depends on people's ability to consider new possibilities.
- Participate - take personal responsibility to be involved in the team's activities.
- Have fun - life is serious, but fun is life. It is important to be successful and make some meaningful change, but don't get too caught up in the moment.

Teams do not need to adopt all of these norms. Select two or three and practice them. This list is not all encompassing but it does list some powerful behaviors that can help teams be successful.

Define Phase Tollgate Questions

The team should pretty much complete its Define phase by the end of day one. Normally, the team's Champion would drop in at the end of day one for a fast, informal briefing of how the day went. Questions to address for the completion of day one include:

1. Does the charter and team goals make sense and does the team fully support them?
2. Has the process walk through been completed?
3. Has the team divided into sub-teams?
4. Are the people in the work area being observed comfortable with the presence of the Kaizen team members?
5. Is the team ready to capture baseline metrics?
6. Did team members agree to follow a few norms?

Define Phase Example - Computer Chip Machine Setup Time Improvement

Pearlie Johnson was charged with reducing the set up time on two high-speed computer chip machines and to improve the overall throughput on the line. She worked with her project Champion and agreed that the improvement target should be a 50% reduction in changeover time and $100,000 in cost savings. She also discussed with the Champion who else to include on the team. They decided to include two operators from the work area (one from the day shift and one from the night), a supervisor from one of the five production lines, a system programmer (person that changes the programs on the high speed insertion machines), a design engineer (because the company was going to be redesigning the base product family next year), a person from the materials department, and a maintenance technician.

The team received a half-day of training on basic *Kaizen* analytical tools and participated in a simulation called the "Ping Pong" factory that highlighted how newly designed processes are sometimes less than a thing of beauty. The Champion met with the team and shared her expectations. Their scope was from the last good part of the old job to the first good part of the new.

When the team went to do their walk-through, a changeover was already taking place. So they just stood back and watched. Pearlie realized that she needed to coordinate better with the scheduling group and the supervisor on the floor, so that the team would be on the floor at the right time for their next observation. Matt (one of the line workers) observed, "These people don't know what they are doing. They should have had the chip sets for the new production run already in place." People also seemed to drift in and out of the workstation while the changeover was taking place.

After the observation was completed, the team members went back to their meeting area and divided into sub-teams.

Define Phase Example - Billing Department Customer Service Improvement Project

A team was commissioned to look at the customer dispute process for a credit card transaction processing center. Kathy Simmons, the team's Champion, had wanted to work on this for some time. The organization was growing so fast that they were having a great deal of difficulty keeping up with the mail backlog of customer requests for information and customer concerns over transactions on their credit card billing statement. Kim Redbill, a supervisor in the billing department, was selected as the team leader.

A team was selected that included two senior customer service representatives, one new customer service representative, one person who answered customer questions on the phone, a person from the information services department, and someone from accounting. People in the department were concerned because they were being required to work excessive overtime and they often had difficulty finding information.

Kathy met with the team to share her expectations. She was looking for a 50% improvement in customer response time and $300,000 in cost savings. The team did an observation walk-through. The mail initially was stored in file drawers and then later bundled out to the workforce. When people started to work on the disputes they would create a paper file that also went to a central storage area in the department.

After the walk-through, the team members looked again at the charter and asked, "How can we possibly come up with $300,000 in cost savings?"

Kim replied, "I'm not sure yet but let's wait and see what we learn."

Charter Section	Customer Dispute Team
Business Case:	Currently the mail is piling up in the files. Backlogs are increasing by 10% per month. And when we do not respond to customers quickly, they overload the operator representative system by calling in to see what is happening with their information request. In the current process, new mail is not even opened for the first 10 days.
Goal Statement:	Reduce the backlog and improve customer response time by 50%.Identify and implement cost savings in excess of $300,000.
Scope:	Start - when a dispute is received in the departmentEnd - when the customer file is closed
Exclusions	No additional staff hiresNo new information systems

Figure 9-6 Portion of Customer Dispute Charter

Measure Phase for the Kaizen Team

The purpose of the Measure phase is to baseline the current state of the process, and to provide factual evidence that a problem or opportunity exists. Basically, it answers the question, "How are we doing?" Six Sigma and Lean teams tend to look for "Critical" to customer or to process metrics. *Kaizen* teams really do not have time to do an in-depth analysis to verify anything to that level of detail. So *Kaizen* team metrics tend to be more process-centric and time-oriented. It is always wise to check with the immediate upstream process (supplier) and downstream process (customer) to make certain there are no obvious conflicts. Like any effective improvement team, *Kaizen* teams need to collect data to verify the baseline.

Deliverables in the Measure Phase

The primary deliverables for the *Kaizen* Team's Measure phase typically include the results for specific metrics that they gather during the course of their work. While Six Sigma and Lean teams need to put together a measurement plan, Kaizen teams simply do it now. Metrics typically include:

The primary deliverables for the *Kaizen* Team's Measure phase typically include the results for specific metrics that they gather during the course of their work. While Six Sigma and Lean teams need to put together a measurement plan, Kaizen teams simply do it now. Metrics typically include:

- ***Cycle time*** - the actual elapsed process time from the completion of one output unit to completion of the next unit.
- ***Lead time*** - the length of time from the beginning point of a process to the completion of a finished output (calculation varies depending on scope; it includes: queue, wait and move times).
- ***Takt time*** - the unit-to-unit pace of production required to meet customer demand. Takt is defined in the Lean chapter. It simply refers to pace.
- ***Processing time*** - the value-added component of lead time
- ***Queue time*** - the lead time that multiple units WIP (Work-In-Progress) wait until they reach the next step in the process (non-value adding time)
- ***Set-up time*** - measured from completion of the last good unit of the completed job to the first "good" unit of the new job.
- ***Travel distance*** - distance a part or person travels in doing their work or a job being completed.
- ***Value-added time*** - the time spent changing the form, fit, function, or information content of a product or service in a way that the customer values and is willing to pay for.
- ***5S*** (sort, set in order, shine, standardize, sustain) - this is an assessment score on the unit's housekeeping and organization.

Baseline metrics are tightly linked to process maps, process steps, flow diagrams, etc. Every *Kaizen* team may not do all of the metrics listed above, but it is a reasonable list of things to consider.

Data Capture Forms

An important step for most *Kaizen* teams is to capture the time it takes to do each step in a process. The information can be pulled together in an Excel spreadsheet or some other format. There are three forms that *Kaizen* teams use to capture process steps: Set up Analysis Form, Process Times Observation Worksheet (for repetitive processes like assembly), and some type of a Process Characteristics Summary (PCS) Form that gathers multiple pieces of information on a process in one place. The listing of the process steps is the same for all three forms, and *Kaizen* Teams would typically just use one of them. They simply have different ways of tracking the time and related data. This section on Measure describes the PCS; the Set Up Analysis form is described under Analyze. The Process Times Observation Worksheet is not included here.

<u>Process Characteristics Summary (PCS) Form</u>

The *Kaizen* Team's version of a Process Characteristics Summary (PCS) is a simple form or spreadsheet to pull together a few related information items. The information that goes into a PCS is process specific. A PCS chart organizes the process data in a useful format that is not possible on a process map or plant layout with material flows. It relates the various process characteristics to each other - within each process element and the process overall.

A *Kaizen* team can build their PCS on flipchart sheets taped to the wall and use post-it notes and hand drawn columns. Or the team can create a simple data collection form using Excel or some other spreadsheet tool. Teams should avoid the temptation to measure secondary data items because the project timeframe is so short that there is not sufficient time available for excessive information gathering.

To build its PCS Form, the team first decides which data items to capture and lists them in column heads at the top of the basic chart. The visual review of the workplace can provide cues to select data items. For example, if floor space reduction is a key goal, the team would probably create a column for just that.

Next, the team lists the steps for the process in the left hand column of the basic chart. The team describes each step with key words telling what activities occur at that step, leaving extra space between steps so that additional steps can be added as they are learned. This is easy to do if you are using post-its.

Then the team goes out to the work area and collects base data to enter into the chart. The team can split up the work in logical groupings for work efficiency - one person or one sub-team does not need to observe the entire process. It is wise to do this in groups of two or three with one person writing the steps, one person observing the steps and saying when the next one happens, and one person running the stopwatch. At the start of each new process step, the timekeeper should state the running time.

Process times should be captured for each step of the process. This is the actual elapsed time for each step in the process. It includes:

- Touch Time (time that the person performing the task actually touches the product, part or document),
- Machine Processing Time (for the automated portion of the process), and
- Inventory Time (time while materials or documents wait in queues).

The team does not need to get overly granular in capturing these times. Whenever someone picks something up, sets something down, goes somewhere, comes back, etc. is a new process step. This is not a classical industrial engineering study where the Kaizen team is looking at micro movements at a workstation. Many process improvement project teams find that by focusing their attention on overall cycle time, the flow of their analysis work naturally bores down to the related factors of elemental process times (inventory, space, equipment utilization, etc.) as they dissect the waste components in each of the process steps.

Other time characteristics that might be important to the team are lead time and cycle time. Lead time refers to the total elapsed time of the elements along the critical path of the process flow. Cycle time is the time per unit of throughput at the limiting station in the process (the step with the lowest throughput capacity). That station paces the system by allowing a net output of one unit per its normal cycle time. The process cannot produce faster than the limiting station.

Other items that could be captured include: inventory levels, wait time as a separate column, space, transportation distances, walking distances, etc. A sample PCS worksheet follows. In this example, each column shows an "as-is" baseline number and an opportunity number. The latter can be filled in during analysis.

#	Step Description	Who	Value Add	Quality Base	Quality Oppty	Setup Time Base	Setup Time Oppty	Process Time Base	Process Time Oppty	Capacity Base	Capacity Oppty	Walking Dist Base	Walking Dist Oppty	Transport Dist Base	Transport Dist Oppty
			X	*%*	*%*	*Min.*	*Min.*	*Sec.*	*Sec.*	*U/Hr.*	*U/Hr.*	*Ft.*	*Ft.*	*In.*	*In.*
A1	*Raw material bin*							*2,240*	*25.7*						
B1	*State "Load ball" every 5 sec.*	*TK*						*30.1*	*0.0*	*120*	*NA*				
C1	*Move ball from bin to range finde*	*LM*						*110.4*	*7.6*					*18.0*	*12.0*
C2	*Aim & release into processor*	*LM*		*25%*	*100%*			*24.0*	*1.0*						
	S.T.							*134.4*	*8.6*	*27*	*419*				
D1	*Processing unit*		*X*			*10.0*	*0.5*	*2.4*	*2.4*	*1,500*					
E1	*Receiving bin*							*224*	*15.4*						
F1	*Wait for ball from processing*	*MA*						*19.2*	*0.0*						
F2	*Transfer ball into inspect. bin*	*MA*						*19.8*	*12.8*					*15.0*	*0.0*
F3	*Pass insp. bin to inspector*	*MA*						*2.6*	*2.6*			*2.0*	*1.0*	*26.0*	*12.0*
	S.T.							*41.6*	*15.4*	*87*	*234*				
G1	*Inspect received pallets*			*0%*	*100%*										
	Totals (from sheet above)			*0%*	*100%*	*10.0*	*0.5*	*2,673*	*67.5*	*27*	*234*	*2*	*1*	*59*	*24*

Figure 9-7 Process Characteristics Chart
Used with permission from the Cumberland Group - Chicago

There are many other characteristics to describe and analyze a process. Again, the team should avoid the temptation to over-measure the process. Too much information can be worse than not enough. The team should try to pick the few measures that help to manage the upstream causal factors (process times, set-up times, etc.) while keeping an eye on the downstream end results (operating costs, inventory investment, etc.). Upstream measures should produce process understanding and real change management. And the overall results measures track that results are what they were expected to be.

Spaghetti Diagram

This is an extremely simple tool used by most *Kaizen* teams. Two team members simply walk around and follow the steps taken by employees in doing their work. The following spaghetti diagram was developed for changeover of a hydraulic pump manufacturing operation. The two operators walked a fair amount.

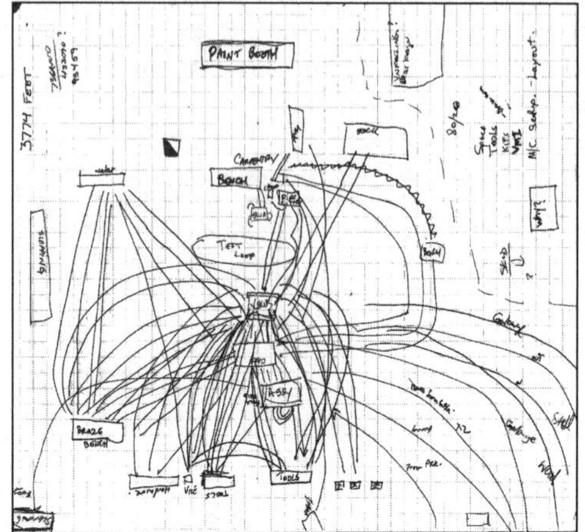

Figure 9-8 Spaghetti Diagram

The diagram works for administrative and service processes as well as manufacturing processes. In administrative and service processes, people walk around looking for information, attending meetings, coming unprepared to events, looking for analytical tools, files, etc. The diagram is easy to do - just grab a blank sheet of paper and a pencil, do a rough sketch of the area under observation, and start following people around. It is generally best for one observer to track one employee. It is hard to track two or more workers at the same time. The Spaghetti Diagram can also be one of the first steps in defining "standard work." Standard Work is the best practice for the way the organization would like for work to be done.

Five S

Another common tool for *Kaizen* teams is 5S - Sort, Set in order, Shine, Standardize, Sustain. Some add "safety" as a step and call the tool 6S. A 5S baseline should be established during the measurement phase.

5S is a simple way to organize a workspace - be it a shop floor, an office, or a research department. It's much more than "good housekeeping". It can reduce aggravations in doing work, instill a discipline for doing standard work, and build flexibility into the workplace.

The 5Ss were originally five Japanese words that all have an 's' sound at the beginning. The typical English language categories used are:

- Sort - to only have what you need, when you need it. Items that are not needed for doing work are removed from the work area, the department, and the operation.
- Set In Order - items are identified in a way that:
  - They are easy to see
  - Anyone can find them
  - People can put them away without a lot of hassle
  - Adequate storage space is provided for storage and disposal of tooling, equipment, reports, etc.

  Visual controls are a critical part of set in order (sign boards, labeling, color coding, etc.)
- Shine - means to remove dirt and grime from the workplace. Shine should help to keep everything clean and working so that when someone needs to use something, it is ready for use. Shine should happen daily, not once a year during spring cleaning.
- Standardize - describes the easiest way of doing the job to achieve quality requirements in a safe way. It fits with "standard work" and prevents set backs in sort, shine, and safety. Standardize focuses on eliminating non-value adding steps and practices and waste in all its forms. Standardize works when:
  - Work instructions are clear and easy to follow
  - Metrics exist and are utilized to reinforce expected practices
  - Visual controls, guides, and focal points are utilized
  - They are kept up-to-date
- Sustain - to properly maintain effective 5S practices. It includes ongoing evaluation of 5S practices, and management and associates cooperate to make it happen.

Pre 5S Practices

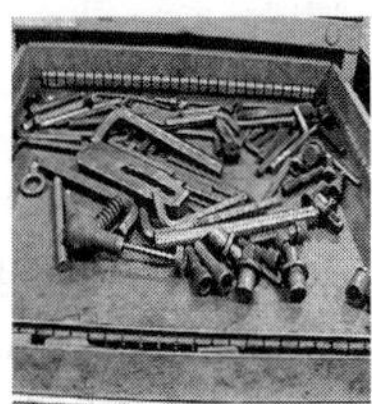

Post 5S Practices

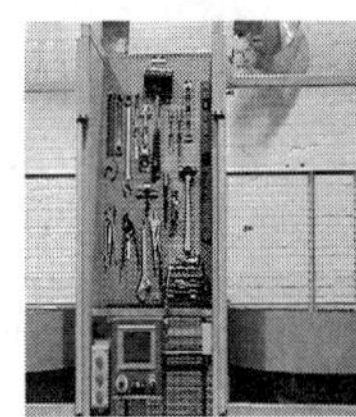

Figure 9-9 Before and After 5S Practices

This is a simple, yet powerful, tool. Some organizations have used 5S to drive an entire process improvement effort because it is practical and low cost. The only problem with using 5S as the sole approach is that the payback from 5S activities take some time to impact the bottom line. The workplace is cleaner and better organized. More work gets done following standardized procedures. And employee associates can become actively involved in the improvement process. But these initial changes rarely uncover major savings opportunities. Over time, as more people follow the 5S practices, the benefits will be realized. But it does not happen overnight or as a result of implementing 5S in one area of an operation.

Some *Kaizen* practitioners have difficulty in scaling the 5S instrument. The example 5S below may be help. There are 20 questions on the example. Use a five point scale with a '5' being World Class and a '1' being significantly less than that. If you think of a totally organized grocery store, the best one you have ever visited in terms of product display and signage, and consider question number five on the 5S Worksheet, "Storage locations are identified….." If the grocery store rated a '4' on labeling in the meat cooler, use that as a benchmark. Now compare the actual work area being examined to that standard. Also, consider places that are not quite so well organized; say for example the average "retail outlet." They are typically not bad, but probably do not rate much higher than a 3 in most categories. Once the team has scaled the instrument (i.e., all team members have some idea of what a '5' is and what a '1' looks like), they can rate the area they are observing. It is best to score the worksheet for the entire area overall, and not for individual sections or areas of the workplace. The final score for each line item is the lowest common denominator, not the average of the area.

The team can reach consensus on a score scale and system by doing the following: Bring the team together and go through the questions one by one. Ask people what their score is. Find the highest number and the lowest number. Ask those raters why they scored it that way. Don't judge their responses; simply seek to understand their thinking. Allow some discussion after each question and then ask the team if they can agree on a score. Sometimes it will be the higher or lower number and sometimes a number in the middle. Usually, people can agree on a number pretty quickly. It is better to stay away from decimals and just work with whole numbers.

The 5S worksheet can be used as one of the tools to help sustain changes after improvements are implemented.

#	Description	Score	Comments
Sort: Get rid of what is not needed (no clutter)			
1	The work area does not contain boxes, containers, tooling, product, raw materials, information, reports, etc. that are not relevant to current work		
2	A log sheet is kept up-to-date and reviewed on a regular basis, unneeded items are tagged and removed from the work area		
3	Team members understand Sort and follow a process to regularly tag and remove items that are not needed		
Set-in-Order: Arranging items so they are easy to see, use and put away			
4	All items needed for the current BOM being worked are present and the work area does not include items for other BOMs or jobs		
5	Storage locations are identified and utilized for tools, products, reports and other work items in the cell		
6	Use of visual tools is high (i.e., labels are clear, easy to read and accurate) and easily seen by anyone visiting the work area		
7	Team information board and other documents are up-to-date and trigger some team actions		
8	Tools, products, reports are put away in their proper place when not in use; it's obvious when tools are in use away from storage locations		
9	Equipment, tooling, supplies, & gages accessible and stored by frequency of use		
Shine: Work area is clean and free of debris			
10	Each workstation has a daily cleaning list, and each team member's responsibilities are clearly defined and done daily		
11	Floors, workplace and equipment are clean and free of debris		
12	Recyclables, process waste and garbage are collected and removed from the work area on a timely basis		
Standardize: Consistency of work processes and practices			
13	Standard operating procedures (instructions) are posted for set-up, operations, cleaning and maintenance at relevant workstations		
14	People understand and follow the standards, operators do their jobs in a consistent way, everyone is involved in the daily routine		
15	Cross-functional departments cooperate in maintaining and practicing 5S		
16	Job check sheets and control screens used and easy to understand		
Sustain: Self-discipline to maintain best practices over time			
17	Areas that have been sorted consistently stay clean		
18	All team members feel responsibility for and ownership of 5S, with rotating responsibilities and cross shift cooperation serving as the norm		
19	Standards are considered targets for improvement with systems to promote on-going improvement at work-site		
20	Visible performance feedback tools are up-to-date & used in the work area including regular 5S reporting		
	Total Score		

CUMBERLAND

Figure 9-4 5S Evaluation Worksheet

Measurement Displays

To the extent time is available, the *Kaizen* team should visually display their measurement results using histograms, Pareto diagrams, and other simple data displays.

Measure Phase Tollgate Questions

The team's Champion should address these questions after the team has completed the Measure Phase:

1. How much data were collected?
2. Has the team established a baseline for the key time related metrics (leadtime, cycle time, value-added time, etc.)?

3. Were the data checked by anyone outside of the team for reasonableness? (A Kaizen team would typically not go through a measurement validation like a Six Sigma team, but they still should do some reasonableness checks.)
4. Are all significant process steps captured?
5. Does the information gathered look useful for analysis and decision-making purposes?
6. If process maps were needed, have they been completed to a sufficient level of detail?

Measure Phase Example - Computer Chip Machine Setup Time Improvement

It was getting late in the afternoon of day one for the Changeover *Kaizen* Team. Pearlie thought they could get one pass at observing a changeover that was scheduled before the end of the shift. So the team talked about how they should divide up the work. They decided to split into four sub-teams of two people each to capture their information. Each person also agreed to get a 5S Form completed by 10 a.m. the next morning. There were two Spaghetti drawing teams and two teams to record process steps and times. One sub-team would observe the front end of the semi-conductor build line where the high-speed chip insertion machines were in operation, and the other sub-team would observe the rest of the production line.

The team members headed out to the floor to start their observations. The setup worksheet breaks process steps into internal and external time, but on their initial observation the team simply listed the steps being done and the time it took for each step. Much of the time seemed to involve getting the new chip reels ready to load. The operator was getting chip reels from a couple of different places. Mary, the system programmer on the team, also noted that the system programmer who should have been loading the new job instruction program was nowhere in sight. Mary took responsibility for writing down the process steps on the Setup Worksheet. Her partner Matt called out the time on the stopwatch as each new step was started. He started the stopwatch running when the last circuit board of the old job moved out of the chip insertion machine.

This is not the complete worksheet, but provides a flavor for the type of information the team captured.

No.	Task / Operation	Run Time Hr:Min:Sec	Current Time Internal	Current Time External	Improvement Opportunity	Proposed Time Internal	Proposed Time External
	Get reel	***3:50***					
	Insert reel in tray	***3:58***					
	Look for job instruction sheet	***4:02***					

Figure 9-11 Portion of Setup Analysis Worksheet

They made certain they captured every separate step, even when Matt said, "but that isn't how we normally do it." If the operator did something, they wrote it down. There were two operators doing the work, so Mary had a separate worksheet for each operator. She decided if they observed this process again, they needed a third person on their sub-team. Overall setups were running about 35 minutes per changeover.

The sub-team members watching the operators work and drawing the Spaghetti Diagram were also pretty busy. They each watched one operator moving around the equipment. When they went back to *Kaizen* team meeting room and merged their drawings, they ended up with a drawing that looked like:

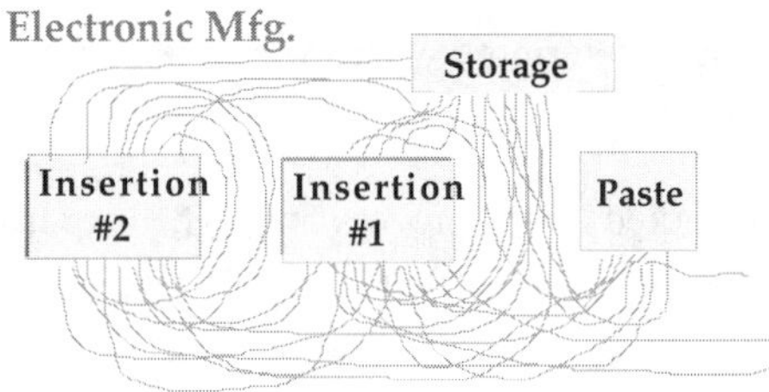

Figure 9-12 Spaghetti Diagram

Their drawing truly did look like a bowl of spaghetti.

The other two sub-teams gathered their information for the rest of the line. The team could calculate transport distances (materials, products, documents) from path measurements on a scale flow layout (for physical processes) or actual floor measurements (for physical processes), using the Spaghetti Diagram or a Standard Worksheet.

Measure Phase Example - Billing Department Customer Service Improvement Project

The Dispute Team decided to use a process map for their initial data and information gathering. They used a simple PCS worksheet to capture the process steps and as a place to store related performance information. They also did a Spaghetti Diagram of people walking around looking for information (customer files, documentation, analysis tools) and looking for supervisors to approve certain things.

To create the process map, the team walked through the process. They then developed a map that highlighted the main steps. They also decided which activities and work volumes needed to be measured. They decided on:

- Lead time - the length of time it took for one dispute document to go through the entire process
- Cycle time - mail was passed out in bundles of 25 letters, so they decided to track how long it took to work one bundle
- Mail volume - determine how much volume was going through the department
- Transaction cost - determine where were the sources of transaction costs, other than staff time

The entire group worked on the process map. The map had more than 75 steps for getting the work done. A condensed version of the map looked like this:

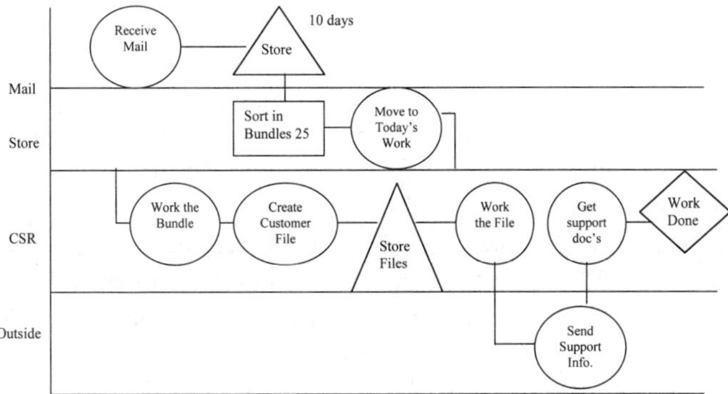

Figure 9-13 Process Map (Condensed version) for Billing Dispute Mail Processing

And these are excerpts from the Process Characteristics Summary (PCS) chart:

Excerpts of Process Steps	Time in Days	Mail Volume	Transaction Costs	Stored Mail
Receive Mail	1	3,000		
File mail in Drawers	1	3,000		12,000
Process Mail Bundle	5	125		
File Dispute WIP folders	1	1,250		8,000
Look for Customer file	2hrs			
Send for proof purchase		1	$5/request	
Wait for proof purchase	5	800		
Etc.				

Figure 9-14 Portion of Process Characteristics Summary Chart

The Disputes team verified that mail did indeed sit there for 10 days before any processing was done, and that it was taking five to ten days to resolve the dispute once a Customer Service Representative began to work it. They also learned that several million dollars were being paid to the credit card parent organization asking for purchase documentation for the customer files.

Analyze Phase for the *Kaizen* Team

The Analyze Phase is where the team determines the key improvement opportunities. Analysis builds on the work started in the Define and Measure Phases. The evidence gathered and the team's skill in pulling the information together in simple ways for people to see can make the implementation road ahead easy or difficult. The team should take advantage of the evidence the data present and use it to drive change. The evidence presented can help people see that new opportunities do exist for implementing change. It's also appropriate to envision how each step might be improved using new methods or technology. At this stage, teams should not try to justify or improve, but merely capture the ideas and use their best judgment to describe potential opportunities.

Deliverables in the Analyze Phase

The primary deliverables for the *Kaizen* team's Analyze phase typically include a list of prioritized opportunities, data analysis that describes a before (a baseline) and a proposed "after state," identification of the problem "root causes," and a few potential solutions. Kaizen team deliverables in Analyze also include:

- A rated 5S score
- Standard work defined for the task(s) being improved
- Projected "opportunity" levels of performance for the areas targeted for improvement
- Takt time, if relevant

Process Analysis

Sub-teams should study the process steps, one at a time, using the *Kaizen* analytical tools. The analysis tools can help to identify the opportunities in the individual process steps. Every step does not need to have an improvement potential. The team should focus on those where meaningful improvements exist. Business process analysis often begins with an identification of the value-added steps or work. Value-added steps typically change the form, fit, function, or information content. These steps tangibly contribute value in the customer's eyes. Non-value-adding steps are waste, all of which should be targeted for potential elimination, if possible.

<u>Waste</u>

Waste comes in many forms. The classic list of waste in a production environment includes:

Manufacturing Waste

1. People (talent, skills underutilized)
2. Delays (waiting, late)
3. Inventory (stock not immediately needed)
4. Facilities (idle machinery and space)
5. Transportation (moving parts, materials)
6. Motion (non-standard, excessive, unsafe)
7. Defects (errors, rework, scrap)
8. Variation (inconsistent process and results)

Most of the items on this list were first described by Shigeo Shingo, one of the pioneers of the Toyota Production System. In a services or administrative environment, the terms might change slightly:

Service Waste

1. Requirements not clear (people don't know or misunderstand customer requirements)
2. Analysis paralysis (an inability to make a decision, when sufficient information is actually available)
3. Wrong sequence (doing work out of order, often resulting in rework)
4. "Turf Wars" (competing against people inside the enterprise, rather than external competition)
5. Unnecessary approvals (unclear or inappropriate accountabilities)
6. Non-value-added overhead (work being done that does not meet an external customer's requirement)
7. Low priority work (working on the wrong or non-important things)
8. Workplace space (idle, excessive, or inappropriate)
9. Variation (inconsistent processes and results)
10. Excessive automation (overkill, more than is needed to get work done, also inconsistent and inappropriate)
11. Information re-handling (rework, re-review, re-inspect/check…)
12. Excessive rework (doing things over again inappropriately)

Source: Cumberland Group - Chicago used with permission

These manufacturing and service waste checklists can help to identify key improvement opportunities. The following questions may help the analysis.

Data Analysis	Opportunities
Value-added	If not, how can it be eliminated?
Requirements	Are they accurate? Do people doing the work understand their customer's requirements? Do unclear requirements cause rework?
Capacity	Which process steps are bottlenecks or capacity constraints?
Lead Time	Would a shorter overall lead time be valuable for customer service?
Cycle Time	Are cycle times comfortably less than planned Takt times at forecasted volume peaks?
Quality	How could defects be reduced to Six Sigma levels?
Material	How can material yields or utilization be increased?
Setups	Add no value? Reduce or eliminate? Run one-unit orders?
Processing	Reduce time, increase rate? More efficient methods? Automation?
Throughput	Where are the bottlenecks? More production hours? More shifts? Reduced maintenance time?
Walking/Handling	How to reduce? Materials closer? Tools closer? Fewer trips?
Inventory	Make one-unit lots? Deliver small lots just-in-time? Flex throughput to exactly meet daily demands? Long set-ups require supply or demand variations?
Floor space	Room for more production? Wasted areas?

Throughput

It is usually worthwhile to calculate the potential throughput rate for work if the team is looking at a multiple workstation process. Throughput Capacity is calculated at each workstation (person or machine) acco rding to the formula:
Capacity = One Hour ÷ Station Process Time Per Unit (or per number of units in the batch, if processing is done in batches)

The Overall Process Capacity is the lowest elemental capacity (through the limiting station). Look for imbalances, this calculation can help to identify underutilized or bottleneck stations.

Takt Time

Takt is a German word for metronome. It's the beat at which the orchestra plays. If an organization is going to move toward flow, it needs to know the Takt time for the operation. Takt time is the rate of customer demand compared to the organization's capability to deliver. In Lean manufacturing, companies try to match the rate of production to the customer's consumption. When they are in balance the customer "pulls" production, and gets what they need only when they need it. Some call this "the customer's pull rate." As product gets made, it gets sent to meet a specific customer order.

Takt time for a *Kaizen* team relates to the individual process being examined. The team must take care not to go too crazy with this number, because *Kaizen* teams are usually just looking at a piece of the overall process. The Takt number is good to know, for determining the ideal rate of work flow through a work center. Takt time is calculated according to the following formula:

Function	Description	Calculation
"#"	Number of shifts per day	
"X"	Number of hours per shift	
"="	Number of hours per day	
"X"	60 minutes per hour	
"="	Number of minutes per day	
"-"	Less: Break (minutes)	
"-"	Less: Wash (minutes)	
"-"	Less: Tool change (minutes)	
"="	Total minutes available per day	
"/"	Unit sales per day	
"="	**Takt Time - Minutes per unit**	
"x"	60 seconds per minute	
"="	**Takt Time -Seconds per unit**	

Figure 9-15 Takt Time Formula

It's a fairly simple concept. How much production or work time is available? So non-work times (time for breaks, lunches, standard meetings, clean-up, sanitation, etc) would be subtracted. Non-work time is any time during the day that is officially not available for work. If that number is subtracted from the regular workday hours and number of shifts, you end up with available "work time". Dividing available work time by the customer demand (often stated as "units per day") yields a production or work rate that equals the customer demand or "pull" rate.

The Takt time concept sounds and is simple. It does not always relate well to the "real world." It is a fantastic concept for a manufacturing company with lines or cells dedicated to a single product. If Takt times change daily (i.e., changing customer demands) or if more than 10,000 variations of products could be made at a work center, the concept may still make sense, but it is not quite as easy to do.

From an analysis standpoint, the *Kaizen* team is seeking to balance the rate of production to the customer's rate of demand. If the rate of demand is not stable for a given product, the team should investigate whether is it stable for a family of products. If not, then Takt time may not be critical to improvement for that particular process.

Setup Time Reduction

A very common target for many *Kaizen* teams is setup time reduction. All setup time is non-value added from a customer perspective. When teams look at setups, they use the Setup Analysis form rather than a PCS to capture the process steps and the process times during the Measure phase. Then, in the Analysze phase, the team members classify all time as "internal" or "external." External is the time it takes to do work that could be done while the old job is still running.

Setups usually include several discrete steps:

- Preparation time to get ready for the setup (this time may start while the old job is still running)
- Remove time to pull out the fixtures, materials, etc. for the old job,
- Install time for putting in new fixtures, materials, etc. for the new job
- Adjust time for setting controls or positioning fixtures
- Test or sample time to check product quality
- Clean time to prepare machinery or the work area for the new production run
- Wait, watch or delay time where nothing seems to be happening

Preparation, adjust, test, and wait are all targets for total elimination from internal time. The other steps can also be improved and usually are by focusing on eliminating all non-essential external time. Setup time reduction projects are good ones to use videotape analysis to facilitate capturing process times.

An example of a setup analysis follows. World-class changeovers would happen in less than 10 minutes. The example form shows the current process on the left and the proposed process times on the right.

Setup Analysis — Work Center: ____________

No.	Task/Operation	Run Time Hr:Min:Sec 0:00:00	Current Time Internal	 External	Start Time - Hr:Min::Sec Improvement Opportunity	Proposed Time Internal	 External
1	*Retrieve tooling items from storage crib - two trips*	*11:30*		*11:30*	*Retrieve standard tools before shut down*		*6:30*
	Turn off machine	*13:30*	*2:00*			*2:00*	
	Adjust fixtures	*15:30*	*2:00*		*Preset fixtures*	*0*	
	Replace cutters	*18:00*	*2:30*		*Easier placement of cutters*	*1:30*	
	Fit tooling to machine	*28:00*	*10:00*		*Add quick change clamps, pins and locator blocks*	*5:00*	
	Airclean cavitities and machine	*35:30*	*7:30*		*Clean cavities first and rest of machine after start-up*	*1:00*	*3:00*
	Remove tooling to cart for transport to tool crib	*40:00*	*4:30*		*Reposition table for easier access*	*1:00*	
	Return items to tool crib	*50:30*	*10:30*		*Return tooling after start-up*		*6:30*
	Run adjust, test and make first piece	*1:30:00*	*39:30*		*Add quick change clamps, pins and locator blocks*	*21:00*	
	Total	*1:30:00*	*1:18:30*	*11:30*		*31:30*	*16:00*

CUMBERLAND

Figure 9-16 Sample Setup Analysis

Setup Reduction in the Office

The setup reduction concept applies to office operations as well as factories. Consider the typical accounting close. How much preparation time is external to the close and how much is internal? External time should include all standard journal entries, information analysis for decision-making, special reports, anything that can be calculated ahead of time. Internal time is waiting for the final information to be input into the system, doing calculations using the full period's numbers, then finishing a minimal number of transactions that can only be done at this time, followed by a final report. Checking accuracy, testing data and information analysis, are all things that are typically internal time, but should be external to the process. A setup reduction type of analysis can be used to shift internal processing times to external, thus improving the cycle time for closing the books.

Standard Work and Workload Balancing

Standard Work is a key foundation for *Kaizen*. Standard work is a defined sequence of activities that are organized for efficiency and effectiveness. It describes a best practice for how work gets done that makes work repeatable and decreases variation during the process. In fact, the primary goal of standard work is to reduce the variation from operator to operator, thus improving quality, eliminating waste, improving productivity and improving safety. It is amazing how much variation exists in the ways people go about doing their work. This is true on a factory floor, and the variation is at least ten fold that for administrative and service processes, including processes like software design.

A key output for most *Kaizen* teams is to define "standard work" and to balance the workload. A team does not need to do this to the point of total inflexibility, but best practices should be defined and practiced on a day-to-day basis. The people who do the work use a couple of analysis tools to determine their own work sequence and develop improved practices. Those tools are the Standard Worksheet and the Workload Balancing Worksheet.

After standard work has been defined, adherence to the standards can be checked and corrective actions can be taken to improve or restore the process when any problems are discovered. This is a key component of the Toyota Production System.

Manufacturing or design engineering usually initially designs standard work. The *Kaizen* team can use that as a starting point, along with the work the team has already completed, for developing improvements. In the Measure Phase, the team recorded each step of the process, mapped the work area, tracked travel distance, recorded all product or information moves, and examined storage locations and workflows. In the Analysis Phase, then the team would look to best practices, review their observations, and develop improved ways to do the work.

The team documents the current practice with a diagram on a Standard Worksheet. Some of the movements identified in the Spaghetti Diagram can be layered on to this sheet, if appropriate.

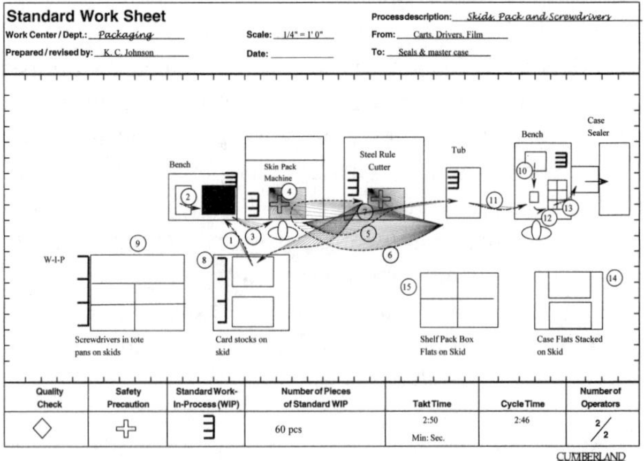

Figure 9-17 Standard Worksheet

The diagram shows where improvements can be made. Even areas that were worked by a *Kaizen* team previously still have improvement opportunities. That is the continuous improvement drive. A member of a *Kaizen* team once said, "There is so much waste in this process, it is difficult to find the work." Very few process steps add real value. Many steps are simply things that are necessary to do because of the way the process has evolved over time.

If the *Kaizen* team is observing a multi-station operation, the team should look at Workload Balancing analysis after the Standard Worksheet is competed. A simple workload balance analysis is shown below. Sometimes Workload Balancing becomes too big a task for a five-day *Kaizen* event, and the project should evolve to more of a Lean or Breakthrough team opportunity. Regardless, Workload Balancing analysis applies to service and administrative teams, as well as the factory floor, and sometimes more so. Just think of the accounting close with the preponderance of work done at the end of the month, the end of the quarter, and the end of the year. They are usually far from balanced.

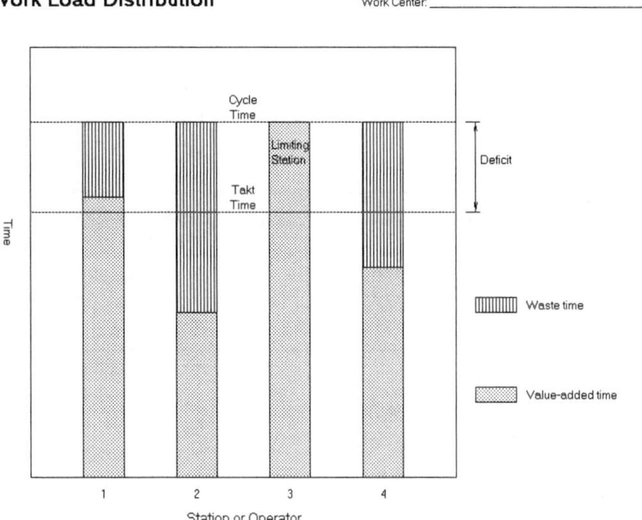

Figure 9-18 Workload Balancing

This diagram shows that the workstations are not balanced and that workstation one and three are taking longer to do than the Takt time requires. The waste time shown includes non-value adding activities and wait time. So when the team brainstorms improvement opportunities, they will need to address these issues. This diagram, with along with the Standard Worksheet, can help to define the new best practice.

<u>Visual Controls</u>

A *Kaizen* team should also focus on analyzing the process for opportunities for a few good visual controls. The visual control opportunities that the *Kaizen* team should be seeking should improve communications, improve the work, and make it easier for people to get their jobs done. The table below shows common types of visual controls.

Visual Controls		
Description	**Purpose**	**Application Example**
Location Markers	Indicates where to locate items.	Tape strips on floor, numbered locations, color coded areas, painted areas, etc.
Standard Methods	Indicates how to perform work.	Method sheets, work sequence charts, flow diagrams, process maps, videotapes of operation, digital photographs.
Tags	Indicates abnormal or special conditions requiring attention.	Red tags (flags) for excess or obsolete items, broken items, maintenance tooling or equipment, scrap and defective materials.
Kanban(s)	Controls production and movement of materials.	Cards, containers, bins, signal flags, etc.
Performance Reporting	Let people know the score, production status, successes and progress toward targets.	Safety, days without accidents, production schedules and quantities, product/service quality, customer satisfaction levels, cost, revenues, profitability, etc.
Product Displays	Show people finished quality products and/or defective products.	Tables, boards, pictures, pictures of customers location using products, tooling, raw materials, etc.

The *Kaizen* team should consider these questions when deciding where visual control improvement opportunities will be most beneficial:

1. Is the work area neat and orderly? - Use the 5S analysis as a guide to cleaning and organizing the work area, visual controls start here.
2. Are people having problems getting work done? - Use signs or designated storage for parts, materials, tools that could be visually identified.

3. Do quality problems happen as a result of incorrectly followed procedures? - Define standard work practices and look for ways to visually communicate the proper procedures.
4. How can communications be improved? - Every organization suffers a communications shortage. Establish better feedback mechanisms on "How are we doing?" This is a primary purpose of schedule boards and posted data diagrams.

Visual controls need to be balanced between providing information to people inside the department and to people from outside the work area. Determining that level of detail is the difficult hard part of developing a visual control system. It is fairly easy to slap up a few signs, or put labels on containers and markings on the floor. But if this is done haphazardly, the credibility of a powerful improvement tool may be undermined. It is a little more challenging to find the right visual controls.

When a team comes up with a great visual controls idea, people will look at it say, "Wow, why didn't we do this sooner?" One company had a lot of pressure valves and meters spread throughout their facility. During analysis, one of the *Kaizen* team members said, "Wouldn't it be great if we could reorient all of the dials so that if the needle was in the 12 o'clock position, the equipment was OK?" They embarked on a mission to see how much change was possible. About 30% of dials were already oriented that way, and the team was able to reorient 70% of the remainder. So a total of about 80% of the dials indicated in control equipment if the needle appeared near the 12 o'clock position.

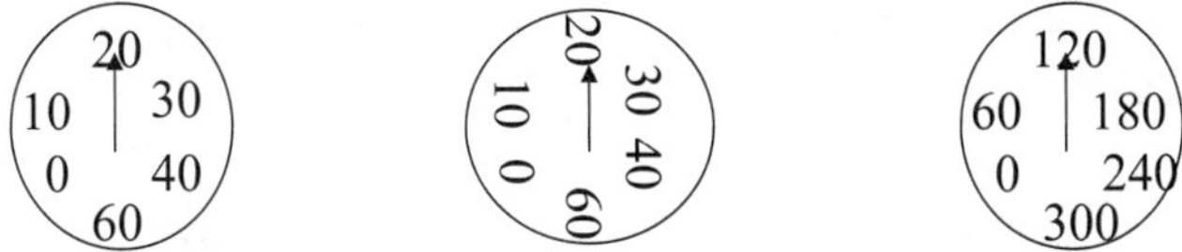

The meter condition can be detected even from a distance by setting standards on the 12:00 position. Multiple meters can be checked. The condition is obvious.

Figure 9-19 Example of Visual Controls

Other applications for visual controls include the use of shadow boards for tooling and cut outs inside drawers so that tools and materials are better organized.

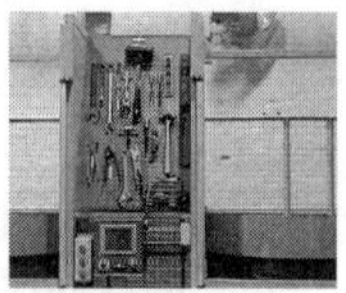

Figure 9-20 Additional Example of Visual Controls

Visual controls should first be tested. If they prove successful, they can be expanded to other parts of the operation, after the test or burn-in period is over.

Poka(e)-Yoke (Mistake Proofing)

The Japanese term Poka-Yoke (pronounced pokeah yokee) comes from the words yokeru - (to avoid) and poka(e) (inadvertent errors). The term poke-yoke is a hybrid word created by Japanese manufacturing engineer Shigeo Shingo to mean avoiding inadvertent errors. Today, most people refer to this as "mistake-proofing," making it impossible to do a task incorrectly. Like most of the *Kaizen* tools, it is a very, very simple concept yet not widely used. It has application in the office as well as the production floor. *Kaizen* teams should analyze processes for opportunities for mistake-proofing.

This is one example of mistake proofing a process: In assembling ball valves, operators would sometimes forget to put the ball onto the valve before installing a cap. Functional test at a subsequent process would uncover valve leakage and the unit would be disassembled, and then reassembled with the ball in place. The operation was improved as shown in the diagram:

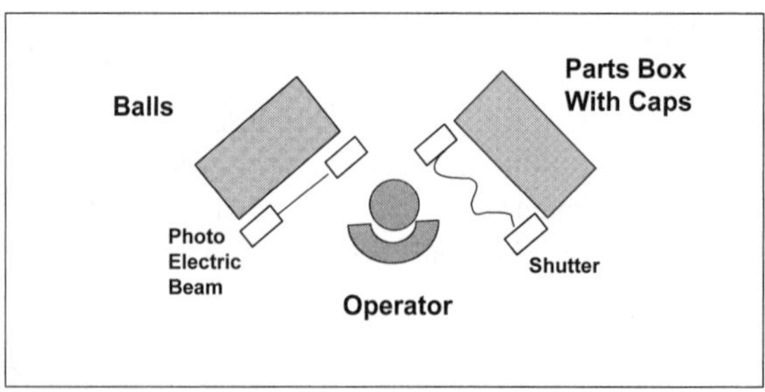

Figure 9-21 Improvement Through Mistake Proofing

- A photoelectric switch was installed in front of box containing ball valves
- A shutter was attached to the front of the box containing caps.
- When an assembler's left hand reached in to take a ball, the photoelectric switch tripped.
- If the assembler reached into the cap box before the ball switch activated, a buzzer sounded.

This made it impossible for the worker to forget putting in the ball valve.

Poka-Yoke fixtures only allow the parts to be loaded correctly and/or alert the operator when a part has been loaded incorrectly so the condition can be corrected. Typical Poka-Yoke devices include:

- Checklists
- Meters and simple control devices
- Error devices for positioning round, square, post rings, etc.
- Temperature gauges, limit switches, shine lights, etc.
- Color coding

Poka-Yoke in Service and Administrative Processes

Teams working on improving service and administrative processes are often not familiar with the Poka-Yoke word, but they do understand mistake proofing. One sales process improvement team was experiencing a constant source of data entry error and missing information from the salespeople. The data collection process was based on spreadsheets and free-form paper forms. Eliminating the free-form paper forms, and collecting all data in structured spreadsheets improved the process. Required information boxes on the spreadsheet were color-coded, so it was obvious what was mandatory to fill-in. For information boxes that required account or activity coding (e.g., size, model number, etc.), a pop-up menu would provide the valid codes. The salesperson could scroll to the correct selection, eliminating coding errors went away.

Analyzing Improvement Opportunities

As the *Kaizen* team identifies improvement opportunities, they should keep a list. A visual way to do this is to write improvement opportunities on a Post-It, and paste them on a flip chart sheet hung on the wall.

The team can use a round-robin session to identify the most obvious opportunities, based on their analysis, and work to surface as many ideas as possible. Each participant takes a few minutes to write down ideas, and then take turns going around the table or room, sharing one idea per round. (Passing is okay.) In any one-week *Kaizen* project, a team of eight or more people should be able to come up with at least 70 ideas.

After all ideas are recorded, the team goes through a clarification and combination discussion. When combining like items, a good rule to follow is if one person wishes to keep an item separate it stays separate and does not get combined. The team should be careful not to create too many "super categories" where so many ideas are combined that the team is actually looking at a super project. At the point, the team also needs to remain focused on the opportunities, and attempt to stay away from solutions.

One way to keep track of opportunities is to number each opportunity on the list. Five ideas that are combined into one opportunity would have one number for the five combined items.

Once all items on the list have been clarified and combined, the team can prioritize the list. There are a number of ways to do this. One way is "weighted voting", where each team member is given one vote for every seven to ten items on the combined list. For example, if there are fifty items on the combined list, each team member gets five votes.

A team member's most powerful vote is the number '5' vote, worth five points; the second-most powerful vote is the number 4 vote, worth four points; then the 3 vote, the 2 vote, the 1 vote. Team members should individually review the list of opportunities, and decide where they wish to put their votes. One way to record votes is to provide each team member with Post-Its with the vote values, like this:

Figure 9-22 Vote Values

Team members can write the line items, or opportunity numbers, on their Post-Its. From the list of fifty opportunities, say one team member selects these and records opportunities like this:

Figure 9-23 Voting on Opportunities

This team member is giving her most important vote (five points) to opportunity #9, the next most powerful vote (four points) to opportunity #21, etc.

Once all team members have determined their prioritized votes, they can put their Post-Its on the flipchart sheets at the same time. (The reason for writing the opportunity number on the Post-It is that the Post-Its will cover the opportunities on the flip chart.) After everyone has put their Post-Its on the list, the leader should tally the point totals. This produces a prioritized list of opportunities. This technique is called the Nominal Group Technique (weighted voting).

Some organizations have an actual opportunity log and opportunity form, which may be appropriate to use after the key opportunities are identified.

Analyze Phase Tollgate Questions

To determine whether the team has successfully analyzed the process, the Champion should address these questions:

Has the team:

1. Completed the analysis forms to a sufficient level of detail?
2. Identified "root causes" of problems?
3. Obtained a reasonable understanding of the improvement opportunity?
4. Communicated with people potentially impacted by changes to obtain their ideas?
5. Prioritized the improvement opportunities?

Analyze Phase Example - Computer Chip Machine Setup Time Improvement

The team pretty much completed their measurements by the end of the morning on Day Two, so they were ready to begin analysis. Matt and the other operator were surprised by what they had learned and by their observations. They both thought everyone did the job in pretty much the same way. But based on their observations, the same people doing the same work didn't do it consistently in the same sequence. And Matt learned that he did some things differently from the way other operators did them.

The team used the Setup Analysis Worksheet to capture most of the process steps and the process times. They were surprised to see how much time was being spent internal to the changeover process that really should have been external. An abbreviated version of their setup analysis worksheet showed:

		Run Time Hr:Min: Sec	Current Time			Proposed Time	
No.	Task / Operation		Internal	External	Improvement Opportunity	Internal	External
11	*Get reels*	***3:50***		*2:50*	*Move all new reels ahead of time near insertion equipment*		*2:50*
12	*Insert reel in tray*	*4:10*	*:20*			*:20*	
13	*Look for job instruction sheet*	*4:40*	*:30*		*Have job instructions available at job start*		*:30*
25	*....wait for programmer*	*15:20*	*5:00*		*Should be zero time, prg. should be there*		
26	*Load new program*	*16:20*	*1:00*			*1:00*	
34	*Look for tooling*		*1:30*		*Should be zero time*		
41	*Make adjust to rails for new boards*		*2:00*		*Simplify rail adjustment*	*:30*	

Figure 9-24 Portion of Setup Analysis Worksheet

Just looking at the Spaghetti Diagram, the team knew there were improvement opportunities. The funny thing was that none of the operators, including Matt, realized they were walking so much during the changeover. They were trying to get their work done, and did not pay too much attention to how much walking it required.

In their 5S assessment, the team agreed on an overall score of 35 points, out of a possible 100. Tools were hard to find, and some of the safety equipment in the area was not well labeled or visible. The job instructions were hard to use. The most spirited debate the team had when they were trying to reach a consensus on their scores was during the discussion of question number seventeen, *"People understand and follow the standards, operators do their jobs in a consistent way."* The operators knew there was some variation, and they felt the overall score was a '3'. One other person said a '4,' and most of the rest of the team graded it a '1 or 2.' They started to get into an argument, when Pearlie intervened, "Who has the highest and who has the lowest scores and why did you grade them that way?" Jim, the design engineer on the team, said, "I graded this question a '4' because we don't have any quality problems with the boards." Sydney the maintenance worker said, "I graded it a '1.' I did not see any standards that showed people where to store their tools or how to get the new reels ready to go, plus they really didn't seem to use the job instructions." When Sydney said that, the team was quiet for a moment. Jim then said, "You are right. How about a '2' for the overall score for this question?" Pearlie then asked if everyone had a thumbs-up on a '2' as the overall score. Everyone raised their thumbs, thus stating they agreed or at least could live with that score.

The team moved into a brainstorming analysis of the improvement opportunities everyone felt existed. They listed over 85 items. When they clarified and combined their list they had 51 different improvement opportunities. Using a weighted voting technique, they prioritized 10 different improvement opportunities that they felt they could do something on this week.

The top five were:

1. Use some type of a setup team, to make set ups happen faster
2. Have all of the chip reels for the new job, set right by the insertion machine, before the old job is finished running.

3. Make certain the system programmer was there when needed.
4. Pull at least 20 minutes of "internal time," out of the change over process
5. Have some type of tracking and recognition system for how fast setups were happening

Analyze Phase Example - Billing Department Customer Service Improvement Project

The Billing Dispute team members were feeling a little anxious. They had finished their measurement work early on the secon day, but they really did not know where the improvement opportunities were to accomplish their goals. Kim, the team leader, told the team to just stay with the process and see where it took them.

They did a diagram of how long the disputes stayed in the department. The diagram looked like this:

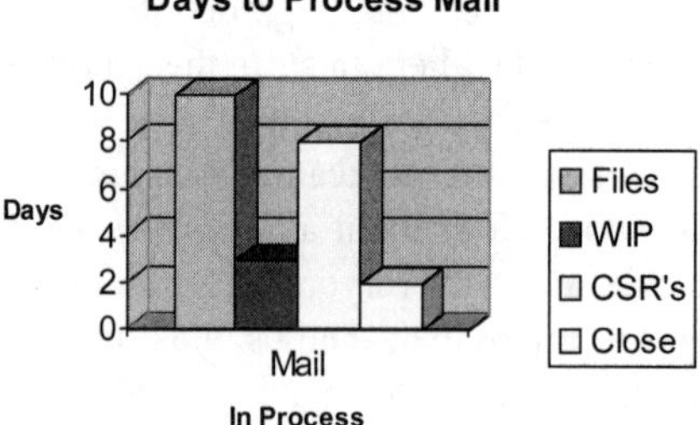

Figure 9-25 Days to Process Mail

Almost all mail sat for ten days before it even started to be processed. It then went to a temporary storage area in the Customer Service Representative's (CSR) room, and was finally given to a CSR to work. They would open a customer file for every mail item. When the CSRs worked the files, sometimes they would have to call and talk with the customer and also talk with the credit card parent organization to get copies of the transaction slips or other proof of purchase information. They were not authorized to talk to the merchants directly, unless they first contacted the credit card parent organization.

The team members decided they needed to dig into the mail to understand just what type of information was coming in. One sub-team took responsibility for this. Another sub-team was commissioned to see how the CSRs spent their time during the day.

The mail sub-team looked at 100 bundles of mail. Each bundle consisted of 25 items. They classified the mail along the following lines.

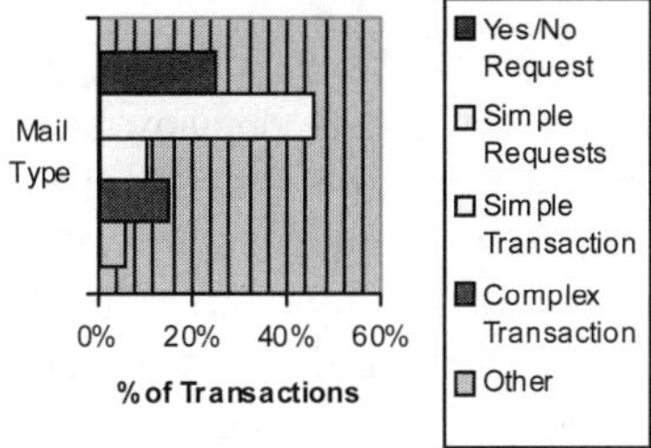

Figure 9-26 Classification of Mail Type

Twenty-five percent of the mail were questions that could be answered "yes or no" (e.g., I should not have been charged interest this month.), 45% were simple requests for changes (e.g., address change), 10% were simple transactions that did require contact with a third party or an information request (e.g., my payment was posted on the wrong date), 15% were complex transactions that truly were a dispute, and 5% fell into the other category.

The second sub-team looking at how CSRs spent their time learned the following:

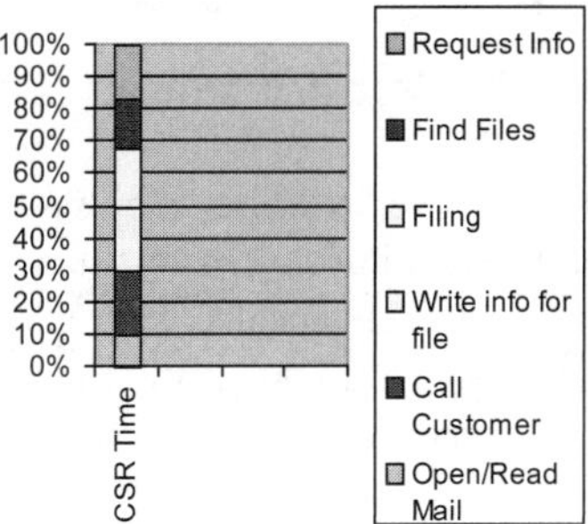

Figure 9-27 CSR Time by Activity

The sub-team was a surprised to learn how little time was spent opening and reading the mail. They thought that required a much greater percentage of time. They also noted all of the time spent filing and looking for files, because it was total non-value adding time from a customer perspective. Sometimes it would be a direct cause of customer dissatisfaction, when customers would call into the center and the CSR could not locate their information.

A third two-person sub-team went to look at the types of information that came back from the credit card parent company to document transactions. They learned that sometimes copies of credit card slips contained no useful or new information. For example, credit card slips from mail order companies would only provide the information "mail order" about a transaction no matter what specific product a customer had ordered. The charge every time one of these slips was ordered was $5. If the transaction was an airline ticket, the credit card slip simply repeated information already on the statement. This team determined that, together, the cost of obtaining these useless credit card slips exceeded $250,000 per year.

When the team came back together and shared the results of their analysis they agreed on three priority improvements.

1. Eliminate the request for information from the parent credit card company that contained no additional explanation.
2. Find a way to process the mail transactions that were "yes/no" request and simple information request much faster.
3. Find a way to reduce the time spent filing information and looking for files.

Improve Phase for the Kaizen Team

This portion of a project is very different for a *Kaizen* team than it is for Lean and Six Sigma process improvement teams. *Kaizen* teams are the ultimate "Do it Now!' team. This is also where the fun begins for *Kaizen* teams because change happens so quickly.

Deliverables in the Improve Phase

The *Kaizen* team's deliverables include:

- Tested improvement solutions
- Documentation on how to do the new process
- Cost benefits for proposed changes
- An implemented improvement

At this point in time, it is totally appropriate for the team to begin brainstorming solutions to the high priority improvement opportunities identified in the Analyze Phase. This can be a whole team activity to promote creativity.

Principles of Brainstorming

The team leader should enable the team to employ sound brainstorming methods:

- Generate as many ideas as possible.
- Encourage freewheeling thoughts
- Don't allow criticism (positive or negative)
- Keep discussion to a minimum
- Allow equal opportunity to participate
- Record all ideas

Use the round robin technique for maximum participation. Then the team should form sub-teams to study and test the brainstormed opportunity solutions. Experiment with several solutions when possible. Don't get locked in on one idea. Make certain to test recommendations for practicality on the job.

Finally, determine the bottom line dollar savings measures including secondary benefits. Give some thought to whether the savings are "hard" or "soft" dollar savings. *Kaizen* teams typically implement improvements that do not require significant capital or information systems expenditures, although it may be necessary to further evaluate more complex opportunities involving capital expenditures for equipment or information systems.

Kaizen Team Presentations

Toward the end of the project the *Kaizen* team will deliver a presentation for the management team. It's usually best to not worry about formality. Use flip charts, overhead copies of worksheets, or videotapes to supplement the presentation. Write overhead copies with large print for easy readability. A minimal presentation should include the following information:

- Project Charter (description of what the team was asked to do)
- Opportunities Log (list of the total number of opportunities found by the team), with the opportunities selected for implementation highlighted
- Opportunity Descriptions (more detail on the selected opportunities)
- Demonstration of Changes Made and Benefits (If at all possible, take management out to the work area and show them the changes that were made. If this is not practical then present a summary of the changes and the benefits.)
- Progress Report (show the baseline starting point and progress to date)
- Next Steps (list of next steps, including who is responsible and commitment dates)

After the management or leadership presentation is completed, it's a good idea to present to the people impacted by the change. This includes people who do the work, and customers of the process (especially if internal). If it's practical, this presentation can be combined with the management presentation.

If the *Kaizen* session runs four and a half days, team members should start pulling together their presentation information at the end of Day Four. Spend the morning of Day Five actually getting the presentation ready. Many teams like to do a practice presentation. This is a fine idea if time is available, but it is not critical. It is more important to get the implemented improvements nailed down.

Improve Phase Tollgate Questions

The project Champion can use these questions to review the team's Improve phase work:

1. Were meaningful improvements implemented?
2. Did the team measure the results of the new methods? Are the benefits documented?
3. Did the team communicate with people in the work area to see if were comfortable with the new solutions?
4. Has documentation been written on any new procedures? (It may not be complete, but something should be written as a guideline.)
5. Have next steps been defined including what is to be done, who is responsible, and completion dates?
6. Did the team's Champion accept responsibility for sustaining the gains?
7. Did the team present to a management or leadership group?
8. Did the team present to the folks in the work area whom will be impacted by the changes and new procedures?

Improve Phase Example - Computer Chip Machine Setup Time Improvement

The setup team broke into sub-teams to work on implementation. One group was assigned to staging and changeover preparation; the second group was responsible for everything that happened after the last good board of the old run was completed and a third group was working the rest of the line. These two sub-teams made significant changes.

The most significant change made by the staging group was to use a kitting process and have all of the reels for the next job sitting right by the chip insertion machine at the start of the changeover. The team believed this would really cut down on the time spent going back and forth to pick up semi-conductor chip reels. They also grabbed a cart with two shelves from the maintenance department to use for staging the reels. They coordinated with the supervisor and scheduling department to try the new arrangement on the next changeover. There were only a few tools used by the operator, so they tied a small plastic container to the side of the tray for tool storage during the changeover. They also made a tool board for storing the tools when not in use.

The new Spaghetti Diagram looked like this:

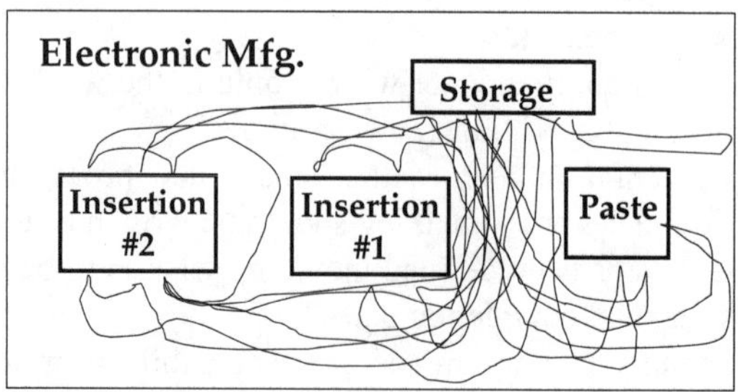

Figure 9-28 New Spaghetti Diagram

An extra operator (a floater available to help out on the lines) came over and adjusted the rails (belts that carry the circuit board down the line). The main operators would load their cart at the storage area, bring the cart over to the insertion machines before they stopped running the last job, and then, after the old job was finished, turn off the high speed insertion machines and exchange the reels. Their preparation time was ten minutes. This time was now all external to the changeover. The team reduced walking distance from over 1,000 each for two operators, to less than 100 feet per operator.

The second team worked on roles and responsibilities. In the old process, the Systems Programmer was not called for until the changeover had started. They actually called the Programmer's office and, if she was not there, left a message. That was part of the reason for the delays. They decided to give each System Programmer a pager. A changeover plan was published each morning, although everyone knew the times could very by as much as 30 minutes. Then, 15 minutes before the changeover, started the line supervisor was responsible for paging the programmer. The Programmer was responsible for being at the line at the 15-minute mark.

The department also had two floaters who assisted with line problems, filled in for breaks, and did some material preparation. They gave the floaters a new responsibility to help with paste-up and to reposition the rails, when the new circuit board was a different size than the previous one.

The actual changeover time using the new procedures, while the machine was not running, was completed in 12 minutes versus 35 minutes following the old procedures. The team completed a new Setup Analysis Worksheet to document the new procedures and to establish the new baseline.

Final results were:

	Before	After
Setup time	35 minutes	12 minutes
5S Score	35 out of 100	67 out of 100
Walking Dist.	1000 feet	100 feet
Capacity		7.5% increase

Figure 9-29 Final Results of Improvement

Improve Phase Example - Billing Department Customer Service Improvement Project

It took a surprisingly simple change to eliminate the request for information slips that contained no useful (additional) information. The *Kaizen* team identified primary mail order houses, airline and hotel vendors and several categories of merchants where the credit card slip would not prove useful, and eliminated the ordering of credit card slips from these merchants. This one change was the source of over $250,000 in annual savings.

The team struggled with how to get to the mail faster because the volume was so large. When the team reviewed the data they had gathered, they noted that CSRs spent less than 5% of their time "Opening and Reading Mail." They ended up brainstorming several possible solutions. The idea that received the most votes, using the Nominal Group Technique, was to add a pre-sort step to the process. Normally, process improvement teams work to eliminate steps in a process, so the team wrestled with this idea for a while. The team leader suggested they try out the idea and evaluate the results.

The team took the mail that came in from the mailroom that morning and began to work it. They quickly realized that 70% of the mail could be handled right away. The action required did not take much longer than the time it took to read the piece of mail. The next challenge was to convince the rest of the CSRs to try the new idea. The CSRs worried that, once the bulk of the mail had been handled, the remainder would be difficult and challenging to handle, and they would spend all day dealing with customers that were upset with their merchandise transactions and were having difficulty dealing with the vendor who made the sale.

After meeting with the team Champion and a few selected managers, the team presented the results of their mail test, and the CSRs agreed to a pilot rollout the following week. The pilot was more of a success than the *Kaizen* team had anticipated. Using the new methods, the team felt they could eliminate 70% of the backlog and process all mail within one to two weeks. In the past, the mail sat for two weeks before anyone even looked at it. The new process more than doubled productivity, resulting in a significant cost savings. Whereas the organization had been in a hiring mode, the new procedures eliminated the need for about ten positions.

As a result of the new pre-sort routine, the department also reduced the number of customer files. Documentation for customer information closed out on day one was scanned to create an electronic copy of the original. The original documents were then put in temporary storage. This reduced the number of work-in-progress files were reduced, and the CSRs found they could easily keep track of fewer files with a simple logging system.

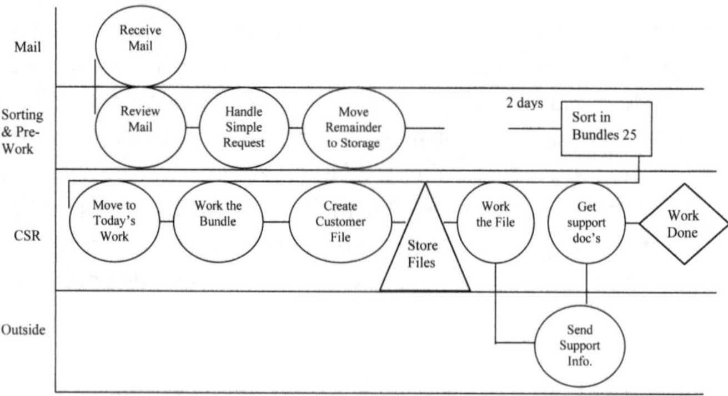

Figure 9-30 New Process Map (Condensed version) for Billing Dispute Mail Processing

Control Phase for the Kaizen Team

A *Kaizen* team typically meets for a one-week period of time and then it disbands after implementing as many improvements as they can practically accomplish. They do not have a great deal of time to spend on control issues. But they still need to give some thought to it or their gains (improvements) are less likely to be sustained.

Kaizen teams don't make recommendations for review; they actually implement the changes they want to see. This helps ensure that changes will remain in place. If a team rearranges equipment, or eliminates of excess equipment, those changes are unlikely to be reversed.

When the team does their management presentation they are largely turning over control to the management team, or at the very least the team's Champion. The Champion is responsible for making certain documentation is sufficient to hand off to the appropriate individuals and that the improvements have actually been tested to make certain they work.

Usually, some *Kaizen* team members (supervisors, managers, etc.) also have a responsibility for the process being improved. They can also have some accountability for ensuring the gains are maintained.

Control Phase Tollgate Questions

The team's Champion should review the team's Control Phase work using these questions:

1. Did the team develop some metrics for monitoring future performance?
2. Does someone have responsibility to check and report on performance for the next 60 to 90 days (or longer)?
3. Was the team recognized and someway and appreciation shown for their performance?
4. Does one person have accountability for following-up on open items (actions items necessary after conclusion of the *Kaizen* event)?
5. Were most of the 'to dos' completed during the course of the *Kaizen* Event?

Control Phase Example - Computer Chip Machine Setup Time Improvement

The *Kaizen* team that improved the setup process on the initial line was given the responsibility to train people on the other four lines. They accomplished their tasks over the next 30 days (after their presentation).

The lead operator was responsible for timing setups. A small whiteboard for each of the five lines was kept on the wall in the production area. At the end of the setup, the net time was posted, along with a short description of any problems enountered (e.g., parts not there, system programmer late, etc.) One of the supervisors agreed to aggregate the results for each line in a spreadsheet, which was posted on an information board in the employees' break area. Data were collected by line, by product, and by shift. Results were posted by day, by week, and by month.

The company also sponsored a recognition program. Over the next year, the lines had a friendly competition to see who could best maintain or improve upon the target times. Every team that maintained the target times more than 90% received a letter of thanks, and each team received a $100 check. The top performing team received special recognition.

The supervisors had a daily morning meeting. Once a week, they discussed the status of the new setup procedures, and shared the changeover problems that were being experienced and what actions were being taken to address them. Problems outside of the supervisor's responsibility area became the plant manager's responsibility to address. Changeover times under 15 minutes were added to the supervisors and managers performance objectives.

Control Phase Example - Billing Department Customer Service Improvement Project

The *Kaizen* team set up a table for doing presort. One senior CSR and two new CSRs were assigned to the pilot team. They planned to test the new system for a two-week period of time, including during a month-end close, where work volumes and pressure to close out open items were more intense.

The pilot team's experience was pretty much the same as the results found in the Kaizen experiment. The CSRs also realized there was still a broad variety of dispute and information request types, so that the CSRs would not burn out just dealing with difficult issues. Once the CSRs accepted this fact, the new procedures were quickly rolled out.

A new procedure was also created for CSRs to identify any future occurrences of information slips with no useful information. The CSR who found the most slips of this type during a month received recognition in the weekly newsletter. All employees who found one of these transactions types received a waste finder star, which they could post in their work area.

The manager of the CSR Department posted a daily information board that described the number of transactions handled in 24 hours or less, and the number processed in one week, two weeks, and longer than two weeks. The distribution for the first two weeks after rollout looked like this:

< 24 Hours	One Week	Two Weeks	Over Two Weeks
45%	15%	35%	10%

Figure 9-31 Transaction Distribution

They also published the number of transactions. The Senior Vice President was so impressed with the work done by the *Kaizen* team that within 6 months of the completion of their project, half of the team members had been promoted to new positions.

Summary of *Kaizen* Teams

Kaizen teams represent a powerful way to involve people who "do the work" in improvement. They tend to be short duration teams, usually meeting for one week or less. They use simple analytical tools to analyze a process. And they *implement* improvements, rather than just *recommend* improvements.

Chapter

10

Lean Teams

Lean teams are different than *Kaizen* teams. *Kaizen* teams tend to work on physical processes. *Kaizen* teams observe work using a few simple tools to identify and implement improvement. Lean teams are more like detectives on a mission. The real requirements are rarely known by all participants in a cross-functional process. They need to be discovered, and the organization needs to seek better alignment for seamless delivery.

Cross-functional improvement teams should be used more often than organizations typically use them today. Just consider information systems which almost always cross functional boundaries. How often do organizations automate what they think is the existing process? Everyone knows the existing process is not the desirable way of doing work, but nobody knows the real requirements of the needed cross functional process. The temptation to get it done quickly usually overwhelms thoughts or desires to understand and fix the current process before automating it. The end result is that the freshly implemented information system rarely meets the users' needs. And that is just one example where cross-functional lean process improvement should be applied more often.

Lean initiatives try to get people to think beyond the "walls" of their department or job function, and to answer the questions:

- How do we work together seamlessly to better serve our customers?
- How do we provide a fair return for the shareholders of this organization?
- How do we make this a desirable place to work?
- And more and more today - How do we provide our products and services in a way that does not harm the environment?

A Lean team is a cross-functional group of people working together to implement significant, meaningful improvement to a key business process. Key business processes can be categorized into two broad categories:

- Core Processes - defined as a series of business activities which cross functional boundaries to create the end product or service that is delivered to external customers. Core processes deliver "value" to customers. Core Processes are almost always cross functional processes. They include, for example:
  - Market and Sell
  - Take Orders
  - Manufacture Product
  - Deliver Product or Service
  - Create New Products or Services
  - Maintain Customers
  - Enabling Processes - defined as a series of tasks and activities that are internal to the business, but contribute to the performance of the core processes. The customers of "enabling" processes are the core business processes. They include:
  - Hiring
  - Training and Development
  - Information Technology
  - Compensation
  - Legal
  - Administration (including finance)

Project Selection and Success Criteria

In the spirit of continuous improvement, every one of the core and enabling processes could be improved. While everything can be improved, only a few things are important to improve at any given point in time. Generally, more business growth and cost savings can be obtained from working core processes rather than enabling processes. The Six Sigma Management System strives to identify the key selection factors. If the leadership team takes the time to prioritize opportunities from a "Voice of the Customer" and a "Voice of the Business" perspective, it gets much easier to identify the important process improvement opportunities.

Characteristics that make for successful Lean projects include:

- First and foremost, these types of projects should be important to the business.
- They need to be Championed by someone with clout - someone who can largely approve a team's recommendations.
- Teams should be launched with clear objectives and a clear project scope so they know the territory they are working.
- Teams should be focused on understanding "real customer" requirements and better aligning the cross-functional players in meeting key stakeholder needs.
- Team goals should relate to time reduction, flow improvement, waste reduction and revenue growth.
- Assignments should be appropriate to tools like Process Mapping and Value Stream Mapping.
- Team members from all functions of the cross-functional process should be available.
- The ideal project should be completed within a 30 to 90 day time frame.
- The team should have 5 to 12 members with at least 20% of their time dedicated to the project.

Generally, if a project is important, the organization has an easier (not to say it is easy) time providing appropriate resources to work the project. Some organizations elect to start their improvement initiatives with projects to simply show a success. If "showing a success" is the only criteria, it can be problematic. First, people are spending time and resources working on something that is probably not important to the business. Secondly, it's easier for people to let things slide in terms of deadlines and project resources because they know their leadership does not really care about their project, thus undermining the credibility of the entire improvement process.

Lean Teams' Strengths and Weaknesses

Strengths	Weaknesses
Moderate Team Training Needs	
Rapid deployment	Critical to have the "right" Champion
Common data gathering tools, easy to learn	Requires 20% or more team member time per week
Implement changes quickly	May move to solutions too quickly
Can develop breakthrough improvements	May get distracted with "quick-hits"
Promotes cross-functional cooperation and understanding	If scope is too broad or narrow, may not accomplish goals

Lean Team Membership

A software Vice President once said, "The most important ingredient on this successful project was having smart people...Very little else matters in my opinion...The most important thing you do for a project is selecting the staff." Jim Collins in his book, "Good to Great," also talks about the importance of "having the right people on the bus." Good people can overcome many obstacles. They will get things done! One could probably ignore just about everything talked about in this book and if "good, bright" people are put on a team, they will get something done. During the course of their work, "good people" would naturally do much of what is talked about on these pages. But they would also encounter many obstacles and barriers. Organizations that follow the thought processes outlined here will make it easier for "good people" to get something meaningful done. They will also decrease the risk of burnout because for the most part, good people get overused. They are truly an organization's most precious resource.

Cross-functional process improvement teams typically have 5 to 12 people as team members that spend a minimum of 20% of their time working the project. It is critical that the key stakeholders in the process have a voice in the work done by the team.

Sometimes key stakeholders will be on the team and sometimes they will be consulted during the course of the project. If a team gets too large, it becomes difficult to get work done. So everyone cannot always be on the team. It is critical to have strong analytical skills and specialist insights on a cross-functional process improvement team. Team members typically include: employees who work within the process, one or two employees who are not familiar with the process (often they ask insightful questions) and possibly a process customer or process supplier. Process improvement teams also tend to have professional staff as team members. Like a Kaizen Team, Lean teams should also include the manager or supervisor with the most vested interest in the team's results to remove barriers, obtain resources as needed, and share background information on the current process, when requested. Again, the supervisor should not dominate a team or try to overly influence recommendations.

Some people wonder if the Champion should be on the team. There is no one right answer to this question. But if the Champion is not on the team, he or she can play a more objective role in guiding the team. If the Champion is not on the team, it also gives an organization three distinct groups driving change: the Champion, the team leader and the team members. Each voice speaks with a different perspective that can help make change happen.

Keep in mind that people can really blossom when given the right opportunity. So maintain an open mind when putting people on teams. There is an amazing number of people that participate on process improvement teams and come out of the experience a better person and a better member of the organization.

Lean Team's Purpose - "Quick Hits" and/or Innovative Changes

The focus of process improvement teams is often to understand the real requirements and to identify innovative solutions for improvement. They should always be looking to solve a major pain or pursue a meaningful opportunity. Lean teams will also have a "get it done now" focus. But their time window is longer than a Kaizen team. Lean teams typically implement a series of quick hit ideas that can be accomplished over a 30 to 90-day timeframe. Sometimes those changes are sufficient and the team's work is completed at that point. If they have a longer-term focus and charter, then the leadership team and Champion need to make certain the Lean team does not get sidetracked with only implementing "quick-hit" ideas.

Lean/Process Improvement Team's Work Plan

The roadmap for a Lean team is a little more varied than a Kaizen team. But there are basic steps that every team should follow. The DMAIC model will also serve as a guide for Lean teams to do their work.

Team Launch

There is no single way to do this. Like most things in life, there are trade-offs. Launching this type of team during a multiple day workshop allows team members to hit the ground running and if the workshop is done well, the team will walk through all of the steps of the DMAIC model in getting organized and beginning to come together as a team.

One member of a Critical Process Improvement Team at Pratt-Whitney stated, "I was against spending one week in a workshop at the start of this project. But after just one week, I can now say that the amount of work we accomplished during this time would have taken us three months to get done without the workshop." People at GE talk about "learning while doing." It's a powerful, but time-consuming, way to get a team off to a successful start. It's difficult to have a group of people away for so long but if the organization can free the resources, they can accomplish quite a bit in a very short period of time. The best workshop launches use Action Learning and have team members working on their project as they learn the analytical tools they may be using. They learn process mapping or value stream mapping, by mapping their process. It is a powerful learning vehicle. Typical launches would run 3 to 5 days.

The DMAIC Project Phases

A Lean team would go into each of these steps more deeply than a Kaizen Team. A Six Sigma team will go even deeper especially in the Measure, Analyze, and Control phases of the DMAIC model.

Define Phase for the Lean Team

The primary purpose of this first step is to develop a clear understanding of the improvement task and to answer the question, "What is important?" Activities during this stage include clarifying the process/project scope, identifying the key stakeholders or players in the process, and beginning to pinpoint which blocks within the process map offer the most leverage for change. The work being done by a Lean team will not differ significantly from a Six Sigma team at this stage.

The Lean team's Define Phase should provide answers to the following questions:

1. What is the current situation?
2. What are the primary symptoms of the opportunity?
3. Why was this opportunity chosen for improvement?
4. What are the critical customer requirements?
5. What are the "quick win" opportunities? (This is most likely to happen if the team is launched in some type of a multi-day work shop; otherwise, it will probably not happen until the Measure or Analyze phases)
6. Who needs to be involved in the project, outside of the team members?

In this section, the actions to accomplish these deliverables are described along with other tools and techniques that are appropriate for Lean teams.

Lean Team Charter

Getting answers begins with a validation of the project charter and improvement goals provided by the team's Champion. The Charter may come from the Champion, but the team needs to agree to the spirit of the document. The team needs to resist jumping into problem resolution at this point. The team's charter would be similar to the one used for a Kaizen team or Six Sigma team. It will include:

- Business Case - describes the purpose and primary benefit for under taking a project. The business case addresses the following questions:
  - Why should we do this?"
  - Does this project align with key business strategies or other initiatives?
  - What impact does the pain or opportunity at hand have on customers, the business, and/or employees?
- Scope: defines the boundaries of the business opportunity. The scope addresses the following questions:
  - What are the boundaries for this opportunity (i.e., starting and ending steps of a process or initiative)?
  - What authority do we have as a team?
  - What processes are we addressing?
  - What is not within scope?
- Goal: the goal defines the objective of the Charter in measurable terms.
  - Defines the improvement objectives and specific targets.
- High-level project plan and team members are also typically listed on the Charter along with the name of the team's Champion.

A good charter provides some stretch in the goals, but not beyond believability. Sometimes the business case and opportunity statement are split into two sections. The Charter provides a team with a starting point. The team members need to accept the Charter. That discussion should take place with the Champion at a kick-off session. If the team feels the Charter needs to be changed for any reason, first they should check their reasoning for doing so. If they still wish to amend it, then they should meet with the Champion and obtain his or her agreement to the amendment(s).

If a number of specific items are excluded from the team's scope that a reasonable person might think are "in the scope," they should be mentioned in the charter, perhaps even creating a separate section for Exclusions from Scope.

Validating the Charter

Team members should answer several questions in validating the Charter:

1. Do we understand it and buy-in to this concept?
2. Does this make sense from our the team's understanding of the business and does it address a real business issue?
3. Is the scope clear, and does it fit with what the team is being asked to accomplish?
4. Is the problem measurable (i.e., can baselines be established and improvement targets set)?
5. Are the right people on the team, to work this project? (this is an important question)
6. Are the goals obtainable and is the project completion date realistic?

Ideally, the project stirs some excitement and some fear (e.g., How are we going to do this?) in the individual team members. It's not a direct question one can really ask because people's responses will typically be to say the right thing. But if people are experiencing excitement and some anxiety at the outset, that is a good sign. One way to tell if the Charter fulfills the spirit of the questions above is if it sparked a healthy debate. Did people on the team question the Charter? Was there some emotion or strong opinions expressed? Did they propose alternative considerations, while at the same time practicing "active listening?" If yes, then the Charter is probably a good one. If "active listening" was not taking place - by the Champion or by the team - then the team is not getting off to a good start and actions should be taken to address the situation. A healthy dialogue can create a better Charter and begin to lay the groundwork for open communications between the team and the Champion. This dialogue might take place between the Champion and the team; it might also, instead, take place with the team leader and the Champion.

Good Charters

There are examples of Charters in the Define section at the end of this chapter. Consider for a moment though just how people "define" what is to be worked (in the Business Case/Opportunity). Would these be good Define Statements to commission a team? The following are all excerpts from actual Charters.

1. We need to outsource this function to.....
2. Sales needs to provide us the information earlier...
3. Customers desire delivery within 48 hours of order......
4. People need more training to.....
5. Our costs are too high relative to.....

Questions 1, 2, and 4 really have a solution buried in the Define statement - need to outsource; Sales needs to provide; need more training. These are all potential solutions to problems and represent one way to address each problem. If the answer is known, then a team is not needed to develop it. Questions 3 and 5 are more statements of fact. They begin to express the business case of why something needs to be done. A team is then commissioned to develop appropriate solutions. Solutions in the Charter stifle innovation!

A Lean team could pursue a variety of next steps. So the exact sequence of the steps that follow is not critical. It is critical that each of these steps receive consideration. In some instances, they may not be relevant to a particular situation, but most Lean teams will want to do something for each of them. Part Four of this book describes a Six Sigma team's approach to understanding and using many of these same tools. There is more than one way to do just about anything. In general, the information outlined in the Part Four will allow a team to become more analytical, and to dig deeper into the layers of understanding. That is one of the key strengths of Six Sigma level analysis. The tools outlined here may be executed more quickly.

Voice of the Customer (VOC)

The team needs to gain some focus on the opportunity. The sooner they can begin to understand key customer needs and concerns the better. This will be a major filter in determining what to improve.

Customer Requirements

Requirements are agreements between customers and suppliers as to what is needed to perform a job properly. The requirements must be:

- Mutually agreed upon

- Attainable
- Well communicated
- Measurable
- Changed officially if they need to be changed.

Below is a model to show the relationship between a customer and supplier regarding their requirements:

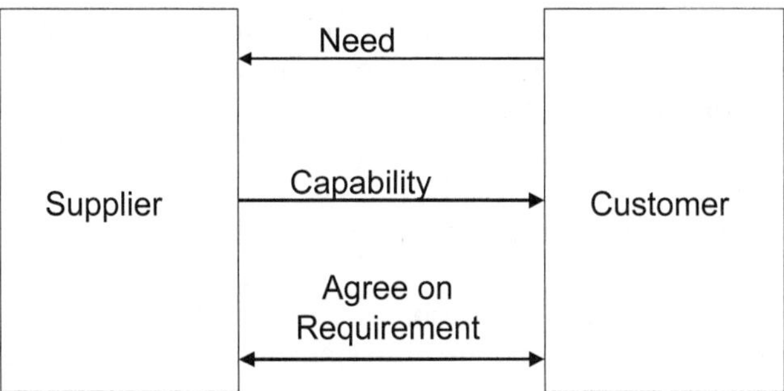

Figure 10-1 Customer and Supplier Requirements

There are levels of Customer Requirements:

- Specifications
- Expectations
- Delighters

Specifications are things that might be written down. Think about reserving a hotel room. Specifications would include: date, bed size, number of beds, smoking or non-smoking, the hotel location, etc.

Expectations are typically not written down. For a hotel room, they might include: the room is ready and available when you arrive, someone will be at the front desk at check in and out times, early check out is available, knowledge at the front desk about places to eat, directions on how to find your way around the area, or where to run, etc.

Delighters are those things organizations offer to try to differentiate their products or services. In our hotel example they might include: newspaper at the door, health club, frequent guest points, special eating or relaxation areas for frequent guests, knowledge about personal preferences of frequent guests, etc.

Competitors quickly copy delighters that work, unless they are done on the people side. Ritz Carlton Hotels pride themselves on the one-to-one guest care that each associate is trained to deliver. This is not as easily duplicated as a newspaper at the door.

Where organizations get in trouble with requirements - assuming they deliver the specifications - is with expectations. These unspoken rules cause frustration. Customers who defect (leave for another supplier) may not say, or even know, specifically why they switched. So when working on improvement, it can be helpful to gain insights to the world of expectations. Understanding here can lead to breakthrough changes.

For an effective relationship, customers and suppliers must be honest with each other. Real needs and real capabilities must be expressed. Once those are on the table, an agreement can be reached. Different customer groups will most likely have differing requirements. This is another reason for knowing which customers are most important now, and which are most likely to be important in the future.

Customer Needs Example

From a customer perspective (customer being someone who pays for the service or product) answer the question, *"In order to meet my needs you must...."* A company that provides the outsourced home delivery service for a large national retailer stated:

Our customers' key expectations are:

- "Provide 7 day delivery with extended hours - after 5pm and on weekends"
- "Accurately communicate delivery schedule and level of service I can expect and answer all my questions"
- "Provide same day service"
- "Be on-time, be complete and be careful"
- "Be flexible enough to accommodate sales fluctuations and changing expectations"

- "Provide friendly, courteous and professional delivery"
- "Provide Delivery service that is a transparent representation of our retail store"

This list can usually be generated using brainstorming techniques (described in the Kaizen chapter). Make certain that some of the people doing this brainstorming really do deal with customers directly.

Segment Customer Groups

In the example above, the distribution company actually had two paying customer groups. The retail outlets for whom they distribute and the customer of the retail outlet awaiting home delivery. When the team identified and prioritized these requirements, it caused them to focus on a different set of improvement opportunities than they were originally considering.

Not all customers create equal value. In order to discover growth opportunities, gain a competitive advantage, and build loyalty into the business strategy, it is helpful to segment customers. Customer segmentation should play a role in Listening to the VOC.

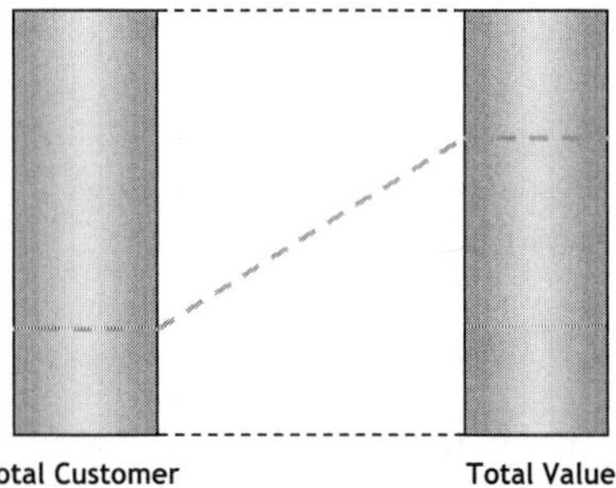

Figure 10-2 Identifying Value

The greatest value might come from a small portion of the customer base. Part of the challenge is to understand how these customers define and prioritize the various needs and expectations they have of the products and services provided by the process. The Value-Added concept can also be applied to internal business operations. It's not exactly the same as an external "customer" view, but there are activities businesses must do in order to operate. Some people call these "Operational or Business Value Added" actions. An activity adds operational or business value if it is:

- Required to sustain the workplace's ability to perform customer value-added activities
- Required by contract or other laws/regulations
- Required for health, safety, environmental or personnel development reasons
- Done right the first time

Validate VOC Information

Depending on the criticality of this information, many different sources of information can be used to check and validate VOC information. In any event, it is a good idea to at least test what the team thinks is a priority with a few customers. Make certain these customers represent the important customer constituencies (based on current business volume or future business volume and profitability).

Figure 10-3 Sources of Customer Information

Voice of the Business or Key Stakeholders

Just behind the customer in terms of importance are the key stakeholders in this business process. First, identify who they are. Then begin to categorize them into common blocks. In the New Product Development process, for example, after customers key stakeholders might include: sales, design engineering, manufacturing engineering, operations, executive leadership and others. A separate Voice of the Customer has already been done. Now it's time for the team to consider the other key groups. It's a good idea to try to pare down the stakeholder classifications to a limited number of groups, such as four groups. In the brief example above, manufacturing engineering and operations might be combined into one group. From a key stakeholder perspective, the requirements for this process should answer the question, "In order to meet my needs you must…."

Usually a team is broken into sub-teams to brainstorm ideas. Assign each sub-team one group of stakeholders, including customers, to further refine. A good way to do this is using wall posters or flipchart sheets. Each group should label a poster with the stakeholder group they have been assigned. The idea here is to identify what this stakeholder group expects from the client. The sub-teams should express their ideas in 'must do' statements as this keeps the discussion on the level of action - things they can actually tackle.

Prioritize Stakeholder Expectations

The way to get leverage in requirements is to align the stakeholder expectations and determine the priority. Expectations will typically fill three or four posters. It's helpful if these are hung while the groups are brainstorming. After brainstorming, each sub-team should consolidate the ideas on their posters into common categories. Then use a Nominal Group Technique (explained in the Kaizen Team chapter) or some other tool for the group to prioritize brainstormed list.

Each sub-team should prioritize their requirements. A good way to identify the overall priorities then, is to have each sub-team share their top requirement for their individual stakeholder view. These priorities can be captured on a list. After every group has shared their top four to six priorities and put them on the common list, the team members can vote on which of the important items are most important.

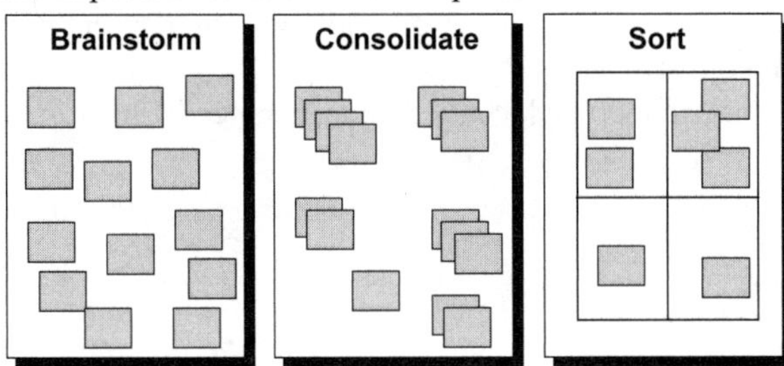

Figure 10-4 Prioritizing Requirements

By moving from broad customer and other key stakeholder expectations, to a list of priority expectations that produce mutual gain, a team has arrived at the key stakeholder requirements for improvement. Any improvements the team ultimately comes up with should be matched against this list of *requirements*. Once the key requirements have been agreed upon the team is in a position to take its first cut at identifying innovation opportunities.

Identify Potential Process Innovation Opportunities

Consider the team's charter and the key process requirements from a customer and key stakeholder perspective. Then answer the question: "What innovations could have a big impact on the organization's performance?" Initially, just get some ideas identified without worrying about the practicality. These are macro level innovations that could have a big impact on the enterprise. They should incorporate normal best practices:

- Eliminate mistakes and miscommunications
- Waste less time and fewer resources.
- Better support customers.
- Eliminate Steps.
- Do things in parallel rather than in sequence.
- Stop gathering worthless information

If there are a lot of ideas, it may be worthwhile to sort them into classifications:

- those that need more money, time or people
- longer-term opportunities
- those that need more decision making authority than the team or Champion have, and
- those that, if approved, could be implemented in the near future.

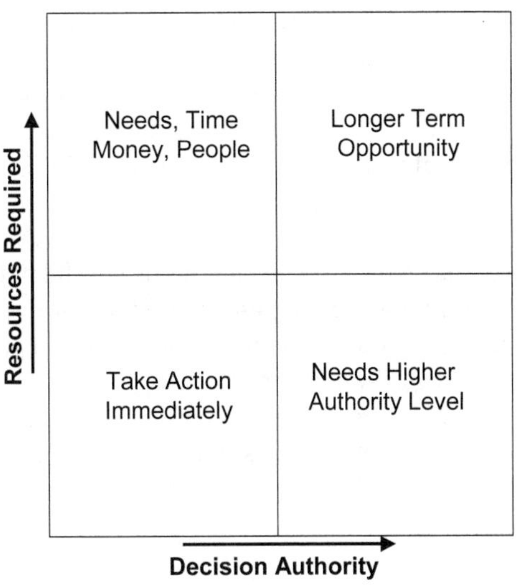

Figure 10-5 Opportunity Sorting Matrix

Teams should pursue opportunities in the lower left quadrant. "Take action immediately" refers to the duration of the team's charter. Typically, these opportunities can be implemented over the next 60 to 90 days - or six months at the longest.

Now, the team could actually begin its first pass at mapping new "could-be" processes. A more traditional approach is to map the "as-is" processes first. There is benefit in that approach, but teams should be cautious not to get too bogged down in the process.

Process Mapping for the "As-Is"

Lean teams start process mapping in the Define phase. A map further validates the scope outlined in the Charter and is often a source for identifying quick hit improvement opportunities. Teams may do a Value Stream Map or a traditional process map (swim lanes) depending on the process and scope.

It is easy to get trapped into doing too much mapping of the "as-is" process. The secret to getting this done quickly is to focus on the leverage factors. Think about it. More often than not, work is being done in the organization. Orders are being processed. Products and services are being delivered. Everything is not broken. It's highly likely that no matter what changes a process improvement team tries to make, at the end of the day, 70% to 80% of what is being done today will still be done tomorrow.

One exception to this would be when a process is being totally recreated. For example, a team implementing one-piece flow (lean manufacturing) is radically changing the way work gets done in a batch processing environment. Those changes impact the sequence of steps and physical layouts. So in this type of project, the percentage of change could be much higher. But the key leverage factors still need identification.

Traditional Process Mapping

If a team is looking for the leverage factors, one way is to start with a 10-step process map. Then go a level deeper based on where the most improvement opportunities exist.

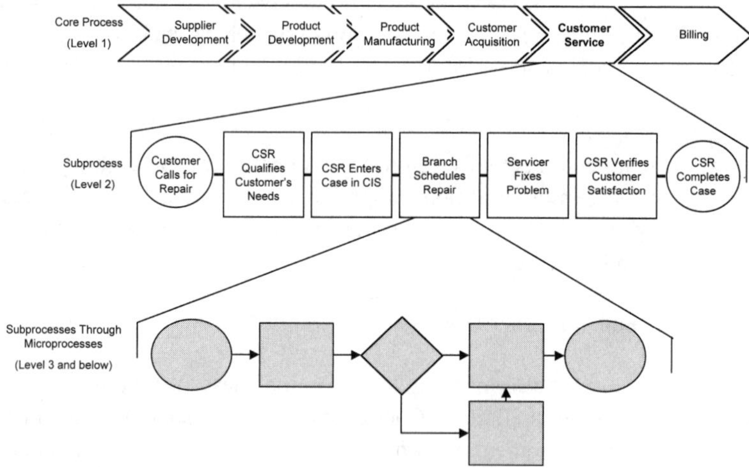

Figure 10-6 Example Top-Down Hierarchy Process Map

The example above shows three hierarchy levels of maps. *Level one* is a description of enterprise level processes. These maps are an example of outputs created by leadership teams in a Motorola Jumpstart Workshop.

In this example, the group felt the high leverage area was "Customer Service." So a *Level Two* map was created for this function. Within that map, "Branch Schedule Repair" was identified as a major improvement target. So a *Level 3* map was created to show those process steps.

Typically, each map is a layer of five to ten steps. One to three steps at each layer usually represent the most likely sources of improvement. Each step identified as a major improvement opportunity source gets mapped down to the next level. By the time a team gets down to the third level, it has normally arrived at a low enough level of detail to identify the specific improvement opportunities.

A traditional process map with swim lanes may amount to more than 150 steps. Using the hierarchy approach a team may have mapped 50 to 70 steps in total. A Level Four map is a desk top level of detail and is sometimes necessary. Later the team may need to come back and document the other steps to a lower level of detail, for information, standardization and training purposes. But it is not a key to do at this point in time. Spending time doing excessive documentation now, delays getting improvements identified and implemented.

Traditional Process Flow Chart with Swim Lanes

A process flowchart can be built both on the wall and on paper. Some people find it easier to write things down first and then transfer them onto the wall diagram. Others work better "off the wall" right from the beginning. Both methods use the same process analysis symbols, so the information can be transferred back and forth quite easily.

The Traditional Process Flow Chart uses common symbols:

Symbol	Represents	Examples
BLUE	**Operation** Person(s) doing something	Type letter Fill out form Post total Sort orders
GREEN	**Transportation** (Physical) movement, requiring work time	Walk to file cabinet Walk to copier Send order downstairs

Symbol	Name / Description	Examples
YELLOW	**Inspection** Person(s) reviewing or inspecting something	Check for completeness Verify totals Approve expenditure Authorize requisition
PINK	**Delay** Waiting for information or for other steps to be done	Wait for approval Wait for mail pick-up Wait for response Wait for end of month
ORANGE	**Decision** Splits work into alternate paths, or signifies the start of a re-do loop	Make yes/no decision Make conditional choice Split multi-part form
TAN	**File** Pertaining to documents	File document Store on microfilm Throw away
WHITE	**Document** Report or form	Customer order Purchase requisition Monthly sales report Telephone message

Figure 10-7 Traditional Process Flow Chart Symbols

Whether working with a worksheet or a wall diagram, here are some tips to complete the activity effectively:

- Don't get lost in detail. A task such as "Walk to the Copier" might be important because it's on another floor and takes time. "Make copies" might also be significant, especially if there's only one copier for dozens of people and getting copies made can be a lengthy ordeal. However, "Turn copier on" and "Change paper in copier" are too fine a cut at the detail needed to evaluate work flow.
- Document, then analyze. Do not be overly concerned with analyzing as the chart gets built. Concentrate on defining what tasks are performed as part of the overall process. Any quick ideas that are generated can be entered in the "Notes" section of the worksheet or on a separate list from the wall chart.
- On the first pass with swim lanes, try to describe the process with 6 stakeholders and in 25 steps or less. Deal only with the major stakeholders (process doers) and stick to symbols which are mostly operational. This is another way to keep you from getting into too much detail at the beginning.
- Get input from all involved. Pass the worksheet around or ask others to look at the wall chart. Get as much input as you can.

- Don't map the exceptions on the first pass. It's acceptable to include key decision points and show a yes or no path, but avoid mapping how exceptions get handled, otherwise the map will become mired in detail.

The following is a sample process:

Process: Processing Customer Orders

Process Begins: Customer order received in Mail Department

Process Ends: Customer order sent to Shipping Department to be filled

Detail:

1. Customer Orders (CS-223, a two part form) are received in the Mail Department. The average daily volume is 50 Customer Order forms.

2. The Mail Clerk sorts them alphabetically and sends them to the Order Process Department.

3. The Order Processing Clerk checks the order for completeness:
   a. customer account number.
   b. quantity information
   c. color information
   d. method of payment

4. If the order is complete, the Order Processing Clerk separates the white and pink copies. The white copy is sent to shipping for order fulfillment and the pink copy is filed.

5. If the order is incomplete (10% of the time), the Order Processing Clerk fills out a Request for Additional Information (Form #CS-217). The two are paper-clipped together and placed in the incomplete orders box. Once per day, the contents of this box are delivered to Customer Service.

6. The Customer Service Clerk calls the customer to get the necessary information and fills out the request form. The completed form (with order attached) is placed in an "out" tray, where it is picked up once per day and delivered to the Order Processing Clerk.

Figure 10-8 Example Process

Process Analysis Worksheet

The Process Analysis Worksheet has been partially filled out below to reflect the example process. It's very similar to a process map but the information is gathered on a few sheets of paper. This was originally an Industrial Engineering tool that predated the use of personal computers. So it is a form that is not used too frequently these days, but you may find it useful in some situations. It probably best lends itself as a tool to have different members of a team go observe pieces of a process in-action. They use it to document their observations of the real "as-is" process versus "the way we think we do it" and then use their observations to build a process map. It can also help to capture "cost of quality" information if the organization is gathering that type of data. Simply write down the process steps in the left hand column and enter the appropriate symbol for the work. If a cost of quality column applies, make note of it, and write down any comments or improvement ideas in the last column.

PROCESS: Processing Customer Orders							
DESCRIPTION: Receipt and processing customer orders, from the time they are received in Mailing until they are sent to Shipping.							
PERSON/DEPARTMENT: Mailing, Order Processing, and Customer Service		**COMPLETED BY:** N.E. Budding					**DATE:** 9/16/XX
		TASK TYPE	**COST OF QUALITY**				**NOTES**
	PRESENT PROCEDURE ___ **PROPOSED PROCEDURE** ___	**Operation Transportation Inspection Decision Delay**	**Preventioin**	**Appraisal**	**Internal Failure**	**External Failure**	
	MAIL CLERK	○⇨□◇D▽					
1	Sort in alphabetical order	●⇨□◇D▽					Form CS-223, 50/day
2	Send to order processing	○➡□◇D▽					
	ORDER PROCESS	○⇨□◇D▽					
3	Check for completeness	○⇨■◇D▽					10% incomplete
4	If incomplete, go to Step 8.	○⇨□◆D▽					
5	Separate Parts	○⇨□◆D▽					
6	White copy to Shipping	○➡□◇D▽					

Figure 10-9 Excerpt of Process Analysis Worksheet

Building a Traditional Process Map (with Swim Lanes)

Building a traditional process map is fairly easy. If a team already did the hierarchy maps, do not do this step, it's redundant. A step-by-step description for developing a traditional process map is outlined below:

1. Attach a section of paper (or mylar) to the wall - 6 feet long should suffice.
2. Divide the paper into segments by drawing horizontal lines across the width of the paper. Each section will deal with one of your major stakeholders. Try to limit yourself to six segments or less.
3. Using the process analysis symbols, chart the beginning and end of the process. Put the symbol for the start of the process in the top left-hand corner. Choose a symbol that represents the end of the process and put it in the bottom right-hand corner.
4. Assign a stakeholder to each horizontal section by writing one name on the left of each segment. With the beginning and end of the process, you should be able to put the stakeholders in the order that they will get involved.
5. Chart the process. Try to use 25 symbols or less. Write a short description of the operation on the symbol to describe the task being performed. Work your way across the chart and down through the stakeholders. Remember to stick to the basics.
6. Do a Status Check. After you have finished a first pass, ask yourselves the following questions:
   - Do we need more detail to analyze the problem?
   - Did we miss any stakeholders in the original list?
   - Do we need more functions represented on the team?
   - How will we get input from people not on the team?

If the team feels a need to get others involved in the charting, develop a plan to do so. You may have to wait for their input before proceeding. If a more detailed process map is needed, a separate diagram for each stakeholder's part of the process can be built.

A sample process map is shown below, though most maps involve more complicated processes than illustrated here.

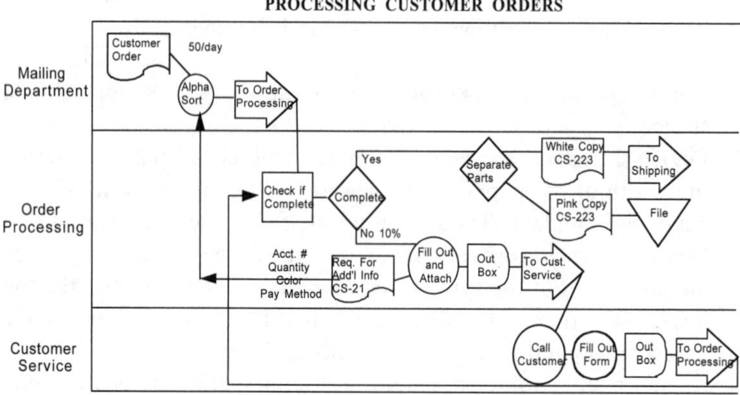

Figure 10-10 Example Process Map

Cross-Functional Roles and Responsibilities Matrix

If a team does not experience any conflict with the organization or on the team when they do the cross-functional process maps, just wait. A Cross-Functional Roles and Responsibilities (R&R) Matrix will usually get people excited. The R&R Matrix simply lays out the process steps along the horizontal axis at the top of the matrix and the individuals departments or functions along the left hand vertical axis.

Then the key job responsibilities for each process step are highlighted for the appropriate department. See the example below.

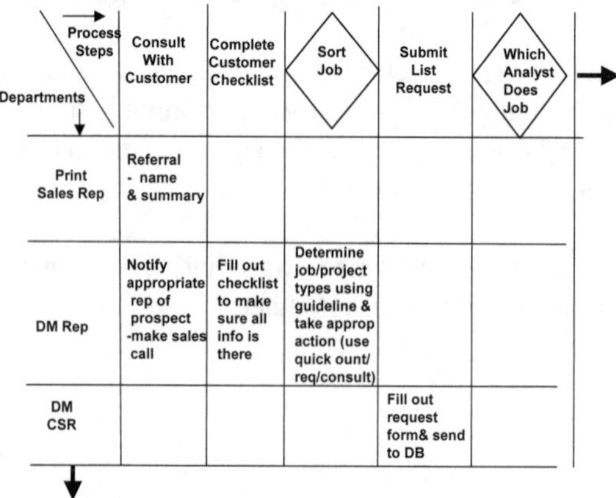

Process Steps → / Departments ↓	Consult With Customer	Complete Customer Checklist	Sort Job	Submit List Request	Which Analyst Does Job
Print Sales Rep	Referral - name & summary				
DM Rep	Notify appropriate rep of prospect -make sales call	Fill out checklist to make sure all info is there	Determine job/project types using guideline & take approp action (use quick ount/ req/consult)		
DM CSR				Fill out request form& send to DB	

Figure 10-11 Example Roles & Responsibilities Matrix

A task shown on the matrix can also indicate responsibility/authority levels. A typical scale is:

R - Who has primary responsibility and accountability for doing this task?
A - Who has authority to say when the process step is completed? There should only be one person or group in a column with authority. It may or may not be the accountable block.
C - Who should be consulted about an action taken? (Consulted usually means a response is expected)
I - Who simply needs to be informed about an activity/action? (But no response is expected)

There could be other categorizations, but the above is a common set usually referred to as a RACI R&R Matrix. The discussions that take place around this matrix can be a powerful means for improvement: determine who should have authority, eliminate multiple or conflicting authority actions, and reduce the number of people needed to coordinate actions to the critical groups that really have something to contribute. It's a simple but powerful tool!

Value Stream Maps

These types of maps are typically used for Lean projects looking to move more toward flow concepts. They were originally used for manufacturing processes but have since expanded to offices, hospitals, and other industry applications. These maps tend to have a strong timeline focus and they show information flows. An example of a manufacturing operation's Value Stream Map (VSM):

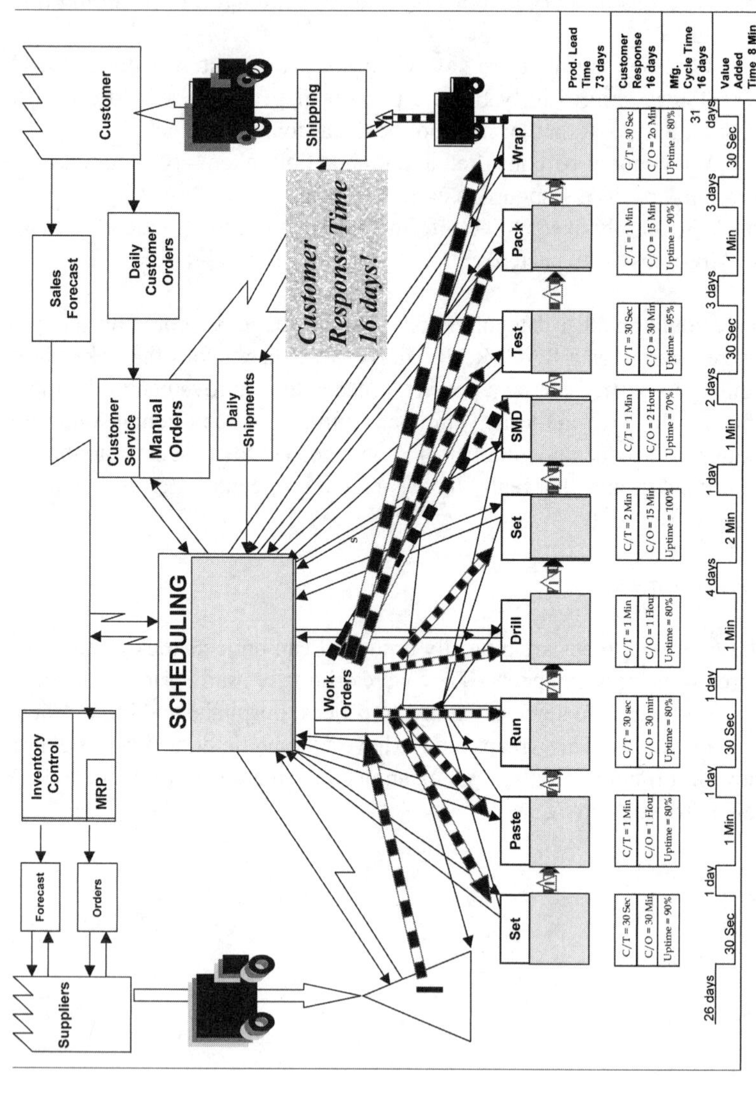

Figure 10-12 Example Value Stream Map

Now this map looks complicated. It is! But it tells a simple story. An order is received from a customer (upper right hand corner). Ultimately, the order needs to make its way into the factory. People need to make the product (the blue boxes across the bottom of the diagram) and the product needs to be shipped. In this real company example, it was taking 16 days to deliver a customer's order. Work-in-Progress inventory existed between each of the blue box steps. It was taking a total lead-time of 73 days for materials from the time they were ordered from a supplier until they were shipped in a finished product. The arrows indicate information flows, so it is pretty easy to see that in addition to the formal scheduling system, there was quite a bit of expediting taking place. The white boxes under the functional steps of the process contain information about up time for that workstation - changeover time and cycle time. These were simply the metrics this company wanted to track. The total "Value Added" time was 8 minutes against 16 days to fulfill a customer order - not a very good ratio.

A VSM tells a different story than a process map with swim lanes or the hierarchy maps (levels). A VSM shows key data with the picture and it typically presents a time line. The data could be captured on a Process Characteristics worksheet, but when it is layered on a picture, it has a more powerful impact. Who could resist making improvements to a process with only 8 minutes of value added work and a cycle time of 16 days?

A VSM is typically done for a product family (since other products most likely go through significantly different processes), a major customer group or a major functional department. However, once an organization builds a half dozen of these for a functional area, a repeating picture will usually emerge. A VSM for different functional departments will yield yet another perspective. The term customer sometimes needs to be converted to internal but if the external customer can remain in the picture, it will usually produce a better map. For example, a VSM for engineering might focus on the new product development process, custom product design or even engineering change notices - all of those have a direct external customer impact. An accounting example might be the receivables process, the payables process, information reporting, or payroll processing. The first two - receivables and payables - clearly have an external component; information reporting might as well depending on accounting's role and responsibilities. Payroll would most likely be a process focused on internal customers only.

Future State Map (or "Could-Be")

It can be powerful to do a future state map at this point. Don't worry about how it is going to happen. Answer the question, "In the ideal world, this process would look like *(fill in the blank)*?" Then create a map that reflects that picture. If there are major exclusions from the team's scope or authority (e.g., no new information systems), those concepts could be excluded from a proposed design. Sometimes, "could-be" maps can be done before the "as-is" process mapping, taking into account the key process requirements and major innovation opportunities identified by the team.

It's powerful to do a future state map at this time because there are few restrictions. People may feel uncertain at first, but after they get over their initial anxiety, they often come up with some exciting innovations. This is another reason for having an external customer view because the question can be, "So now that we know our current state, what would be the ideal way to serve this customer group?" And then create a future state map. The future state map for the above example looked like this:

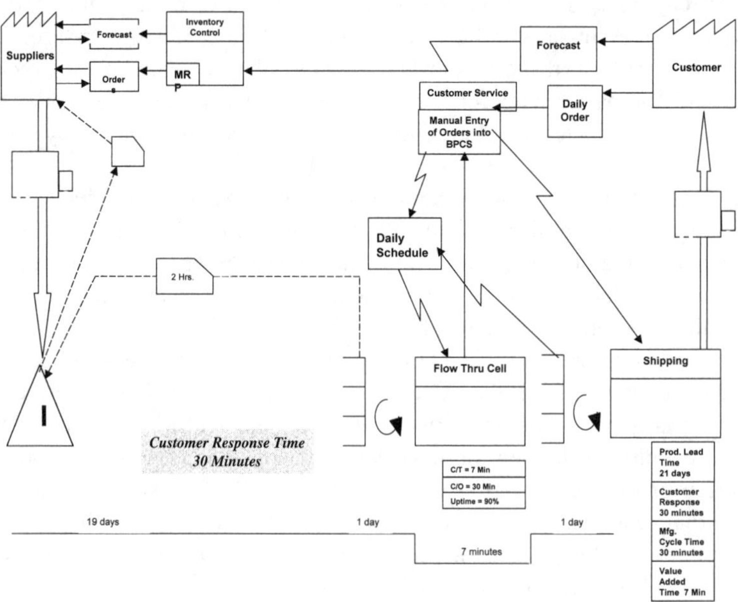

Figure 10-13 Example Future State Map

In this map, the organization would have moved from inline assembly to a series of product related cells. Scheduling would be much simpler and customer response time to an order would drop from 16 days to 30 minutes, plus shipping. Value added time only dropped from 8 minutes to 7 minutes. This is rarely a significant change relative to the overall timeframes. This, in fact, is the direction this organization moved, and they hit these targets.

Quick Wins

Processes that have not been reviewed for some time usually have easy and obvious "Quick Win" improvement opportunities that can be seen from the process maps and initial observations. These "low hanging fruit" opportunities sometimes have high Return on Investment (ROI). Criteria for defining an Improvement Opportunity as a quick win:

1. Easy to implement and making the change does not require a great deal of coordination and planning.
2. It's largely within the team's (including the Champion's) control
3. Fast to implement
4. Cheap to implement
5. The change is easily reversible

Obviously, the team needs to guard against the danger of losing sight of the overall goal for improvement, and to avoid the distraction of solely focusing on quick wins.

Sustaining the Gains

Consideration needs to be given to sustaining the gains and managing the change right at the beginning of the project. The top reasons for employee resistance are a lack of awareness about the change or why the change is necessary. Fear of the unknown is also a major concern. Middle managers tend to have greater concerns about losing control, an overload of task and clear roles and responsibilities. Effective process improvement teams begin to address these issues early in the project with their communications to people outside of the team. Keep in mind that communication is a two way street and requires active listening.

Define Phase Tollgate Questions

As the team progresses through the Define phase, the team leader should schedule meetings with the Champion to review all that has been accomplished. The following are a list of questions that can prompt discussion in these meetings and to monitor the team's progress through the Define phase.

1. What are the primary business goals that will be influenced by this project?
2. What Voice of Customer data were used to establish Critical Customer Requirements? How were the data validated?
3. What Voice of Business data were used to establish Critical Business Requirements? How were the data validated?
4. What are the boundaries of the process to be improved?
5. What is the specific problem being addressed?
6. Has this problem been tackled before? What was learned from that attempt?
7. What are the goals, in measurable terms, of the project? Are they achievable in the timeframe established?
8. Were team norms established? How are violations of the norms handled?
9. Has the team created a detailed project plan with milestones and associated activities?
10. How detailed were the process maps? How were the maps validated? Did the team ensure that they were "as-is" maps showing the actual state of the process?
11. Was a "future state" map created? Did it guide the team toward improvement opportunities?
12. Who are the stakeholders that will be affected by this project? What level of communication or involvement is necessary for each stakeholder group?
13. What concerns may the stakeholders have? How will the team prevent these concerns from becoming obstacles (during the project and after change implementations)?
14. What quick wins were identified? What is the plan for implementing quick wins? What are the plans for ensuring that the quick wins work? What effect will the quick wins have on the goal?

At the completion of the Define phase, the team members, team leader, and the Champion should feel comfortable with the answers to all these questions and any others that might be specific to the organization.

Define Phase Example - Publishing Business

The Vice President of Sales for a major U.S. Publication Company was looking for opportunities for revenue growth. She met with Kathryn Ireland, the manager of their Direct Marketing group, to explore opportunities. The Direct Marketing Group sold marketing information, customer analysis, and demographic data.

They both felt the current process was rife with exceptions, the administrative process was cumbersome, and opportunities for revenue growth were being missed. So a team was commissioned to work this process with Kathryn as the team leader.

They were not certain how much time the sales force spent with customers or planning for customer visits, but they felt a disproportionate amount of time was spent filling out administrative forms and preparing client's paperwork for internal processing at the publishing company. There also seemed to be an excessive number of errors that were occurring. Most of the errors were caught so the client never saw them but they caused chaos inside the publishing company, especially just before publication.

So three goals were established for the team:

1. Increase sales face time with customers to 75% (included planning time)
2. 100% accuracy in information processing
3. No decrease in profit margins

They discussed who to include on the team. In addition to Kathryn, they decided on the manager of database information processing, a salesperson, two publishing operations personnel, two customer service representatives and one person from accounting.

The primary purpose of the team was to streamline the direct marketing process in order to increase sales, decrease costs, improve efficiency and improve effectiveness. They would do this by giving sales people more time to sell. The team's charter was as follows:

Charter Section	**Direct Marketing Project**
Business Case:	Streamlining the direct Marketing job workflow will increase sales and planning time, improve profitability and efficiency, and increase revenue
Opportunity Statement:	Standardizing processes for the Direct Marketing jobs gives the opportunity to increase internal and external customer satisfaction, accuracy, productivity and revenue generated by the direct marketing sales team.
Goal Statement:	- Increase sales and planning time to 75% - Increase accuracy of paperwork to 100% - Increase profit margin to a 35% average
Scope:	Process: Direct Marketing job Start - Customer Inquiry End - Customer Bill Sent
Project Plan:	Task — Start — End Date Evaluate current process flow — 10/21/03 — 11/20/03 Identify Quick Wins — 11/20/03 — 12/8/03 Define roles and responsibilities — 11/18/03 — 12/16/03 Form Consolidation and redesign — 11/25/03 — 3/31/04 Data Collection — 11/12/03 — 3/11/04 Pilot CSR — 11/13/03 — 5/3/04 Data Analysis — 1/5/04 — 4/19/04 Final Recommendations/Control — 5/1/04 — 5/1/04
Team Members:	Champion: VP of Sales Team Leader: Kathryn Ireland Team Member: Mgr of Database Information Processing Team Member: Salesperson Team Member: 2 Publishing Operations rep's Team Member: 2 CSRs Team Member: Accounting

Figure 10-14 Team Charter

Kathryn and the Vice President of Sales first met one-on-one with the supervisors and managers of the people they hoped to have on the team to get their approval. Everyone gave a thumbs-up and agreed with the importance of this project. Then they called a meeting to launch the project. They launched it in a three-day working session.

At the beginning of the meeting, the Vice President of Sales stressed the importance of this project. They also discussed how new competitors were entering the scene and that the Internet was a constant threat to their business. The team members were excited about the project and were comfortable with the scope that had been established.

The team prepared a process map using the hierarchy method. Their level one map included 10 steps. They then looked into the Customer Service Representative's task blocks and the Salesperson's administrative work task blocks. Ultimately, they made a Level Three map. An excerpt from this map follows:

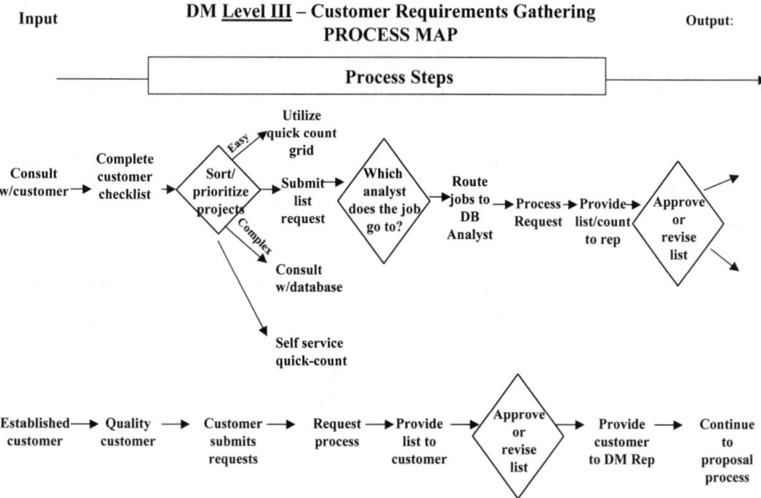

Figure 10-15 Excerpt from Process Map

Define Phase Example - Telecommunications Company

A large telecommunications company commissioned seven teams to look at the end-to-end business processes for their overall enterprise. They had 15 to 20 different regional processes currently in existence. Customers were becoming very frustrated in dealing with the company. The leadership team set very specific goals for the process improvement teams to accomplish over the next three years to increase customer satisfaction, employee satisfaction, contribution margins and cash flow. The first year improvement goals were:

1. Increase revenue $30 million
2. Decrease Cost of Sales $15 million
3. Decrease other costs $40 million
4. Improve customer retention 10%
5. Improve operating margins 1%

The business was experiencing considerable competition. A number of their traditional competitors had been acquired and the new owners were investing in this market. The Internet was giving existing customers new alternatives to the organization's services. And a number of the organization's large customers had been acquired and were no longer using their services.

The organization had purchased a series of regional businesses, had recently gone through a downsizing and leadership was now looking to stabilize employment, improve financial performance, improve customer service and get new products into the marketplace much more quickly.

The organization hired a consulting firm to get the project off to a quick start. A Senior Vice President championed each team. The seven teams needed to accomplish the above five goals, working together. The teams were working on seven different but related processes. The key four processes were:

1. Customer Solutions
2. Customer Acquisition
3. Production through Fulfillment
4. Customer Care and Billing

The teams included over 100 full and part-time team members. They started their project by defining the "as-is" business processes, practices and high-level procedures.

They wanted to make certain they were coordinating their efforts. So they began with a series of executive, team leadership and key personnel meetings to define their project responsibilities. In addition to the above goals, the team leaders agreed to a joint charter:

The following elements were within scope of the project:

- Recommendations for end-to-end processes supporting print, electronic, and national products for the following functions: Marketing Distribution; Sales; Billing; Publishing; Credit Management; Printing; Collections; & Customer Care

The scope of this project did not include:

- International processes and organization structures
- Recommendations for change of underlying systems / platforms

The team leaders conducted a meeting with leadership to organize the scope and responsibilities for each process team. In this meeting, they:

1. Overviewed each project and gained insights from the Executive team
2. Further clarified the scope and laid out a plan on how to work together
3. Defined the deliverables from this workshop including:
    a. Resources, Project and sub-team Scope and Expectations Reviewed & Developed
    b. Team Roles, Norms and draft Individual Process team scopes
    c. Top Issues Discussed
    d. Customer market segmentation capabilities matrix developed

The project team leaders then followed this meeting with a set of individual key process team meetings. Each process improvement team went through a final clarification of the scope, defined how to work together, reviewed the market segment study, and created a communication plan to guide each team's efforts. The deliverables for each individual process improvement team included:

1. Definitions
    a. Team Norms and Roles Developed

b. Market Segment Capabilities and Matrix Reviewed
c. Team Vision for Outcome of Effort Developed

2. High Level Process Maps
Further develop high level process maps to include next-level sub-process overviews, identify process linkages and gaps to ensure that the teams are aligned.
Deliverables:
a. Sub Team Processes Mapped to Critical Steps
b. Team Leaders Met and Reviewed & Aligned Maps

3. Best In Class (BIC) Assessments Initiated
Identify processes that could benefit from best in class research.
Deliverables:
a. Best in Class (BIC) Opportunities Identified
b. BIC Sub-Teams Formed

The teams gathered a lot of information - actually, too much. The teams mapped ten different processes in the first several weeks and planned to map another 30 to 40 more. This caused a few project problems over the next several months. There was also a high degree of concern amongst team members about each of the regional processes and how they differed. The members from each region were being pressured by their regional leadership to avoid making changes to the processes for their region. This information was not initially shared with the other team members.

Measure Phase for the Lean Team

The primary purpose of the measurement phase is to answer the questions, "How are we doing?" and "How far do we have to go?" The team needs to establish a baseline of the current performance level. Lean and Six Sigma Teams begin to differentiate in this step. Lean Teams take a horizontal view across the process. Typically, Lean teams try to understand the overall process requirements and then to identify the waste elements currently inside the process. Classic Six Sigma teams typically use metrics to dig deeply into the process (more vertically than horizontally) to identify and eliminate sources of variation. This is one of the key reasons why it is important to stabilize the process first. There is no need to reduce variation for process steps that are waste. They should be eliminated.

Generally, there is no shortage of opinions about what the problem is and what needs to be done to fix it. It is critical to call a "halt!" to this type of thinking and get the team focused on gathering evidence to describe the current reality, not their assumptions about that reality.

Deliverables from the Measure phase include:

1. Collected data
2. Selection of what measures to use
3. Baseline of the "current state"
4. Data collection plan

One of the first actions in this step is to determine what to measure. This is also one of the most important steps. There is no one right answer as to how far to go with this idea. Obviously, if a team measures too much, it becomes bogged down in trivial pursuits and risks losing sight of the overall game plan: identify the best improvement opportunities and implement them. The right answer is to collect a sufficient level of information to have something that is actionable. So how do you do that? The team probably needs to consider collecting a couple different types of metrics. This is a useful thing to brainstorm first and then do some initial data gathering and check early results to see if the numbers look like they might yield the types of information the team is seeking to learn.

Measurement Plan

Data collection is an important deliverable for any process improvement team. Traditionally, this is a weak area for process improvement teams (but it is truly a strength of the Six Sigma approach). The measurement plan should take into account:

- What is the measurement information to be captured? What is the metric?
- How is the information going to be collected? Does a form need to be developed? Can it come from an existing source?
- When will it be collected? How granular does the team need to be? Is continuous information needed? Will a week-end or month-end number suffice (a point in time number versus continuous)?

- Where will the collection take place? Where does the data come from? What is a satisfactory source?
- Who is responsible for doing it? Who is accountable? If this is some one outside of the team, do they understand what they need to do, do they agree to do it and have they been given permission to do it?

Some of the information will already exist. This is especially true for production related processes. For administrative processes, often the information does not exist. The team will need to develop a simple data gathering plan that does prolong the project nor bring the organization to its knees. It's also worthwhile to give some thought now as to how this information will be displayed. If the team plans to use histograms or Pareto Charts, the information needs to be collected in a format that fits the display.

Selecting Metrics

When measuring a process, the team should not get trapped into just looking at financial metrics. What a customer wants is often measured in other ways. The team should give consideration to a balance of perspectives. Often the solutions will result in trade-offs between the different views. A balanced set of metrics helps to better weigh the alternatives. There are literally thousands of measures that can be chosen. The trick is to pick the few that will be most significant to those involved in the process.

One set of metrics should focus on the degree of the problem. How severe is it? If the team cannot act on the data, they are most likely collecting the wrong information or they have not gotten to a sufficient level of detail. It is one thing to say, "we have a lot of rework," a little more specific to say, "we have problems with the #3034 disk drive," and better yet to say, "we have fracture problems with the #3034 disk drive." The latter statement starts to be actionable. A team may still need to dig deeper into the world of fractures if there are a variety of types. Pareto analysis, cause and effect analysis, frequency diagrams, all are ways to look at the level of the problem.

Another type of metrics addresses the health of the process. These metrics may not be the most important at the front end of a project, but the team should start looking for these types of metrics in this step. They will become important during the Improve and Control phases.

Baseball enthusiasts are legendary for their fascination with data. Every action, activity, and result in every game is carefully tracked and documented - at both an individual and team level. This information is then used to make game decisions and to compare and contrast players, teams and ERAs. But some metrics are better indicators of overall performance than others. Routine baseball metrics include:

Hits	Runs	Errors	Home Runs	Triples	Walks	Saves	Left on Base	On Base Percentage
Doubles	Singles	Innings Pitched	Sacrifice Bunts	Ground Outs	Total Bases	Lead Off Hits	Full Counts	Slugging Percentage
Sacrifice Fly's	Fly Outs	Extra Base Hits	Strikes	Balls	Total At Bats	Complete Games	Pitches Thrown	Fielding Percentage
Team Batting Average	Team Pitching Average	Stolen Bases	Walks to Hits Ratio	Strikes to Balls Ratio	Night vs Day Record	Won-Loss Record	Strike Outs	First Pitch Swings

One way of thinking about measures is in terms of what they look at: inputs, process (activities), and outcomes (outputs). When the process is underway, process measures are where the greatest leverage exist because they give the best predictive information. The key is being able to find the right predictive metrics.

1. Input Indicators - Measures that evaluate the degree to which the inputs to a process (provided by suppliers) are consistent with what the process needs to efficiently and effectively convert inputs into customer satisfying outputs. In the baseball example, inputs represent the number of opportunities. Examples of input indicators include:
   a. # of customer inquires
   b. Type of customer inquires
   c. # of orders
   d. # of positions open
   e. Accuracy of the analysis
   f. Timeliness

2. Output Indicators - Focus on the end result. Measures that evaluate dimensions of the output - may focus on the performance of the business as well as that associated with the delivery of products and services to customers. In the baseball example, they focus at an individual and team level. Examples of input indicators include:

a. Retention rates
b. Total # done, sold, made, etc.
c. On-time
d. Complete

3. Process Indicators - Focus turning opportunities into desired results. Measures that evaluate the efficiency, effectiveness and quality of the transformation processes (i.e., the steps and activities used to convert inputs into customer satisfying outputs.) Examples of process metrics include:
   a. Availability
   b. Time to do something, timeliness
   c. # of non-standard request
   d. Yield (first time through)
   e. # of exceptions (e.g., non standard approvals)
   f. Quality level (could also be an output metric)

Input Measures	Process Measures	Output Measures
• Total At Bats	• Walks to Hits Ratio	• Home Runs
• Pitches Thrown	• Errors / Fielding Percentage	• Won-Loss Record
• Innings Pitched	• On Base Percentage	• Complete Games

Figure 10-16 Baseball Input, Process and Output Measures

Leading Metrics

The measures that give the best predictive information in the game of baseball are those that measure the process. They are the leading metrics that can be monitored during the course of a game and people can try to execute directly linked actions. As Yogi Berra (former Hall of Fame catcher for the New York Yankees) once said, "it ain't over till it's over." A win versus a loss is certainly important but it is the end result of the game. You don't know if you won or you lost until the game is over; at that point in time, it is too late to act. In the game of baseball, the team that has the best walks to hits ratio, fewer errors per fielding opportunity and high on-base percentages will have more wins.

A Lean team looking to implement improvements to a process - whether it is Lean manufacturing across the enterprise or improving the order entry process - needs to identify a set of leading indicator process metrics that reflects the health of their process.

Time Metrics

Lean process improvement teams often look at time metrics. They include:

1. Takt time - the unit-to-unit pace of production *required* to meet customer demand. This is the ideal pace for doing work. At this rate production and demand are in balance.
2. Cycle time - the *actual* elapsed process time from the completion of one output unit to completion of the next unit. Cycle time for a process is the length of time for the longest step within the process. So if a four step process has three steps of two minutes each, and one step of five minutes. The cycle time is five minutes.
3. Lead Time - the length of time from the beginning point of a process to the completion of a finished output (calculation varies depending on scope; it includes: queue, wait and move times).

Time metrics can also include frequency, degree of impact, response time, etc. Teams will typically look at the time that it takes to do the process steps. The Process and Value Stream Map will also provide sources of time measurement data.

Data Gathering Tools

There are multiple tools used to gather and organize measurement data: Cause and Effect diagrams, Cost of Poor Quality, and more Process Step Analysis worksheets are typically key parts of process measurement. If the team does not gather this evidence in a meaningful way, they increase their chance of doing rework during analysis, and of developing solutions that people will not support due to a lack of hard evidence.

A Process Characteristics Chart is another way to gather measurement information about the various steps of the process. Typically, this chart shows the steps of the process down the left hand column and various attributes of the process that the team wants to review across the horizontal axis. An example of the form looks like:

									Process:		Ping Pong Factor			
			Value	Quality		Setup Time		Process Time		Capacity		Walking Dist		Transport Dist	
			Add	Base	Oppty	Base	Oppty	Base	Oppty	Base	Oppty	Base	Oppty	Base	Oppty
#	Step Description	Who	X	%	%	Min.	Min.	Sec.	Sec.	U/Hr.	U/Hr.	Ft.	Ft.	In.	In.
A1	Raw material bin							2,240	25.7						
B1	State "Load ball" every 5 se	TK						30.1	0.0	120	NA				
C1	Move ball from bin to range	LM						110.4	7.6					18.0	12.0
C2	Aim & release into process	LM		25%	100%			24.0	1.0						
	S.T.							134.4	8.6	27	419				
D1	Processing unit		X			10.0	0.5	2.4	2.4	1,500					
E1	Receiving bin							224	15.4						
F1	Wait for ball from processin	MA						19.2	0.0						
F2	Transfer ball into inspect. b	MA						19.8	12.8					15.0	0.0
F3	Pass insp. bin to inspector	MA						2.6	2.6			2.0	1.0	26.0	12.0
	S.T.							41.6	15.4	87	233.8				
G1	Inspect received pallets			0%	100%										
	Totals (from sheet above)			0%	100%	10.0	0.5	2,673	67.5	27	234	2	1	59	24

CUMBERLAND GROUP

Figure 10-17 Process Characteristics Chart

It is the team's responsibility to gather enough sufficient information to move down to a root cause level of analysis in the next phase.

Measure Phase Tollgate Questions

1. Do the metrics link to the team's improvement goals?
2. Do the metrics have a balance of viewpoints (customer, supplier, process, productivity, quality, etc.)?
3. Was a plan developed for collecting the information?
4. Was a baseline established?
5. Does the data collected look like it will provide the necessary information for analysis and decision making purposes?
6. Are all significant process steps captured?
7. Was the data validated for reasonableness by anyone outside of the team?
8. Which metrics may be useful in sustaining improvement after implementation of the changes?

Measure Phase Example - Publishing Business

The team wanted to first establish baselines for their key goals. There was plenty of information available but they really did not know how the sales people spent their time. So they designed a simple survey to be filled out for a short period of time by the sales staff. Based on the survey, observations of work and experience, they learned that the sales staff only spent 27.5% of their time planning for sales calls and with customers. The rest of the time was spent coordination work, filling out forms and miscellaneous tasks.

On the administrative side, they learned that direct marketing projects have many players: clients, artists, printers, the mail house, the database analyst, and representatives. They had a lot of things to juggle and not much time to do it. This resulted in a problem with accuracy. They learned that 50% of the transactions were being reworked in some fashion.

The initial baseline information compared to goals looked like:

CATEGORY	GOAL	METRIC	BASE
SALES & PLANNING (Face Time)	75%	Time use survey, selling time vs. other	27.5%
ACCURACY	100%	% of revisions through database requests and art tickets / production orders	50%
PROFIT MARGIN	xx% avg.	Total monthly average margin by account executive.	+XX%

Figure 10-18 Baseline Measures

They also laid out a Cross-Functional Roles and Responsibilities Matrix where they tracked each step from their process map to the person or function that had a responsibility for doing this task or making this decision.

Departments \ Process Steps	Consult With Customer	Complete Customer Checklist	Sort Job Com-plexity	Submit List Request	Which Analyst Does Job?	Route Request to DB Analyst
Class/Retail/ Print Sales Rep	**Referral** -contact name -conversation summary					
DM Rep	**Notify** appropriate rep of prospect -make sales call	**Fill out checklist** to make sure all info is there	**Determine job/project types** using guideline & take approp action (use quick count/ req/consult)			
DM CSR				**Fill out request form** & send request to DB		

Figure 10-19 Cross-Functional Roles and Responsibilities Matrix

Measure Phase Example - Telecommunications Company

The team members gathered information on measures that were already in existence. Each process team operated independently for this phase. Next, they listed the measures on a matrix to see what types of metrics they had.

Classification	Input/Supplier	Process	Output/Customer
Financial			Total Costs
Productivity		Units per hour	
Learning/Development		Multi-skills	
Timeliness	On-time	On-time	Complete

Measures deemed to be leading indicators (metrics that allowed early action to be taken) were shown in blue. Measures deemed to be lagging indicators (after the fact) were shown in green. If the team members could not agree, or if the metrics were not clearly leading or lagging, they were shown in black. Each team determined if they had a balanced set of metrics that represented a variety of perspectives including customers, suppliers, employees, effectiveness and efficiency.

The team members began to feel somewhat overwhelmed by all of the different measures being used. They found that different regions used the same metrics, but they called them by different names. They solved this problem by giving a description to the metric, keeping a log of the calculations and in the log attributing the measures to the regions that used it. They also noted the source of the data in the log. This proved to be a valuable tool later when they were trying to standardize the metrics.

Once they had a picture of the existing metrics, they determined what was missing. The customer services team quickly realized that no metric existed for a number of important business goals. For example, no real metrics existed for customer loyalty. They had aspects of it (e.g., customer cancellations) but they had no clear idea of how many customers were retained year-to-year, how many provided referrals or how many purchased new products and services. In their benchmarking visits, they also learned that other companies had 50% less paperwork and time in signing up new accounts. The other process sub-teams gained similar insights.

The teams validated their metrics in a number of ways. One of the simplest ways was to check the metrics with people who were not on the team. They asked how the metric was used, what were the benefits and what were the shortcomings of the metric. They learned that most of the metrics were lagging measures. The operational measures that were used had no clear linkages into financial performance.

Analyze Phase for the Lean Team

The primary purpose of the Analyze phase is to answer the question, "What is wrong?" This is where Lean teams use their detective skills. Brian McKibben of the Cumberland Group - Chicago likes to say, "Reasonable people, equally well informed, seldom disagree." It's still important for the team to withhold their judgment at this point in the process. It is far easier to become equally informed if people are not being harassed when they ask a question or float an idea. If team members practice active listening where people spend a moment seeking to understand, then all team members will learn more.

The primary deliverables for a Lean team's Analyze Phase include:

1. Documented analysis of the opportunity/problem
2. A reasonable identification of the root cause
3. A prioritized list improvement opportunities
4. Identification of potential solutions for key opportunities
5. Estimated costs/benefits of any proposed solutions
6. Recommendations for action
7. On-going communications to key parties impacted by potential changes

In this section, the actions to accomplish these deliverables are described along with other tools and techniques that are appropriate for Lean teams.

Root Causes

A key objective for any type of improvement team is to eliminate the root causes of problems. No one wants to go back and solve the same problem over again. For a team working on a Six Sigma problem, this may require statistical analysis and validation. For other types of problems, it's simply drilling down to a root level. The 5 Why's is one common technique. It's described in the Six Sigma section of this book. It works pretty simply. One of the keys to success is how the word "why" is stated. One could ask, "WHY in the heck did you do that?" Or they could ask, "Why do you think this happened?" Same word, but not the same question! A quick true story of how asking "Why" can get to a root cause.

Public Warehouse Mystery

Sam, a superintendent at a Public Warehouse, came up to the Warren, the General Manager, and said, "We need more fork lift trucks. Here is a request for $75,000." Warren asked, "Why?" "Well we never seem to have enough trucks available at peak times….." Warren again asked, "Well why not? I seem to remember buying several trucks last year." Sam said he was not certain and he would check it out. Sam came back the next day and said, "We have too many trucks in the repair shop so we need some new ones for the floor." Warren (a patient man) asked, "Why are the trucks in the repair shop?" Sam, getting a little nervous said, "I'm not sure." So Warren and Sam headed out to the repair shop where they asked Lynn why there were so many trucks in the shop.

Lynn said, "We keep getting all of these trucks in with flat tires. It's not difficult to fix, but it takes time." So Warren and Sam went for a walk around the warehouse. They saw a driver named Mary and asked her about the problem. She indicated that people seemed to get flats when they were driving over in quadrant four. So they headed to that territory. As soon as they arrived they saw the problem. There were nails all over the floor. When they looked up at the racks in that area, they saw that two of the boxes on top had torn open with nails spilling out. The ceiling above the boxes had a slight brown smudge. Sam said, "OK, I will get the roof fixed." Warren further thought, "We need to revisit plant cleanliness."

That is what root cause analysis is all about - drilling down below the symptoms of the problem to the root level. If the root level gets fixed, all of the symptoms disappear. Sometimes this requires sophisticated analysis, but often it simply requires inquiry and an open mind.

Analytical Tools

Many of the analytical tools started in the earlier phases of the project continue to be useful at this stage. This is where people become equally well informed. If a team does its homework, it should be able to sell its recommendations for improvement. Lean teams could use the Waste List in the Kaizen section as a checklist to guide analysis. Teams should study the process steps, one at a time, and identify the most important improvement opportunities in the overall process, using the appropriate analytical tools:

- Process maps - where are the delays, where are the redo's, what are the customer's requirements (internal and external), where is the waste?
- Time studies and process steps - same as above

- Value Stream maps - where are the non-value added steps? Can they be eliminated, combined, sequenced, simplified, etc.?
- Pareto analysis and other data gathering - what are the "vital few" things that need to change? What story does the data tell?
- Universal best practices:
  a. Eliminate mistakes and miscommunications
  b. Waste less time and resources
  c. Make the process work to support customers
  d. Eliminate steps
  e. Have fewer people involved
  f. Do things in parallel
  g. Stop gathering useless information
  h. Push decision making downward
  i. Define standardized 'ideal' practices for doing work

The team then determines the priority ranking of improvement opportunities and begins to develop solutions for the key opportunities.

Key Questions for Interviews

Sometimes the analytical tools will provide sufficient information. In other situations, a team may need to interview people inside or outside the business about the process. If this is true, a set of Key Questions should be created. A good set of Key Questions maximizes the collection of relevant information and helps to avoid collecting trivia and irrelevant data. A Key Question is an information probe to verify, reject or modify a hypothesis. The questions could be relevant to the team's Charter or to a significant recommendation the team is considering.

There are three types of questions people can ask:

1. Open-ended questions generate the highest level of information. They also serve as the basis for developing new questions to further understand the situation.
   a. What are your current needs?
   b. How has this impacted your customers?

Open ended questions allow the interviewee discretion in deciding what (or how much) to say. These types of questions require careful planning before an interview.

2. Close-ended questions are the type most commonly used by inexperienced interviewers. They can be answered with a one to three word response. A few of them are necessary for factual information, but they should be kept to a minimum.

    a. How much did this cost?
    b. When did this happen?
    c. Did this impact your customers?

An interviewer using close-ended questions talks about 80% of the time. This is exactly the opposite of the ideal.

3. Leading questions have an answer embedded within the question. This is usually a poor way to ask a question, and it may make the interviewee defensive. Leading questions typically reveal the interviewer's bias.

    a. How did you feel about causing this problem?
    b. What did you say when they caught you with your hand in the cookie jar?

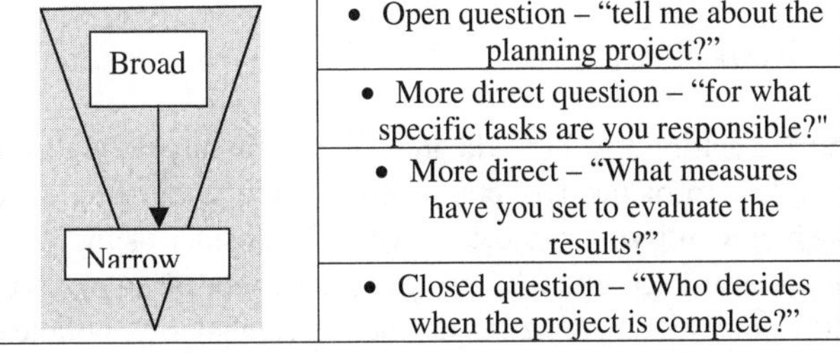

Broad
Narrow
• Open question – "tell me about the planning project?"
• More direct question – "for what specific tasks are you responsible?"
• More direct – "What measures have you set to evaluate the results?"
• Closed question – "Who decides when the project is complete?"

Funneling questions can be an effective way to gather information, moving from the general to more specific.

This type of questioning and probing for more information can yield much information. Interviewers without much experience interviewing should probably do them in pairs, with one person asking the questions and the other observing what is happening and taking notes. Interviewers should listen for ideas, not just facts. The interviewer should also have some enthusiasm and avoid jumping to conclusions - concentrate, wait, listen and keep an open mind. Sometimes use the power of silence. Don't jump into the quiet space - wait and see if the interviewee says more.

At the end of the interview, evaluate the facts.

- Are they from a reliable source?
- Do they make sense?
- Are their any conflicts or holes in the data?
- Do you have sufficient information?

Finally, it's always a good idea at the end of the interview to thank them for their time and ask if it is OK to come back for more information if you have another question. Keep the door friendly and open.

Focus the Analysis

At this point in the process teams can be overwhelmed by the amount of information they have gathered. It's important to gain a focus and not get lost in working on trivia. Remember to look for the leverage factors.

Performance Driver Analysis

If the team is getting lost in trivia and has not already done this, consider identifying "Performance Drivers." Perhaps, the leadership team already did this when identifying projects. Even so, consider using it to help the team gain focus. Performance Drivers cause a process to perform well or to perform poorly. Organizational practices, behaviors, policies, timeliness, work activities can all impact process performance. Look at the following example:

Categories	Internal Business	Financial	Customer & Market	Learning & Growth
Key Metric (What to measure?) Big "Y's" & Little "Y's"	1a. Cycle Time 1b. First Pass Yield (FPY) 1c. Defects per million oppty's (DPMO)	2a. Gross Margin 2b. Revenue 2c. Cash Flow	3a. Performance to Customer Req'd Date 3b. Market Share 3c. Customer Returns	4a. Employee Satisfaction Index 4b. Retention in critical positions
Stretch Goal (How much?)	1a. 20% reduction in application processing 1b. 90% FPY	2a. 10% margin growth 2b. 20% increase in revenue	3a. 80% on time to customer request 3b. 10% increase 3c. 15% reduction in returns	4a. 80% Very Satisfied 4b. <2% turnover in critical positions
Performance Drivers (What factors cause the metric to increase or decrease?)	1a. Data accuracy 1b. Quality of incoming parts	2a. Materials management 2b. New product intro's 2c. Managing Receivables	3a. Supplier lead time 3b. Customer service 3c. Product reliability	4a. Opportunities for advancement 4a. Clear communication

Figure 10-20 Example of Performance Drivers

The Performance Drivers in the above example are factors that cause the metrics and the performance of the process to go up or down. In Six Sigma language, those metrics are the little "x's" that impact the little "y's" and Big Y's. If the team knows the most important Performance Drivers for the process under examination, they can do a better job of focusing on the vital few factors. The team can generate these drivers using the brainstorming and prioritizing approaches described earlier.

Overcoming Barriers in the Measure Phase

Quick hits can be okay, but random changes made to the process before a complete analysis is risky. Changing one part of a process may unknowingly impact another process. It's important to know the process requirements. But excessive debating over them adds no real value. The team should concentrate on improving the process to meet the requirements set out by customers and the process owner.

Sometimes a team comes to a point where they are unable to be creative in the change process. The team's Champion can help here by suggesting "off the wall" ideas or bring in other third parties or stakeholders to get their ideas.

The team needs to avoid "analysis paralysis." This happens when teams are looking for the perfect solution. Recognize that a perfect solution does not exist. All solutions will have trade-offs. This does not mean a great solution can't be found - it can. But don't go crazy over it. Make a decision, move forward and don't go back and revisit unless significant new evidence is discovered that, if known, would have caused you to make a different initial decision.

While it's important to uncover the root cause, it's also important to avoid working down in the weeds. The whole idea of process maps is to find the leverage targets. Save the "nitty-gritty" details until the last step when specific change opportunities are addressed. When there is a problem moving on to the next step, it's usually because the team has delved into too much detail. Obviously, there is some degree of art here. Sometimes, it's important to make a point, just make certain it's important enough in the larger context of what the team is trying to accomplish. Teams and team members will only learn how far to push this with experience.

Analyze Phase Tollgate Questions

At the conclusion of the Analyze phase, the team should have addressed the following:

1. Are the improvement opportunities documented?
2. Have root causes been identified?
3. Does the team have a prioritized list of improvement opportunities?
4. Have potential solutions been created for key opportunities?
5. Have the estimated costs and benefits for proposed solutions been documented?
6. Does the team have specific recommendations for action?
7. Has the team communicated with the key parties impacted by potential changes to obtain their inputs and buy-in to proposed changes?

Analyze Phase Example - Publishing Business

The team quickly discovered that most of the salesperson's administrative responsibilities could be done by the Customer Service Representative (CSR) if the process was simplified and some additional training was given to the CSR.

The team also noted that the salesperson and the CSR filled out 12 separate forms for direct marketing services for a customer. Much of the information on these forms was redundant. There were also many opportunities to make an error on the manual forms and when entering technical information because of all the data codes. The largest sources of errors were simple transposition errors and incorrect coding.

The sales process was also discovered to be cumbersome. When a salesperson called on the customer, they needed to lug several large books into the meeting. These books contained basic demographic information and pricing information for varying data searches. The size of these books was intimidating - some people had trouble carrying the books - and they were difficult to use. The salesperson would never have all the information to quote a job. They would always have to go back to the office, do some research, check with the database analyst and finally get back to the customer with a quote. Quite often, the opportunity had passed, and the customer's interest had moved on elsewhere. Therefore, it was difficult to close the sale.

Sales people were spending less than 30% of their time planning customer meetings and getting together with customers. Fifty percent of all Direct Marketing transactions were requiring some type of rework.

Analyze Phase Example - Telecommunications Company

The teams kept a running list of improvement opportunities from the very start of the project. But as they started the Analyze phase, the teams were becoming bogged down by the many processes they had mapped and were now trying to gather related performance and improvement information.

The organization spent a full six months working on "as-is" analysis. Detailed process maps were spread throughout the corporate offices. Three ring binders and spreadsheets contained tons of information, but little of it was being pulled together. The organization decided to quit working with the outside consulting organization that was helping to co-manage the project because they felt they were spinning their wheels and not moving forward.

In total, the teams had identified over 400 separate improvement opportunities. They were trying to decide where to start. They kept looking at all of the detail that had been mapped and thinking, "How can we possibly create a Future State Map to this level of detail?"

One of the team members attended a Motorola University presentation on Six Sigma as a Management System. She went back and met with her leadership team. They decided to use part of Motorola's methodology. The teams were already launched. So the work normally done by executive leadership had already been accomplished. Their primary problem was the teams had stalled, some team members were becoming frustrated with the process and their lack of accomplishment, and the executive group was perceived by some team members as having moved their attention elsewhere.

A Motorola senior consultant worked with the leadership team to design a three-day Breakthrough Workshop that would result in Future State process maps for each of the sub-teams, along with a list of prioritized improvement opportunities. It was also clear that the connection (or handoff) points between the processes had not been clearly defined. A Process Interconnections Matrix was designed to facilitate discussions between the teams.

All 100 plus team members attended the event. No executive member was available to launch the workshop. The event started off with a current status update that gave each of the sub-teams a feeling for progress to date, an understanding of the work being done by other teams and a chance to explore inter-process connections. They were also permitted to share any improvement ideas they might have for other teams to consider.

Each process team brought along a copy of their charter and "as-is" macro process flow diagram. They were encouraged to use just one map. They were also asked to state what one or two process metrics they felt best stated the overall health of their process.

At the end of the team presentations, a breakthrough happened that probably changed the outcome of the entire three-day event. One of the team members stood up and said, "Many of us are being pressured to protect the processes in our region. If we all keep operating that way, we will never accomplish our goal of making major improvement happen in our company. We need to let this go!" This generated quite a discussion in the room. After a short period of time, a consensus emerged that the statement was true. A team norm was added as a behavior for everyone to follow over the next three days: to be open and to let go of these imposed restraints.

A visible wave seemed to cross the room and the energy level rose. People seemed to sit more upright in their seats and the sidebar conversations ceased all together.

All of the work that the teams had done over the previous six months was recognized and appreciated. Then the facilitator asked everyone to put that work aside for a short period of time.

Each sub-team was given an assignment to identify what the Voice of the Customer said about that process. They had to answer the question from a customer's perspective, "In order to meet my expectations you must (fill in the blank)." Other stakeholders existed but an attempt was being made to go back to a customer focus rather than an internal stakeholder emphasis. They were also asked to identify the key Performance Drivers that caused the key metrics to go up or down. Each team brainstormed a list for their process area and then prioritized the list, identifying the top five items.

Customer requirements included:

- Deliver on time and complete
- Provide new products that decrease my costs
- Don't take too much of my time in developing solutions

Performance Drivers included:

- New product revenue
- Clear understanding of customer (external) requirements
- Number of product offerings

Each team prioritized a list of customer requirements and performance drivers. One of the things they were surprised to discover was how much they had in common with one or more of the other sub-teams.

Next, the teams worked on defining the key interconnections between their process and the other processes being worked. Since this had not been done, they initially needed to focus on the "as-is" process. Each team reviewed their Interconnections Matrix with the other sub-teams and received feedback and suggestions from the other groups on key connections, requirements and improvement opportunities. An example of an Interconnections Matrix is shown below.

Process Name: Customer Service

Supplier **Who supplies critical input?**	**Input** **What do we need?**	**Requirement** **Expectation or critical specification**	**Customer** **Who receives our outputs?**	**Output** **What is it?**
Finance & Accounting	Customer Profitability Report	Variable costs (direct costs) Customer Contribution Margin		
		Key Customers' Needs - Prioritized	New Product Development Grp	Quality Function Deployment Matrix
Sales	Market Share Analysis Customer Total Spending $$$	% of Customer $$$ with us vs. with Competitors		
		Customer Retention	Sales	Reports on: Customer defection Customer reduced purchase % % Key Cust. Buying New Products

Figure 10-21 Example Process Interconnections Matrix

Based on all the work the teams had done, they were asked to identify what they felt were the most important innovations to adopt in the new process and to prioritize the key innovation opportunities. This synthesized the overall list of hundreds of improvement opportunities down to 30 key items. One idea that dealt with differing services to each customer segment took on a life of its own that crossed several of the sub-teams' responsibilities areas. Many of the innovations identified were incorporated in the teams' final recommendations.

The teams were then asked to do Future State process maps. In order to avoid drowning in detail, they were asked to use the Hierarchy Mapping method (explained earlier in this chapter) to create their maps. Each team started with a macro level future state map of ten or less steps.

Figure 10-22 Customer Service Level One Future State Map

The above example is a generic version of a customer service map. The teams would then look at their map to decide which boxes (limit of three) offered the most improvement opportunity (or leverage for change). Then they created a second level future state map.

Customer Service (second level - future state map)

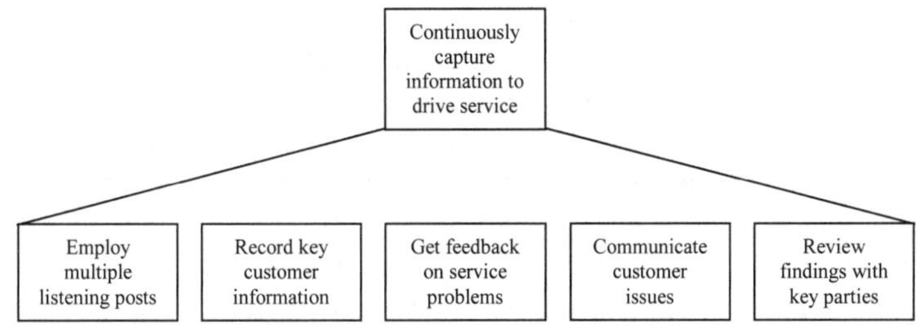

Figure 10-23 Customer Service Level Two Future State Map

Once they developed their Future State maps, the teams needed to develop a Roles and Responsibilities (R&R) Matrix for the future state process. When the teams came back to review their R&R Matrix, it also generated some heated discussion. After the teams reached agreement, they had to go through the same conversation with other key players in the organization to reach a consensus on the changing roles.

Improve Phase for the Lean Team

The primary purpose of the Improve Phase is to answer the question, "So what needs to be done?" The improvement opportunities have been prioritized, benefits have been determined and now its time to "Do it!"

The primary deliverables for a Lean team's Improve phase include:

1. Solutions to the selected improvement opportunities/problems

2. Alternative solutions for key opportunities that could also work
3. Process maps and documentation for the new practices
4. An action plan with key implementation milestones
5. Identification of potential solutions for key opportunities
6. Estimated costs/benefits of any proposed solutions
7. Recommendations for action
8. On-going communications to key parties impacted by potential changes

In this section, the actions to accomplish these deliverables are described along with other tools and techniques that are appropriate for Lean teams.

If asked, "Where are you most creative or innovative?" most people would not answer, "In the conference room where our team normally meets." And yet, when it comes to the point in the project where the team must generate new ideas and new ways to run the process, the team leader generally schedules a meeting in the same conference room with the same brainstorming technique with the hope that this will produce something spectacular. Unfortunately, it does not usually work that way. Look at some of the suggestions elsewhere in this book for ways to generate meaningful new ideas. The Kaizen Analyze section talks about the importance of going to where the action is, and to avoid doing most of the analysis and improvements locked in a conference room. While a cross-functional process improvement tends to be more analytical than the physical observation analysis by Kaizen teams, the same idea still applies of going to where the action is.

In the Analyze phase, the expression "reasonable people, equally well informed, seldom disagree" was used. That also plays out in the Improve Phase. If the team has done their homework gathering evidence, maintained an open mind, practiced active listening and if they have been communicating with key parties along the way, it will be much easier to gain buy-in to improvement ideas.

Consultative Problem-Solving Model

There are a variety of other ways an improvement team can create recommendations. One method is to follow a classic consulting model:

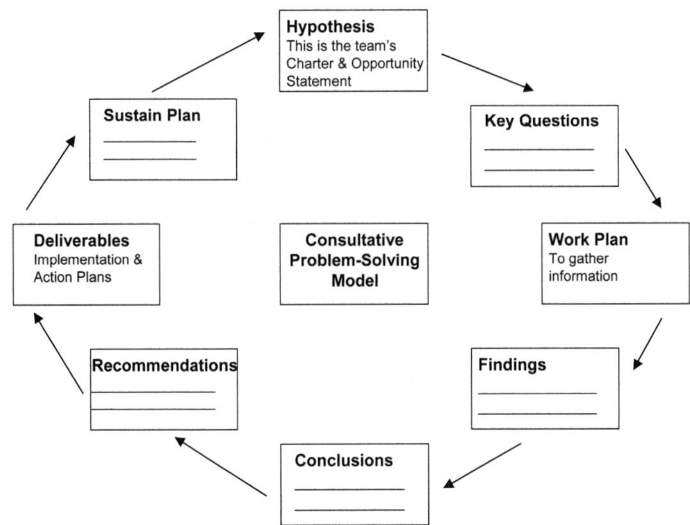

Figure 10-24 Consultative Problem-Solving Model

- The Hypothesis is a tentative conclusion about an opportunity. It is stated in the team's Charter.
- The team tests the Charter by creating a set of Key Questions to go and interview people and to test the hypothesis. These are information probes to verify, reject or modify the hypothesis.
- A Work Plan of some type is usually put together to guide team member activities and accountabilities for gathering data or evidence that either supports or refutes the hypothesis.
- Findings represent the data facts. They are linked to the key questions. There is little interpretation of a finding. Findings fill in knowledge gaps and may further refine the hypothesis.
- Ultimately, Findings provide proof of the current situation and Conclusions can be drawn about the situation.
- After the current situation is understood, Recommendations can be made about what actions to take. Recommendations are prescriptions for action. They specify the actions needed to resolve the current situation.

This analytical consulting model fits with being equally well informed. People can see how a team reached its recommendations. A team could actually map these relationships in an Excel Spreadsheet.

Hypothesis Stated near the top.							
Number	**Findings**	**Source of Findings**	**Linkded Finding Numbers**	**No.**	**Conclusions**	**Linked Conclusion Numbers**	**Recommendations**
1	done right		1,3,33, 45	1	interesting to note...	1, 3, 14, 21	Recommend one
2	lost						
3	momentum						
4	wow		23,17, 65	2	we learned that	2,6,9,15	Recommend two
5	winter slide						
.....list to							
150	late		56,3,14, 37	3	we were surprised about	19, 20	Recommend three
				....list to			
				22			

Figure 10-25 Excel Version of Consultative Problem-Solving Model

Rule of Three

As a team moves to end of the Improve phase, sometimes team members get impatient and simply want the project to finish. If this is happening, revisit the project's purpose, have the Champion meet with the team to reiterate his or her support and the importance of the team's mission. Find some way to maintain energy. Another way to do this is to let team members know ahead of time that just one solution is not acceptable for key recommendations or sensitive recommendations (those recommendations where team members know some conflict is likely to arise.) There is no hard and fast rule, but think about Mike Bremer's "Rule of Three!"

There are always alternative solutions to sensitive areas or key issues. Once a team has developed a first solution, have them develop a second, workable solution. Usually, this ends up being a modification of the first solution. Then push them to develop a third. This is where breakthroughs will often take place. Team members would be (or should be) embarrassed to only modify the first solution one more time, so they will often seriously begin to consider an alternate approach at this point. It is surprising how often this becomes a significant change. Stay with it and develop them for the hot items.

Team Presentation

Lean teams also typically wrap up a project with a presentation to management. Suggested ways of doing this are outlined in the Kaizen and Six Sigma Improve sections. Those ideas also apply here. Often these presentations are just to the leadership team. It can be very powerful and informative to invite the people from the work areas impacted by the changes and deliver the same presentation to that group. Keep the presentation simple. Where practical, include a tour of the work area(s), especially if some changes have already been implemented.

Pilot Project

If practical, a pilot is a good way to test proposed solutions and see if they work as planned. This is not always possible but should be tried whenever time and risk will permit. A pilot program can yield a wealth of information regarding the practicality and reasonableness of improvement ideas.

Barriers to Successfully Completing the Improve Step

The recommendations are the team's opportunity for innovation. Some companies have reputations for innovation, but it is people who innovate, not companies. In the end, it's not so much about innovation as it is selling ideas and making them happen. If a team fails to get its recommendations approved, a couple things may have happened:

- The team's ideas are illogical, or dumb. This is possible, but rare. If the team uses a logical approach similar to the Consulting Problem-Solving Model described previously, they should be able to address the logic issue.
- The team failed to gain the Champion's support along the way. This is a frequent occurrence. Champions do not like surprises. If a team is proposing a recommendation that is outside of the Champion's or the organization's comfort zone, the team has the responsibility to sell their idea ahead of time.
- Sometimes a "no" is too easily accepted. What are other ways the recommendation might still move forward with out jeopardizing someone's career? The Rule of Three talks more about this concept.

As people consider the team's recommendations. they are going to be asking themselves several questions:

1. Is this idea going to be successful?

2. Will this idea move us to where we need to go?
3. Is it worth doing?
4. Does the team have credibility and capability to see the project through (or if not the team, the people they are suggesting take responsibility for the idea)?

If the answer to any of these questions is, "NO!" the team has a problem. The team should make certain that they have as much alignment as practical on the idea before they formally recommend it. That does not mean the team needs to back down in the face of resistance. But the team should not surprise people that have the clout to say no to the above four questions. If conflict exists, bring it out beforehand to allow people to save face, and try to agree on an acceptable approach for resolving the conflict. Pilot test the idea, vote on it, consider alternatives, create a simple prototype of a working model that people can touch and hold or a working demo if you are looking at information management.

Improve Phase Tollgate Questions

1. Has the team effectively communicated with the key parties impacted by potential changes?
2. Has the team developed meaningful solutions to the selected improvement opportunities/problems?
3. Have alternative solutions been developed for the most important and politically sensitive opportunities?
4. Have process maps and documentation for the new practices been created?
5. Does an action plan exist with key implementation milestones for moving forward?
6. Has the team documented estimated costs/benefits for any proposed solutions?

Improve Phase Example - Publishing Business

The team decided on several changes. First, they decided to simplify the paperwork process. The team took the 12 information forms and distributed them among the team members. Each team member was responsible for defining the purpose of the form, recipients of the information, and whether the information comes from any of the other 11 forms. They were able to consolidate the 12 forms to eight forms. They also created a simple automated solution. The remaining eight forms were put on an Excel spreadsheet. Each form was a separate worksheet in one Excel file.

Any information that was redundant (same information on another worksheet) was automatically populated on all worksheets after it was filled in one time. The worksheets were also color-coded. Any block that needed to be filled in on a worksheet was a different color from the rest of the worksheet.

Wherever they could, the team used scroll down menus to make coding easier. When multiple codes existed for one information cell, the person entering the information would simply move the computer cursor to that cell and a list of choices (scroll down menu) would appear. The person would use his or her mouse to scroll to the right selection and the data were automatically entered.

The team redefined the roles and responsibilities along the following lines:

- Direct Marketing Salesperson - Shift responsibility of administration and production from sales to CSR. Transform sales time into strategic, consultative selling.
- Customer Service Representative (CSR) - Existing personnel with strong customer service skills, added training for database, mail specifications, to follow with art production and billing training.
- Database Traffic Coordinator - Redefined and reinforced role of gatekeeper and quality check responsibility.
- Database Analyst - Redefined roles based on expertise of the analyst as well as project type.

They developed a pilot program to test the new responsibilities and procedures. They looked at two different variations.

1. 100% Model - where one CSR worked with one salesperson
    - One Direct Marketing (DM) Sales representative was paired with one CSR.
    - Sales was responsible for customer contact, relaying information to CSR, approval stages of proposal, creative, etc, and customer follow-up...selling.

- CSR was responsible for all request forms to data base, printer, mail house, scheduling, proof relays, vendor contact, etc.

2. 50% Model - where one CSR could support more than one salesperson
   - One DM Sales Representative was paired with one CSR, for half the time.
   - Same as above, but on a limited number of projects.

The results of the changes were significant. Sales and planning time went from 20% of the time to over 40% customer face time and sales call planning.

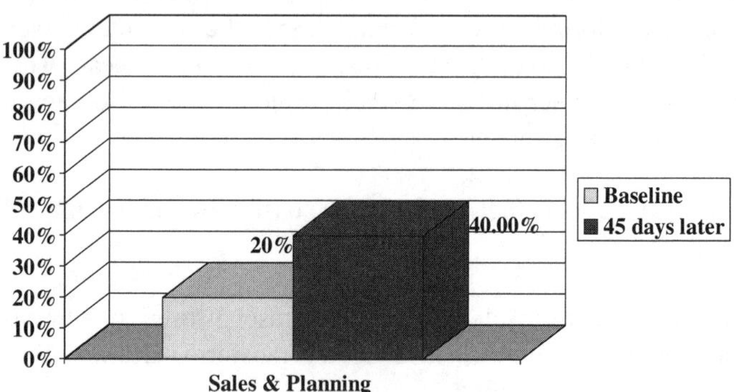

Figure 10-26 Improvement in Sales and Planning Time

Information accuracy also showed a significant improvement.

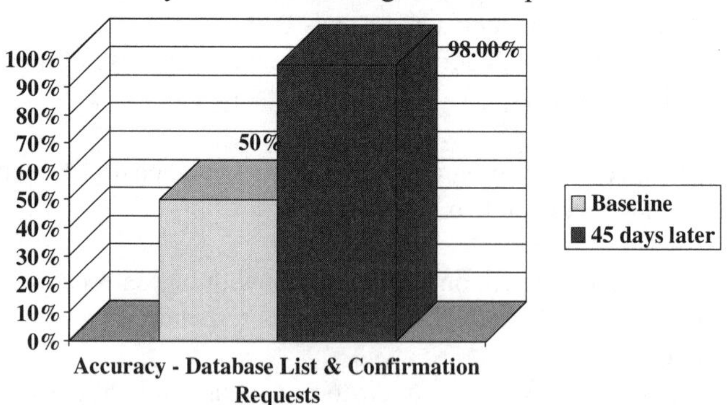

Figure 10-27 Improvement in Information Accuracy

Accuracy went from 50% of the transactions requiring rework or correction to a 98% level of accuracy. This freed up a considerable amount of time for people in the departments.

Improve Phase Example - Telecommunications Company

After the Breakthrough Workshop, the teams finalized their "should-be" processes over the next few weeks, and high-level business cases were developed for 32 individual projects that emerged from the seven process teams. A subsequent two-day follow-on event was held to further define the interconnections (or touch points) among the processes and to identify the critical metrics.

The teams had done a fair amount of communication with people in the organization so there was broad support at a grass roots level for the changes. Many team members, though, had concerns about top management's support. A number of people were feeling like the teams' Champions had abandoned them. This turned out to be more of a communication problem than reality.

All along, the team leaders and the overall project manager were interacting with the executive leadership group but the executive group was not directly interacting with very many team members. After this issue surfaced, each executive champion met with his or her team individually. Then, the entire executive group participated in a Project Walk Through where all of the teams came together. They each took a separate space in a room used for very large meetings and posted visual charts and graphics on the walls. As the executive group toured the room with the project manager, each team presented their conclusions and recommendations.

Of the 32 projects presented to management, 31 were approved to move to implementation planning. Process improvements included:

1. Pre-canvas sales process steps reduced by 35%
2. Sales close process steps reduced by 62%
3. Customer Service call routing changed from random to segmented
4. Graphics process steps reduced by 387%
5. Account Management Process Steps reduced by 50%

The above changes resulted in a significant increase in revenues and margins that the company has asked remain confidential. They also developed significantly different procedures for dealing with each customer niche.

Control Phase for the Lean Team

The primary purpose of the Control Phase is to answer the question, "How do we guarantee performance?" Improvements have been implemented. Now the emphasis is on finding ways to sustain the gains.

The primary deliverables for a Lean team's Control phase include:

1. Pilot project evaluation
2. Implementation rollout plan beyond the pilot
3. Identification of metrics to monitor the changes
4. Determination of whom is accountable for shepherding and maintaining the gains
5. Report out plan to the Leadership team by the responsible executive
6. Post project implementation evaluation plan

In this section, the actions to accomplish these deliverables are described along with other tools and techniques that are appropriate for Lean teams.

Numerous studies have shown that up to 70% of the gains received from improvement projects get lost in the next three to six months. Why is that? Is it true that people are simply resistant to change? Or could there be another less sinister reason? Just think about it for a moment. Let's assume a new change has been implemented, people have received training and everyone is excited about the new procedure. What happens during the first two weeks if a major crisis comes along? Could there be a tendency for someone to say, "Wait a minute! We are under a crunch here and we have to get this task done. How about if, just this one time, we do it in the old way, so we can get it done faster?" Or if the employee is under the gun to complete a task, is it not less difficult to do it in the old way? This is not to say that resistance to change does not exist. It does! But more often then not, the improvement simply slips between our fingertips. No one is doing anything evil. It just slowly melts away, until six months later people are largely doing their work the old way.

So a key part of Control is the first six months after implementation. The changes need to be monitored more closely and aggressively during that period if they are to be sustained. Activities that happen several times a day or even several times a week are much easier to monitor than activities that happen one or fewer times a month.

Sustaining the Gains

What gets monitored is what gets done. This may not be true 100% of the time, but this is part of maintaining any gain. The team should develop a few, visual metrics that people can use to see if the gains are being maintained. Ideally, these metrics already exist. If it's an operational process, they probably do exist; if it is an administrative process, they probably do not. Remember the discussion in Measure about leading and lagging indicators? Leading is best! The best rule of thumb for improvement type metrics is: keep them simple, keep them visual and avoid "extra" work.

Answers to the following questions may help people and organization to learn and apply new behaviors, skills and work practices:

1. What is the new behavior skill or work practice that needs to improve? (There is power in naming it.)
2. In what situations will this new behavior, skill or practice get used?
3. What impediments does the team anticipate might arise?
4. How can each impediment get minimized?
5. What will prepare and encourage people to accept the new practices?
6. During the next week after implementation, what specific actions will people take to foster use of the new behaviors?
7. How will progress be assessed, what metrics would reflect positive action?
8. What concrete results are reasonable to expect?
9. How can these results be visually displayed? How often should this happen?
10. What results should be reported in a daily, weekly or monthly management meeting, and has this been made part of the calendar?
11. Who has accountability and responsibility for reporting the results?
12. Who has authority to take action to amend the process if the results are less than expected? Does this authority align with responsibility for doing the work?

If the project Champion and the project team give some thought to answering these questions, they should be able to develop a control plan that has a reasonable chance to succeed.

Change Management

Take a grizzled, 50-year-old cop who's been patrolling the streets for decades and has grown quite comfortable filling out five-ply carbon forms to process all his arrests and casework. Then you suddenly order him to start doing his reports on a computer, an alien-looking object he may have never laid a finger on in his life. Indeed, one veteran cop on his first day of system training on a PC picked up the mouse without logging on, pointed it at the screen and started clicking away. When nothing happened, he asked why the damn thing wasn't working. That's the whopper of a change management task that faced the Chicago Police Department. Their initial roll-out of a new information system was an absolute disaster.

The information systems group and leadership of the police department went back to the beginning and took a look at their change management process as well as the new system. "They began by sending a team of their best programmers out to the field for six weeks to document all the issues users had with the system. The group came back with 200 specific requests for changes, ranging from implementing an easier approval process to changing the format of how the reports printed. IS leaders also made it a point to glean user input throughout the development process. They instituted JAD (joint application design) sessions, which involved teams made up of management, users and technical staff. They formed focus groups from all ranks to gather input. Teams of officers went out to the 25 districts to field-test new apps and train officers. Having officers-not civilians-be the trainers, made a huge difference to the cop on the street." [2] The newly revised system was much more successful in its roll-out.

Change management actually works on two levels, the individual level and the organizational level. Considering both views is critical to sustaining the gain. The number one contributor to success is strong, visible, effective project Championship. Employees want to hear messages about change from 'C' level executives and from their immediate supervisor. Supervisors can speak to the individual view, and 'C' level executives

---

[2] CIO, "No Small Change," Todd Datz; http://cio-asia.com/pcio.nsf/0/6AA156FC628887F748256E770024B27E?OpenDocument

should speak from a customer perspective. The number two contributor to success is exemplified in the Chicago Police Department story. People designing the change need to communicate with the people impacted by the change and clearly understand their needs (Voice of the Customer).

One of the simplest methods to stay focused on sustaining the gains is to have the appropriate manager over the work area impacted by the change to report out in management meetings. This needs to be proactive. If the numbers start to slip, action needs to be taken to address the situation. If the leadership team does not see fit to do this, then one must question whether or not this was an important endeavor. The leadership team should take care not have too many change improvement opportunities underway. If their capacity to pay attention and follow up when action is needed is exceeded, the entire credibility of an improvement initiative is undermined and rendered less effective.

Control Phase Tollgate Questions

1. Does a plan exist with clear roles and responsibilities for Pilot Project Evaluation?
2. Does an implementation rollout plan exist for beyond the pilot?
3. Have key metrics been identified to monitor the changes?
4. Do clear roles and responsibilities exist for shepherding and maintaining the gains?
5. Will the Leadership team receive a report by the responsible executive (peer)?
6. Has the team planned for a post project implementation assessment?

Control Phase Example - Publishing Business

The team planned to repeat their measurements at 45 days, 90 days and six months after implementation of the new procedures to make certain the gains were sustained. Roles and responsibilities of the players were also formally changed after the results of the test pilot.

Kathryn had already committed to a 15% increase in her sales target for the coming year. There were new markets and an additional salesperson already approved for the sales staff. As a result of the team's improvements, Kathryn increased her sales targets for the coming year further to 25%.

Control Phase Example - Telecommunications Company

Most projects were rolled out as pilot projects. At the end of each pilot, there was an evaluation by a core set of the project team, the line managers involved and the executive team. Once these three groups approved and validated the changes, they were further rolled out to the organization. Most pilot projects met expectations and were rolled out largely unchanged. In a few instances, modifications were made.

The teams also realized the organization was going to experience a culture change as a result of the greater focus on external customers. A desired "Business Cultural Model" was developed that identified skills needed and the skills and cultural gaps could be addressed. Procedures were put in place to adjust reward, recognition, accountability and work planning systems.

The executive team held itself accountable for sustaining the gains by using the Corporate Scorecard. Revenues, Cost of Sales, Margin Growth, etc. had all been increased from the original plan to incorporate the accelerated growth and benefits expected as a result of the project teams' work.

Transition responsibilities were planned by an Implementation Team, with controls incorporated in new processes to ensure that the original issues did not recur and that the processes continually improved moving forward.

The organization also used the Roles and Responsibilities Matrix (Future State) as the basis for a new assessment tool to see how effectively people were adopting the new policies, procedures and spirit of the new processes.

Summary

Lean teams represent another powerful way to involve people in improvement. Most of the actions taken by this type of team also apply to Six Sigma teams.

If the organization involves good people in improvement, good things are likely to happen, but there is no point in making their life more difficult. Leadership has an obligation to do its part out of respect for the people they are trusting to develop major new improvements for tomorrow.

Cross-functional teams work on improvement. When organizations that have difficulty sustaining gains are asked what they would do differently the next time, they say, "Begin change management activities earlier in their next project, rather than waiting until the start of implementation." Change management activities include many of the actions described in the steps of the DMAIC process:

1. Appropriate Champion
2. Clear reasoning on why change is needed (business case/team charter)
3. Time tables to implement the change
4. Clear accountabilities
5. Two way communications with people impacted by changes
6. Clear priorities
7. Action plan and milestones
8. Metrics to monitor progress, including baseline established
9. Documentation
10. Solution test (pilot projects)
11. Assessment of results

This is not complex. Despite what people say, a lot of good has come from improvement initiatives over the last twenty years. Just about everyone realizes it could have been done better. Realize that the competition is also launching improvement teams. To truly jump ahead of the pack, use the tools, methods and concepts outlined in this chapter and elsewhere in the book, that offer the most leverage for your organization.

Part Four:

Six Sigma Teams, Methodology and Tools

Part Three illustrated the power of *Kaizen* and Lean teams in process improvement initiatives. This section is dedicated to the power of Six Sigma teams that apply a deeper level of analysis and statistical rigor to reduce variation, stabilize and optimize business processes - all for the purpose of bottom-line impact for the organization.

The section begins with an introduction to the DMAIC methodology for process improvement in the context of Six Sigma Teams. The subsequent chapters provide a comprehensive view of the activities, tools and deliverables in each phase of the DMAIC model. In addition, two case studies flow through the chapters to provide real examples of the concepts presented. Chapter 17 summarizes some tips for Six Sigma teams to use to ensure their project is on track.

Finally, Chapter 18 discusses the DMADV methodology for new process/product design.

Chapter

11

Introduction to the DMAIC Process Improvement Methodology

Define, Measure, Analyze, Improve and Control (DMAIC) comprise the major phases of a process improvement project. Each phase consists of a set of tools and deliverables. In the previous section of this book, DMAIC was illustrated in the context of *Kaizen* and Lean teams. In the history of process improvement, DMAIC is just one of a variety of proposed methodologies. In grade school, most students learn the scientific method based on observe, formulate a hypothesis, collect data and form a conclusion. Shewhart, a principal figure in the history of quality control, suggested the well-known Plan, Do, Check and Act (PDCA) cycle for improvement. Although DMAIC looks different than these methods, it really encompasses both approaches. It focuses on using data to make decisions and then verifying those decisions before committing business resources.

The Five DMAIC Phases

The advantage of the DMAIC approach is *not* the top-level phases themselves but what is contained in each phase. The contents provide a common, structured approach to solving a problem. For each phase, there are some primary activities and an associated overall question to answer.

Define

The Black Belt team determines the boundaries of the process area to improve and the requirements for the output of that process. The team answers the question "What is important to the business?"

Measure

The team determines how the current process is performing compared to the requirements. The team answers the question "How are we doing with the current process?"

Analyze

The team determines what is wrong with the process. The team answers the question "What is wrong with the current process?"

Improve

The team finds solutions to the problem and conducts a pilot on the selected solution to determine feasibility. The team answers the question "What needs to be done to improve the process?"

Control

The team implements the solution and transfers the ownership of the new improved process to the responsible owner. The team answers the question "How do we guarantee performance so that the improvements are sustained over time?"

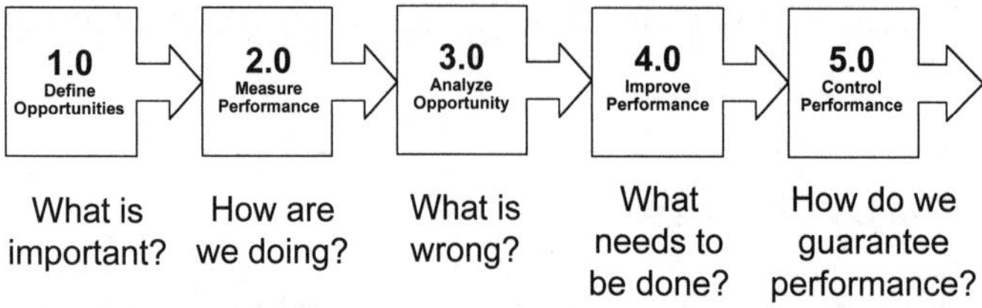

Figure 11-1 DMAIC Flow and Associated Questions

A team may not complete all five phases for several reasons, including:

- The team finds some obvious problems that can be fixed easily, called Quick Wins, and these quick fixes bring the process to the desired performance levels. Management reassigns the team to a more pressing business opportunity.
- The team discovers that a DMADV (Define, Measure, Analyze, Design and Verify) approach, used for designing a new process or

major redesign of a process, may be more appropriate. The team changes the focus from process improvement to process creation or re-creation.

DMAIC Project Timing

Most DMAIC projects should be carried out in four to six months, from the start of the Define phase to the validation of the proposed solution based on a pilot in the Improve phase. Since the team does not know what the solution will be in the Control phase, the time required for this phase varies from project to project. Strict time management is critical to keep the project on track. If the project takes too long, many things may happen. The team may lose focus, the priorities of the business may change, or the business may be losing money during that time. The Black Belt and the Champion should ensure that the project meets the planned deadlines.

DMAIC Leadership Roles

Through all the phases, the Black Belt should communicate progress of the project to the Champion. The Champion should help the Black Belt if the team runs into conflict or resource allocation problems. The more involved the Champion is, the more successful the Black Belt and the team will be. The Six Sigma Weekly Review process and the Tollgate Review at the completion of each DMAIC phase are critical to keeping teams on track. Leaders should thoroughly understand the DMAIC structured approach and the primary objective of each phase and use their reviews to ensure that teams are on the proper path.

For inexperienced teams, the beginning of the project may cause feelings of trepidation, and people may have a sense that what they are about to take on is too much or impossible. The Champion needs to ensure the project has been scoped so that it is doable but challenging. Champions should also support the team as it hits any lows in the process. The payoff will be at the end of the project when the team proves that what might have seemed impossible is, in fact, possible.

Chapter

12

Define Phase

The primary purpose of the Define phase is to ensure the team is focusing on the right thing. The Define phase seeks to answer the question, "What is important?" That is, what is important for the business? The team should work on something that will impact the Big Y's - the key metrics of the business. And if Six Sigma is not driven from the top, a Black Belt may not see the big picture and the selection of a project may not address something of criticality to the organization.

The primary deliverables from the Define phase are:

- Team Roles and Procedures Established
- Team Charter
- Voice of the Customer and Voice of the Business Analysis
- Critical measures of Quality and Process Chosen
- Process Maps Drafted
- Quick Wins Identified

In this chapter, these deliverables are discussed as well as other tools and considerations to be successful in this phase.

Project Selection

The Define phase is the most critical phase of the five. As Aristotle said, "Well begun is half done." For example, a great project on decorating the break room may have some impact on morale but probably won't make a difference to the bottom line. Even if the team receives the project from management, it still may require validation and refinement. Leadership can see the direction required but may not be able to define exactly the boundaries of the project. Therefore, the team's first step is to take the opportunity as given to them and to refine it to make it their own. There are some criteria to consider when selecting a Six Sigma project. A good Six Sigma project should:

- Impact a key business goal
- Require analysis to uncover the root cause of the problem
- Benefit from the application of statistical tools (e.g. data is available or can be made available for decision-making purposes)
- Have cross-function or cross-business impact
- Address a source of customer pain or dissatisfaction
- Focus on improving a key business process
- Produce quantifiable results (e.g. financial savings, customer satisfaction)
- Be able to be completed in time to make a difference to the business goal
- Be scoped so that results can be achieved in 4-6 months.

The team should keep these points in mind as they review the opportunity given them and complete their Define phase assignment to determine, "What is important?"

Developing the Project Team

Selecting the Team

Developing a good project definition requires having a good team with good teamwork. The Black Belt and Champion must take care when selecting the members. Team members must understand the DMAIC process, want to be on the project, and be prepared to show up to meetings in spite of other commitments. Anything less will impede progress. In addition, the Black Belt selected to be the team leader should have a good understanding of teamwork and associated team tools.

Gathering the correct people is essential to building an efficient and effective team. The best way to select team members is to identify the areas of expertise that are required and then fill the team with people who can meet those requirements. As the team refines (or maybe changes) the charter, the team membership may need to be re-assessed.

Team Management Tools

The Define phase is typically when the team is in the Forming/Storming Stages of development (described in Chapter 2). Strong leadership is needed to introduce the team to its assignment and the roles that member are to play.

Meeting Agendas and Objectives

Before conducting the first team meeting, it should be well planned. The agenda is the planning tool for the meeting. To develop the agenda, first, the objectives to be accomplished need to be defined. The Black Belt will traditionally set the objectives for the first meeting and then the objectives for subsequent meetings will be set with the team at the end of the previous meeting. A list of activities for the meeting, the leader of the activity and the timing of each activity should be included on the agenda. The Black Belt should send the agenda out one to five days prior to the meeting. This allows the team members to arrive at the meeting prepared. The figure below shows a partial example of an agenda.

Task	Activity Type	Duration	Leader	Outcome
Cause and Effect Matrix	Brainstorming	50 min	Mike	X data to collect
Wrap Up / Process Check	Discussion	10 min	Tom	Next meeting's agenda

Figure 12-1 Portion of Sample Agenda

Conflict Resolution and Behavior Guidelines

One of the most common early sticking points in a Six Sigma project is conflict amongst the team members. Most teams experience interpersonal conflicts as the team goes through the storming stage of team development. To help head off and handle conflicts, the team leader should recommend establishing team guidelines for behavior. The team develops this list of rules and everyone should agree to them. This list can then help the leader if behavior problems occur. The figure below shows some example guidelines. Another consideration is to bring in a facilitator to the team. A trained facilitator can help the team work through problems of behavior as well as impasses in making decisions. Conflict will usually happen in a team. The important issue is to address the conflict immediately. Conflict amongst the team members can cause time delays or failures in the project if not handled correctly.

1. Meetings will start on time.
2. We will have one conversation at a time.
3. Cell phones will be on vibrate.
4. Action items will be done or we will notify the team leader if we can't finish them.

Figure 12-2 Team Guidelines

Process Checks and Charting Results Trends

To ensure that the team is operating effectively, the leader or facilitator may conduct a team process check at the end of each meeting. The leader may establish a few categories that are measures of good teamwork such as listening, results, and participation. The leader would ask each person to rate the meeting in these categories using a scale of 1 to 10. The leader would then track this information over time as exhibited in the figure below. If any one score sticks out, such as a low score by one person in the listening category, the leader would address this issue with the team member and the team. These scores should improve as the project progresses and the team learns to work together effectively.

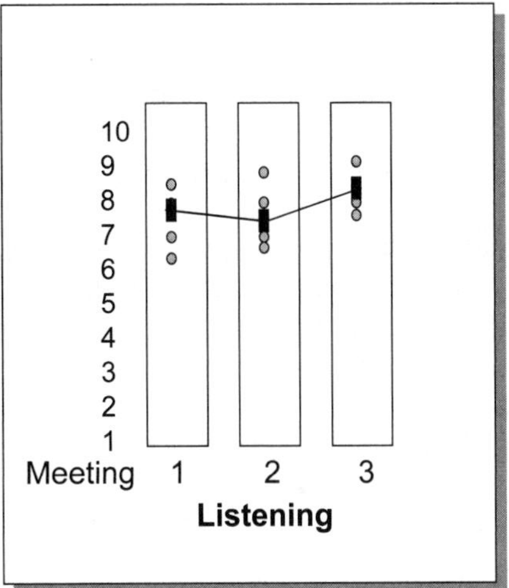

Figure 12-3 Process Check

Brainstorming

Brainstorming is probably the most used team tool and most people know the rules of brainstorming:

- The topic of the brainstorming session should be clearly stated.
- Individuals are allowed to complete suggestions or thoughts without interruption or critique.
- Suggestions should be kept as brief as possible to maintain a fast pace.
- The primary goal is to generate a *quantity* of ideas.
- The focus is building on suggestions of others, as well as generating new ideas.
- An environment of creativity and innovation is encouraged.

However, even though the team may know the rules, the team leader should review them with the members since people often violate the rules and start to judge ideas.

Once the team has completed the brainstorming, the team must evaluate and revise the list. Any unclear ideas should be clarified, common ideas can be combined, and additional ideas can be added.

Multi-voting

Another commonly used team tool is multi-voting. Multi-voting is a structured way to help the team narrow a list of items. There are many ways to run a multi-voting session but the steps for a typical session are:

1. Clarify the meaning of the items on the list. Combine any common ideas or delete any that may not be applicable.
2. Give each team member a number of votes. The number is half the number of items on the list. (If there are 20 distinct items on the list, each team member gets 10 votes.)
3. Members then vote for their top ten items. Usually, each person goes to a flip chart and makes a tic mark by each of their selections. They are not allowed to vote for any item more than once.
4. After voting, generally some items won't have received any votes. These are taken off the list.
5. If the team would like to narrow the list even further, the process can be repeated with fewer votes allowed for each person.

The Pareto Chart

Another powerful teaming tool is the Pareto chart. The Pareto chart is based on the Pareto principle. Pareto was an economist in the early 1900's who discovered that 80% of all the wealth was held by 20% of all the people. This became known as the 80/20 rule and it was found to be applicable to more than the economy. Eighty percent of the warehouse space is taken up by 20% of the part numbers. Eighty percent of the defects are caused by 20% of the defect types.

The Pareto chart is a bar chart. The height of the bars indicates the count, or frequency, of occurrence. The bars represent one grouping of the data, such as defect type. The idea motivating this chart is that 80% of the count will be due to 20% of the categories. The bars are arranged in descending order, therefore the dominant group can be determined and it will be the first bar on the left. This chart can be used in a variety of places in a Six Sigma project. One place where it is useful is at the start of a project to help define the scope. The team may create a Pareto chart and choose the tallest bar as the area of opportunity to be addressed on their project. The figure below shows an example of a Pareto chart.

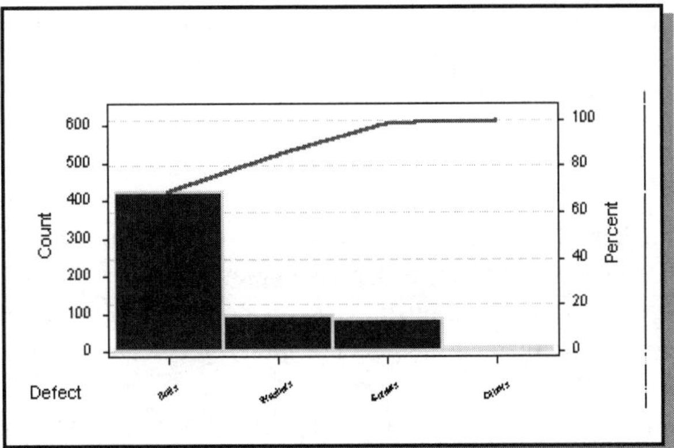

Figure 12-4 Example Pareto Chart

DMAIC or DMADV?

The first decision that the team must make as they validate the opportunity is if there *really* is a current process. DMAIC is designed to be a process improvement technique for current operational process. One project had a goal to improve a process but the company had just merged several smaller companies and each of the companies did the process steps differently. This team did not have one process to improve; they had several!

If a process does not already exist, the team will want to follow a DMADV (Define, Measure, Analyze, Design and Verify) approach to create a new process. Or, if the opportunity is to create or improve a new product, this type of situation would also be better accomplished with DMADV.

Define Phase Deliverables

Once the team is confident that the DMAIC approach is the correct approach, they need begin work on the Define phase deliverables. There is no prescribed order of tasks for teams to follow in the Define phase. Teams will work on preparing the deliverables in different orders depending on the process, the opportunity and the information at hand. Some of the same tasks, including data gathering and research, are required in preparation of different deliverables. However, the Team Charter is usually considered the most important document used assess the opportunity in the Define phase.

The Team Charter

The team charter is used to establish a clear understanding of the project amongst the team, the team leader, Champion, Sponsor and stakeholders. It includes a documented business case, opportunity for improvement, goals, scope, timeline and members of the project team.

The Team Charter is the key document that defines the scope and purpose of any project. The Charter functions as the communication vehicle for the team, the Sponsor, Process Owner and Champion, the team leader, and all other team members involved. It is used to assure that the team sees the vision of leadership, and understands what the project opportunity and performance improvement goals are.

Often, leadership gives an initial draft of the charter to the team. That draft generally needs refinement based on the team's intimate knowledge of the process. The Black Belt and the team will need to spend time on this document until they are sure that they can support it.

One common misconception is that the team charter is written and then not touched again. However, the team should refer to the charter often to ensure that the project is staying on track. In addition, it should be considered a living document that may be revised as the team learns more in the Define and the Measure phase.

The team charter can take on many forms. To further expand on the Charter information provided in the *Kaizen* and Lean team chapters, the charter used by Motorola includes six sections:

- Business Case
- Opportunity Statement
- Goal statement
- Scope
- Project Plan
- Team Selection

Business Case - Why should we do this?	**Opportunity Statement** - What "pain" are we experiencing? - What is wrong?
Goal Statement - What are our improvement objectives and targets?	**Project Scope** - What authority do we have? - What processes are we addressing? - What is not within scope?
Project Plan - How are we going to get this done? - When are we going to complete the work?	**Team Selection** - Who are the team members? - What responsibilities will they have?

Figure 12-5 Elements of a Team Charter

Each of these sections is discussed in more detail below.

Business Case

The business case contains a description of the Big Y, which is the reason for taking on the project. In other words, why does it make strategic sense to take on the project? This allows the team to see the clear application of the project to something that is important to the business. A de-motivator for the team would be if the members think that top management doesn't support the project. With a well-defined business case, the team can see how this project is important to the organization. This is the one part of the Charter that probably won't be updated over time. If the team Sponsor, Champion and the leadership don't see a strategic Big Y impact in the team Charter, the project will likely be abandoned.

Opportunity Statement

The opportunity statement discusses the pain related to the project under consideration. The pain is usually the dollars that are being lost. The pain may also be related to customer needs not being met. Often, the business doesn't know this information at the start of the project. People may have a feel for the pain but there hasn't been an exact measurement of the dollars due to lost opportunities, costs, etc. As the team collects more data, they may need to update the opportunity statement.

Goal Statement

The goal statement describes what success looks like. The goal statement should contain the metric to improve and the desired level for the metric. The goal statement should be focused on the little Y's. By stating a clear goal, the team knows when the project is done. In the table below, two example goal statements are shown. The first states the metric to improve, but it isn't explicit about the measured goal value to achieve at the end of a successful project. The second example gives a detailed statement of the results to achieve if the project is to be successful.

Example Goal Statement	Quality of Statement
Decrease average cycle time.	Incomplete. Does not state desired level. Decreasing the average may not meet all customers' need. Must also consider the variability of the cycle times.
Decrease average cycle time to 3 days with a standard deviation of 0.5 days.	Good. Clear desired level stated and both the target and the variability are addressed.

Figure 12-6 Goal Statement Evaluation

A common mistake made in the goal statement is to list a solution rather than a goal. For example, suppose the goal section contains the statement "Implement an Enterprise Requirements Planning software module for order entry." This statement implies that some analysis has been done and a software module will fix the problem. This is a solution; whether it is a good or bad solution remains to be seen. If the business already knows the solution, the project is not a Six Sigma project.

Project Scope

The scope describes the start and end steps of the process under consideration, as well as any other boundaries on the project. An example scope statement might look like:

Starting Step: Customer calls
End Step: Order is complete

Boundary: The team cannot change the information technology system used in the order entry area.

By defining the scope well, the team can avoid "scope creep." Scope creep occurs when the boundaries of the project keep changing until it becomes too large given the timeframe and resources allocated. Since Six Sigma projects should last four to six months and most projects have limited resources, the scope must be carefully managed. Timing, resources required and scope should be balanced.

Project Plan

The next section of the Charter is the project plan. This section usually contains a chart with the major milestones of the project. Often, the chart will contain the steps of DMAIC as a start. The team will need to follow the chart with a more detailed project plan. The figure below shows a typical chart included in the Team Charter.

Task / Phase	Start Date	End Date	Actual End
Define	1-Feb	20-Feb	
Measure	20-Feb	30-Mar	
Analyze	1-Apr	30-Apr	
Improve	1-May	1-Jun	
Control	1-Jun	15-Jul	

Figure 12-7 Example Project Plan

Team Selection

The last section, team selection, lists the people participating on the team, the Champion, the designated Black Belt and the Master Black Belt, if applicable, supporting the team. This section may also contain the responsibility of each person on the team. If the scope is refined during the course of the project, the team composition may also need to change.

Assessing the Team Charter

As the team finishes the first draft of the Charter, they need to assess whether it is ready for review by the champion and other people involved. One technique that can be used to evaluate the charter is SMART. SMART is an acronym for Specific, Measurable, Attainable, Relevant, and Time Bound. This five-word checklist is used to ensure that the charter is effective, thorough, and actionable. For instance,

- Is the project addressing a *specific* opportunity?
- Does the goal address a *measurable y*?
- Is the project *relevant* to a Big Y that is important to the business?
- Has the team set some milestones *time bound* with dates attached?
- Is the project goal *attainable*?

The team should use SMART to finalize the Charter. Remember, the Charter is a communication vehicle. It doesn't have to be perfect - just understood and supported.

Analysis of the Customer's Voice and Voice of the Business

In the Define phase, the team must clearly understand what the process requirements are. This involves listening to Voice of the Customer and Voice of the Business data, and then translating these data into measures called Critical to Quality and Critical to Process measures. Part of this translation process is recognizing what makes a good metric.

Defining the Voice of the Customer

To ensure the team is working on what is important, the Voice of Customer (VOC) needs to be determined. The VOC data are what the customer wants and needs. Defining the VOC is a leadership responsibility as part of the Align Mode (Chapter 3). In most DMAIC projects, the VOC will be relatively well defined. Usually, the team just needs to validate that the VOC for the process has not changed. Most DMADV projects include a much larger emphasis on VOC.

Sources of Information

The VOC information can be obtained from three major sources: reference databases, listening posts, and research methods. Reference databases may be internal or external. Internal databases may be maintained by a marketing research group. External databases may be maintained by a recognized authority or organization in a particular business arena (e.g. J.D. Powers). Listening posts are anywhere in the business where an employee touches the customer. Common examples of listening posts are billing, sales, customer service, or technical support personnel. Research methods include surveys, interviews, focus groups and observational methods. The team must decide which of these sources will be appropriate for the project.

The team may hear from many different customers. As they contemplate the customer data, the team must determine which customers to listen to. Not all customers bring equal value to the business. The team should consider the VOC data for the customers that bring the most value. The team should understand the possible stratification of the customer base, identify those customers that are important to the business and to the project, and then focus on listening to these customers. A primary example of stratification in the airline industry is the grouping of customers into frequent fliers and casual fliers.

Kano Analysis

After the team has identified the primary customers for their project, the next step is to classify the VOC data for these customers. When listening to Voice of the Customer, the team might classify the data into three categories. The categories are: must-be's, primary satisfiers, and delighters. This classification is called Kano Analysis, named after the developer Noriaki Kano. A must-be requirement can dissatisfy, but cannot increase satisfaction. It is a basic requirement of the customer; they will not do business without it. A primary satisfier requirement is one where the more of these requirements that are met, the more the customer is satisfied. A delighter does not cause dissatisfaction, but it will delight clients if present. The team must ensure that the process meets the must-be's, increases the primary satisfiers, and potentially adds delighters. The table below shows an example of Kano Analysis for the airline industry.

Category	Example for Airline Industry
Must-Be	Bin space, luggage delivered with person
Primary Satisfiers	Leg space, arrival time
Delighters	Extra miles for web based reservations

Figure 12-8 Sample Kano Analysis of Airline Industry

Unfortunately, not all customer data are delineated so clearly. Sometimes, the customers are unhappy but have trouble articulating the reason for their unhappiness. This is often a problem with data from listening posts. The company may have plenty of this data but it may not be filtered correctly. For example, a customer may call the complaint department and say that he is unhappy. The customer service representative needs to clarify and classify correctly what exactly is making the customer unhappy. Was he unhappy with a product characteristic, with the service he received, with the packaging, etc?

Critical Customer Requirements

After the team has the pertinent VOC data, they then need to translate this information into Critical Customer Requirements (CCR's). A CCR is a specific characteristic of the product or service desired by and important to the customer. The CCR should be measurable with a target and an allowable range. The team would take all the VOC data, identify the key common customer issues, and then define the associated CCR's. The table below shows an example of the translation of VOC to a key issue to a CCR.

Voice of the Customer	Key Customer Issue	Critical Customer Requirement
Your product is bad. Your parts are not matching requirements.	Part Diameter	Diameter must meet specification of 2" with acceptable range of 1.98 inches to 2.02 inches.

Figure 12-9 Translating VOC to CCR

Defining the Voice of the Business

Although the VOC is important, the Voice of the Business (VOB) is equally important. The customer may want everything yesterday, with all the trimmings, for free. Unfortunately, a company wouldn't stay in business long making the customers extremely happy with this strategy. Although the sources may be different for VOB data, the concept is similar. The team must determine what is important to the business. This information is then translated into Critical Business Requirements (CBR's). Like a CCR, the business requirement should have a target and an allowable range. The Sponsors, Process Owners and Champions should help clarify the VOB and the CBR's.

Sometimes the customer requirements and the business requirements may conflict. The customer may have a requirement around the price per unit. The business requirement may be around profit per unit. Although the customer requirement might be met by lowering the price, this would have a negative effect on the requirement of making a good profit per unit. The solution the team develops in the Improve phase should take both customer and business requirements into account. Leadership must help resolve these conflicts and establish priorities for the team.

Critical to Quality and Critical to Process Measures

The Critical Customer Requirements (CCR's) often are at too high a level to be directly useful to the team. So the next step is to take the applicable CCR's and translate them into Critical to Quality (CTQ) measures. A CTQ is a measure on the output of the process. It is a measure that is important to meeting the defined CCR's. Often it does not equate exactly to a CCR, though at times it may.

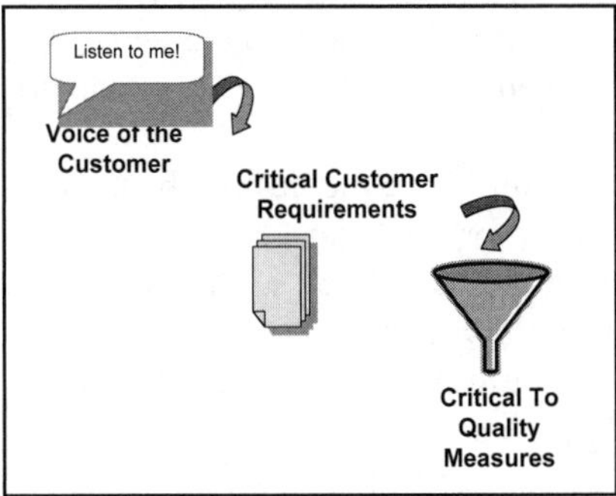

Figure 12-10 VOC to CCR to CTQ

For instance, the customer may require the cycle time from order to delivery to be one month. Since the team may only be working on one piece of the total process for that cycle time, their CTQ may be the cycle time for their process, e.g. the order entry process. The Critical Business Requirements (CBR's) should be converted into Critical to Process (CTP) measures as well. The CTQ's and the CTP's are the little y's that will be in goal statement of the Charter. The table below shows the relationship of a Big Y to little y's for a sales process.

	Example
Big Y:	Increase Sales
Voice of the Customer:	Sales quotes take too long.
Key Customer Issues:	Get quote in a timely manner.
CCR:	Quote should be received by customer in 14 days or less.
CTQ or CTP:	Cycle time for quote.
Goal Statement from Charter:	95% of all quotes should be delivered to customer 14 days from customer's request for quote.

Figure 12-11 Relationship of Big Y to Little y's

What Makes a Good Metric?

"What gets measured is what gets done." If a process is being measured on a certain metric, people will modify their behavior to meet that metric. In one company, the metric that management considered important was how many units started in assembly each day. Everyone worked toward making that metric look good. Unfortunately, what the company got was a lot of started, but unfinished, units and a lot of Work in Progress. But the metric looked good. So what makes a good metric?

1. The metric should drive the desired behavior of the organization and people in the organization.
2. The metric should be defined so that it is understood by multiple levels in the organization.
3. The metric should be clearly related to a Critical Customer or Business Requirement.
4. It should be continuous data - as described in the Measure phase - if possible.
5. It should be collectable.

The little y's for the project, listed in the goal statement of the charter, should follow these guidelines.

Process Maps

Another important set of deliverables of the Define phase are process maps. The maps should be based on the actual state of the process, or "as-is" maps. They should *not* show the desired state at this point in the project. The team should also avoid the temptation to fix the process during mapping. In addition, since the mapping process can be difficult or time consuming, the team may approach the mapping in stages as presented below.

Stages of Process Mapping

Developing process maps is a critical part of the team activities. Even if a process map has been developed previously, the team should review the map to ensure that it is still correct and that everyone on the team has a complete understanding of the process. Often, at the beginning of the project, each person has a unique perspective of the process steps. The mapping can be a team-building activity as well as an important deliverable for the team. However, mapping can also be a frustrating and time-consuming activity if not conducted well. One method to minimize these risks is to build the map in stages.

The SIPOC Map

The first stage would be to create a SIPOC. SIPOC stands for Suppliers, Inputs, Process, Outputs and Customers. The SIPOC is a very top-level view of the process to be improved. By starting the mapping process at this level, it allows the team to quickly develop a common understanding of the process to improve and the key customers and suppliers. The figure below shows an example SIPOC in a manufacturing process making metal parts - the base of a pedestal that is part of a larger piece of equipment.

Form Base Part

Start: Raw metal **End**: Finished part

SUPPLIERS	INPUTS	PROCESS	OUTPUTS	CUSTOMERS
Stock Vendors	Metal Stock	1) Lathe Operation 2) Drill Operation 3) Inspect 4) Heat Treat 5) Degrease	Base Part	Assembly

Figure 12-11 Example SIPOC Map

The steps to create a SIPOC are:

1. Establish a name for the process.
2. Define the starting point and the ending point of the process to be improved. These should already be listed in the scope section of the team Charter.
3. List the key outputs of the process. Usually, this list includes up to three or four main outputs even though the process may produce more.
4. Define who receives those outputs, i.e. the customers. These customers may be internal (part of the business) or external.
5. State the top-level process steps of the process. Keep the list to four to eight main steps. These steps do not contain any decision points or feedback loops.
6. List the inputs to process. Stick with one to four main inputs.
7. Define who supplies the inputs to the process.

Most teams can create a SIPOC in a half-hour or hour session. Remember, the focus should be establishing the boundaries of the process improvement; that is, defining the first step of the process and the last step. This helps to keep the scope manageable. In addition, the SIPOC defines the high-level steps of the process. However, the SIPOC is probably not detailed enough to find opportunities for making the process better. More detailed mapping is required.

The Top-Down Chart

The next stage of mapping may be a top-down chart. The top-down chart takes the information from the SIPOC and adds a second level of information. For each step listed in the SIPOC, the team defines the associated sub-steps. There are still no feedback loops or decision points.

The steps for creating a top-down chart are:

1. Agree on the start point and end point for the process map. These are already listed in the SIPOC.
2. Identify four to eight major steps that describe the process from beginning to end. List those steps horizontally across a flipchart page. These are also taken from the SIPOC.
3. Break each major step into three to seven sub-steps. List the sub-steps under the corresponding major step.
4. Review the map and make corrections as necessary. Rearrange steps, combine sub-steps, or revise the descriptions of the major steps or sub-steps so they accurately describe the process.
5. Agree on a presentation format for the process map.

The figure shows a top-down chart associated with the example SIPOC presented earlier.

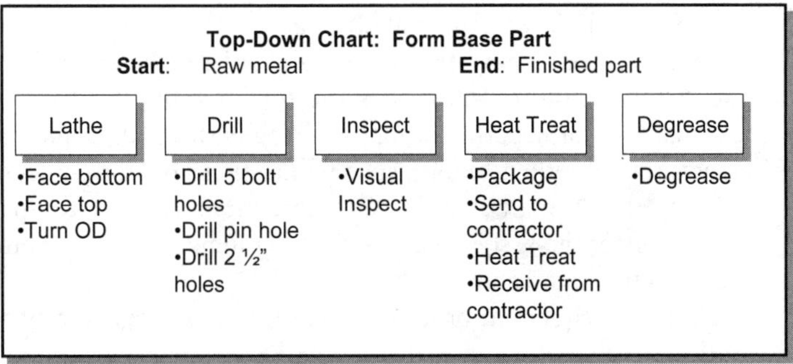

Figure 12-12 Example Top-Down Chart

The Functional Deployment Map

The third stage of mapping is to create a more detailed "as-is" picture. An example of a detailed process map is called a functional deployment map. The functional deployment map displays the steps of a process in sequential order. The functional deployment process map also illustrates what function performs the process step. Symbols are used to illustrate the flow, decision points, and activities performed.

Symbol	Name	Description
	Start & End Points	Identify the boundaries of the process.
	Activity	What is being done. Indicates necessary and unnecessary activities performed in the process.
No / Yes	Decision	Illustrates decision points and where loops occur in the process. Also used to accept, reject, approve, etc.
	Arrow	Represents a process path/flow.
	Input or Output	Shows important inputs or outputs without describing in detail.
	Process Connectors	Connect flow to another page or process.
A#	Activity Number	Shows the activity in the sequence performed.
D#	Decision Number	Shows the decision points in the sequence performed.

Figure 12-13 Commonly Used Symbols for a Functional Deployment Map

The steps for creating a functional deployment map are:

1. Review the top-down chart with the team.
2. List each of the process steps in sequential order in the first column.
3. Use the horizontal axis across the top to show the location/responsibility or department for each step performed. Depict individuals (by job title/position), specific locations, or work functions.
4. Indicate the steps, activities, and decisions that make up the process under the associated functional column. Use the symbols shown in the figure above.
5. Identify the sequential order in which the steps are actually performed.
6. Use arrows to indicate the direction of the process flow
7. Review the final map and correct as necessary.

The figure below shows the functional deployment map for the example presented in the SIPOC and top-down chart.

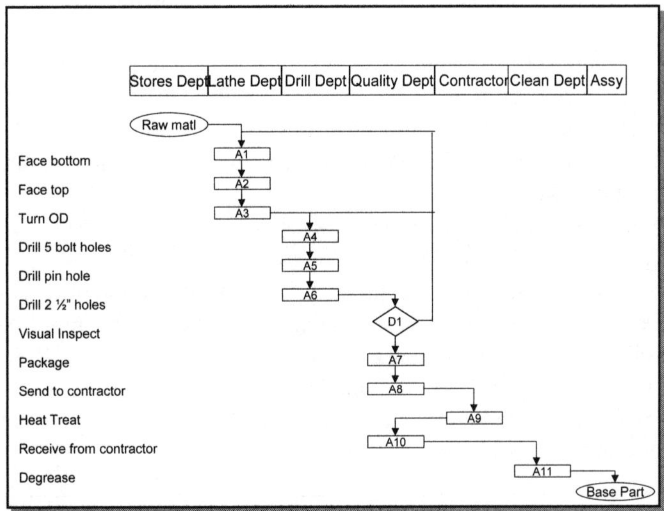

Figure 12-14 Example Functional Deployment Map

Annotating Process Maps

After the map is complete, additional information can be added depending on the project goals. For instance, if the goal is to reduce cycle time, the time for each step may be added. If the goal is to reduce defects, yield information may be added. This may help the team identify the areas on which to focus. However, this type of information may not be readily available and, often, after the data are collected in the Measure phase, the maps are updated.

Identifying Customer Value-Added Activities

Another helpful analysis is to take each activity on the map and identify whether it is customer-value added, operational-value added or non-value added. Customer-value added has *all* the following characteristics:

- The customer recognizes the value.
- It changes the product or service toward something the customer expects.
- It is done right the first time.

An operational-value added activity has *all* the following characteristics:

- It is required to sustain the workplace ability to perform customer value-added activities.
- It is required by contract or other laws and regulations.
- It is required for health, safety, environmental, or personnel development reasons.
- It is done right the first time.

A non-value added activity is one that does not fit into the other two categories. Examples of non-value added activities are: proofreading, inspection and checking, logging information, checking calculations, reviewing and approving, moving and set-up, monitoring work, and rework. This table shows an example analysis from a manufacturing process making metal parts.

Activity	**Customer-Value Added**	**Operational-Value Added**	**Non-Value Added**
Cut stock to length	X		
Check length to make sure it is correct			X
Form part	X		
Check dimension			X

Figure 12-15 Example Value-Added Analysis

The team would focus on the non-value added activities to see if they could be eliminated or minimized. An important clarification is that a non-value added activity doesn't automatically make it an unnecessary activity. The team needs to be cautious when doing this type of analysis. People on the team may have all the activities they do for their job defined to be non-value added. This may cause conflict on the team and the Black Belt needs to be sensitive to this.

Identifying Quick Wins

For many processes that haven't been under continuous improvement, some fixes may be obvious. The team should address these obvious fixes;

they may add to the team's motivational level and may help convince others of the importance of the project. These fixes are called Quick Wins.

Criteria for Good Quick Wins

After the mapping activity is done, the team should then use the maps to assess if there are some obvious opportunities for Quick Wins. Quick Wins should be changes that are easy, fast, and cheap to implement, and that fall within the team's responsibility. In addition, the team may want to ensure that the changes are easy to reverse since these ideas may not be validated with data. The table below shows an example evaluation of a Quick Win idea.

Quick Win Idea	Easy to Implement	Fast to Implement	Cheap to Implement	Within the Team's Responsibility	Reversible
Have workers stagger break and lunch time to help increase capacity	X	X	X	X	X

Figure 12-16 Example of Quick Win Evaluation

If the team identifies and implements a Quick Win, they must make a decision. The Quick Win may be significant enough to reach the project goals. The team's next step would then be to collect data to verify that they reached their goal and to set up a strategy for monitoring the process to ensure that the goal continues to be met. Management must decide if they want to continue the project with a revised goal or to use the resources in another area that has a more critical need for improvement.

Define Phase Tollgate Questions

As the team progresses through the Define phase, the team leader should schedule meetings with the Champion and Master Black Belt to review all that has been accomplished.

The following is a list of questions, called Tollgate Questions, which the Champion and the Master Black Belt can use to prompt discussion in these meetings. Also, the team can use these questions as they progress through the Define phase to be sure that they have completed all the important items.

1. What is the Big Y that will be influenced by this project?
2. What Voice of Customer data were used to establish Critical Customer Requirements? How were the data validated?
3. What Voice of Business data were used to establish Critical Business Requirements? How were the data validated?
4. What are the boundaries of the process to be improved?
5. What is the specific problem being addressed?
6. Has this problem been tackled before? What was learned from that attempt?
7. How do the little y's directly or indirectly influence the Big Y?
8. What are the goals, in measurable terms, of the project? Are they achievable in the timeframe established?
9. How were the members chosen for the team? How did the Black Belt ensure that the members understood their roles and responsibilities?
10. Were team guidelines established? How are violations of the guidelines handled?
11. Has the team created a detailed project plan with milestones and associated activities?
12. How detailed were the process maps? How were the maps validated? Did the team ensure that they were "as-is" maps showing the actual state of the process (not the desired state)?
13. Who are the stakeholders that will be affected by this project? What level of communication or involvement is necessary for each stake holder group?
14. What concerns may the stakeholders have? How will the team prevent these concerns from becoming obstacles?
15. What quick wins have been identified? What is the plan for implementing quick wins? What are the plans for ensuring that the quick wins work? What effect will the quick wins have on the goal?

At the completion of the Define phase, the team members, team leader, Master Black Belt, Sponsor, and Champion should feel comfortable with the answers to all these questions and any others that might be specific to the organization.

Case Study One

Sara Six-Sigma has been commissioned with reducing the level of scrap in production. She worked with her Champion to develop an initial Charter for the project. She accessed the IT system where scrap information was recorded and generated several Pareto charts of the scrap. The first chart she generated was a Pareto by shift of all scrapped items over the last three months. The figure shows this Pareto Chart of Shift.

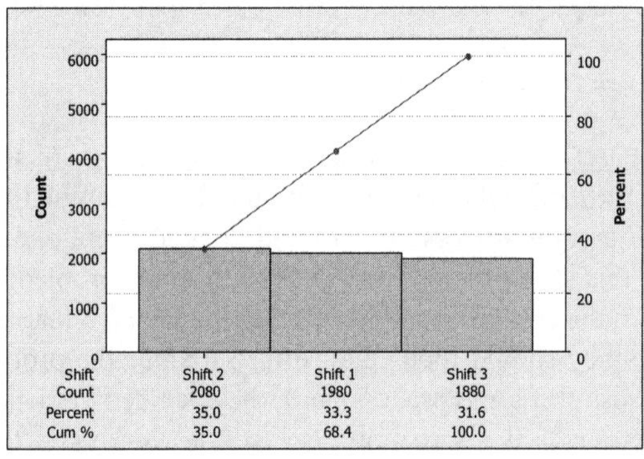

Figure 12-17 Pareto Chart of Shift

The next Pareto Chart she generated shows scrapped items by product line.

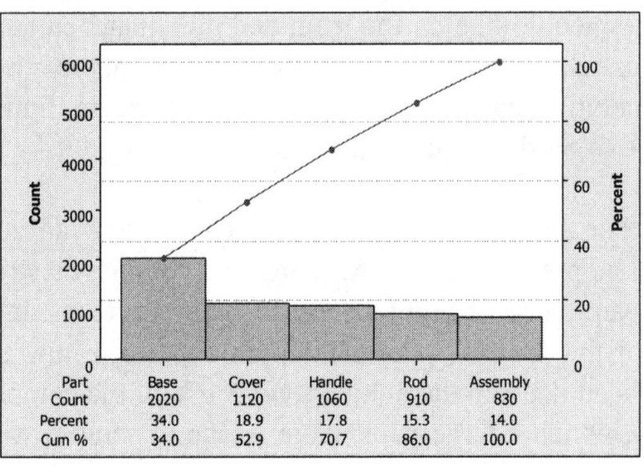

Figure 12-18 Pareto Chart of Scrapped Items by Product Line

She also generated a Pareto based on associated scrap dollars instead of a count of scrapped items, shown in the figure below.

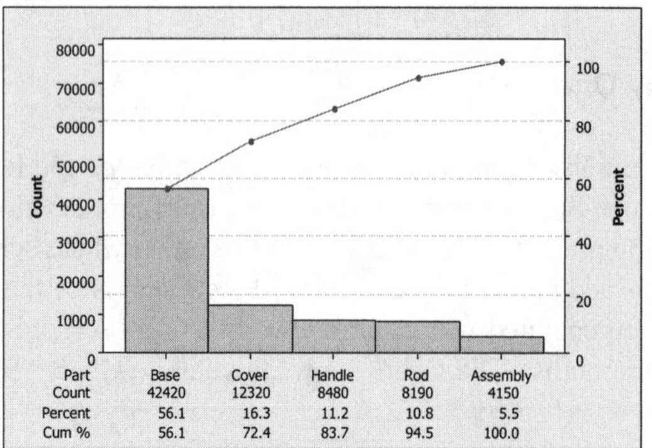

Figure 12-19 Pareto Chart of Associated Dollars

From these Pareto Charts, it appeared that "Bases" were the part type scrapped the most by either count or dollars. She examined the scrapped bases by shift and by process line but both looked like a picket fence. In other words, this stratification didn't help define the problem; no one bar stood out from the rest. She and her Champion decided that base production is where they should focus their efforts. Since the problem seemed consistent across shifts and process lines, they decided to focus on Process Line A. They made this selection because the workers in that area had been making some suggestions about improvements and they seemed willing and eager to accept changes.

In next step, Sara and the Champion selected a team. They listed the characteristics they would like on the team and then matched names to those characteristics. Sara wanted people who knew the process but also had a mixture of analytical and creative skills. And, of course, finding someone who had time to work on the project was important, too.

Sara called her first meeting with the team. She arranged for her Champion to be present as well. Although he wouldn't be a regular member of the team, Sara thought he could help the team understand the importance of the project. The Champion explained how Cost of Poor Quality had been deemed the number one Big Y of the company. Cost of Poor Quality, which a large percentage is due to scrap in the plant, had become 28% of operating costs. He also presented the Pareto charts and explained how they selected the area and the team. Many of the team members had questions for the Champion and almost all of the meeting time was taken with answering these questions. At the end of the meeting, Sara presented the draft Charter that she had developed. She asked the team members to review it for the next meeting, with the understanding that the team and the Champion would need to approve the final version of the Charter. The agenda and time for the next meeting was set and Team Scrap (as they liked to be called) was officially on its way.

Case Study Two

Joe Black-Belt is in charge of a team that is going to reduce the cycle time of customer applications. Joe works in an electronic data interchange company where the primary service provided is processing insurance claims. Currently, the customer application cycle time is anywhere from eight to ten weeks. The customers must go through an application process to be set up electronically and to sign up with the proper insurers. The faster the company can get the customers registered in the system, the faster the company can bring in revenue.

The team set a goal of reducing cycle time by 50%. In one of the first team meetings, Joe led a session to refine the team Charter provided by management. The team agreed on a final version of the Charter. The team's charter is shown here.

Charter Section	Customer Application Cycle Time Team
Business Case:	There is a significant opportunity to add new clients for our electronic processing of insurance claims. New clients will add to our market share and profit margins for the year.
Opportunity Statement:	The current time to register new clients is anywhere from 8 to 10 weeks, causing us to miss out on some revenue. In addition, customers are frustrated with the process and have left our process before getting registered, causing us to lose clients.
Goal Statement:	Reduce cycle time of the registration process by 50% by December 15.

Scope:	Start - Receipt of application form. End - Notification of status to client.
Project Plan:	<u>Activity</u> <u>End Date</u> Define 7/1 Measure 9/1 Analyze 9/20 Improve 11/1 Control 12/15
Team Members:	Champion: Ajit Black Belt: Joe Customer Registration, 1st Shift: Ernesto Customer Registration, 2nd Shift: Vince IT Support: Cathy Vendor Relations: Linda

Figure 12-19 Team Charter

After reviewing the Charter with their Champion and interested stakeholders, Joe's team decided to create a top-level process map to set the bounds of their project. This is the team's SIPOC.

Start: Receipt of Application Form **End:** Notification to Client

Suppliers	Inputs	Process	Outputs	Customers
• Insurers • Clients	• Application Form	• Review of application • IT Set up • Application information to Insurers • Final IT Set up • Notification to Client	• Notification to Client • Notification to Insurer • Notification to Billing	• Client • Insurer • Billing Department

Figure 12-20 SIPOC Map

The team continued with mapping techniques. They completed a top-down chart and a functional deployment map. Since their goal was to reduce cycle time, they wanted to include timing for each process step listed in the functional deployment map. However, since the process is paper-based and manual, this information was not available. They planned to collect the data in the Measure phase.

Summary

In the Define phase of DMAIC, the team is established and the purpose of the project is validated. The Team Charter is created, the VOC and VOB are reviewed, the Process is mapped, and Quick Wins are identified when possible. The Define phase consumes a small portion of the total effort and resource expenditure that will be required to complete an entire DMAIC process improvement project. The Define phase must tie the achievement of immediate, doable project goals to key strategic objectives. If this is accomplished and the Champion and leadership approve, then the team efforts on the rest of the project will be worthwhile.

Chapter

13

Measure Phase

The primary purpose of the Measure phase is to answer the question, "How are we doing?" In other words, the team must baseline the current state of each CTQ or CTP. Many times in a project, the critical measure identified by the team is not an already reported measure. Although the team may not think the measure is meeting the goal, they need to collect data to verify the current performance level.

The deliverables from the Measure phase include:

- x data to be collected
- Operational definitions
- Data collection plan
- Measurement system analysis
- Baselined data

This chapter discusses these deliverables as well as other tools and important for the Measure phase.

Determining x Variables

The team will have already picked the process outputs - CTQ's and CTP's, or the little y's - in the Define phase. The x's are process and input variables that affect the CTQ's and CTP's. The first consideration in the Measure phase is to identify the x data to collect while baselining these y variables. The y data are needed to establish a baseline of the performance of the process. The x data are collected concurrently with the y's in this phase so that the relationships between the x's and y's can be studied in the Analyze phase.

There are a variety of ways to identify the other data to collect (the x's). One of the best ways is to use a combination of the Process Maps from the Define phase, Cause and Effect Diagrams, and Cause and Effect Matrices.

Cause and Effect Diagram

The Cause and Effect Diagram was originally developed as a brainstorming tool to identify potential causes of a particular problem (the effect). This might be how the tool is used in the Analyze phase. However, in the Measure phase, it is used with a slightly different twist. It is used to brainstorm potential x data. It is important to recognize this difference and to have the team use the tool appropriately.

The Cause and Effect Diagram is sometimes called an Ishikawa Diagram after the inventor of the tool. It is also known as the Fishbone Diagram since it resembles the head and bones of a fish.

The CTQ or CTP is placed in the head of the fish. Each bone is labeled with a category. The typical categories are People, Machine, Materials, Environment, and Methods. In the traditional use of the tool, another category called Measurement is added. Since the team should verify the measurement system before collecting data, any issues relating to the measurement system should be addressed first, so this category is not part of Cause and Effect Analysis in the Measure phase of a Six Sigma project.

One of the first steps the team may do is to review these categories and consider refining them to fit their process. For instance, if the team is looking at a bank project, they may change "People" to "Teller" and "Machine" to "IT System".

After the team defines the major categories, they will brainstorm input and process data that should be collected. The steps for developing a Cause and Effect Diagram are:

1. Review the process maps developed in the Define phase. Make sure each team member is able to see the maps during the brain storming session. If appropriate, post the process map on a wall during the meeting.

2. Review the five generic categories with the team. Revise the categories to fit the process.
3. Remind the team that the focus is on x data to collect.
4. Write the CTQ or CTP under consideration in the head of the fishbone outline. Use only one CTQ or CTP at a time. In other words, each y for the team will have its own Cause and Effect Diagram.
5. Review the rules of brainstorming.
6. Each person will suggest some x data to collect and state under which bone category it should placed.
7. When the brainstorming session is complete, clarify any items on the list as necessary.

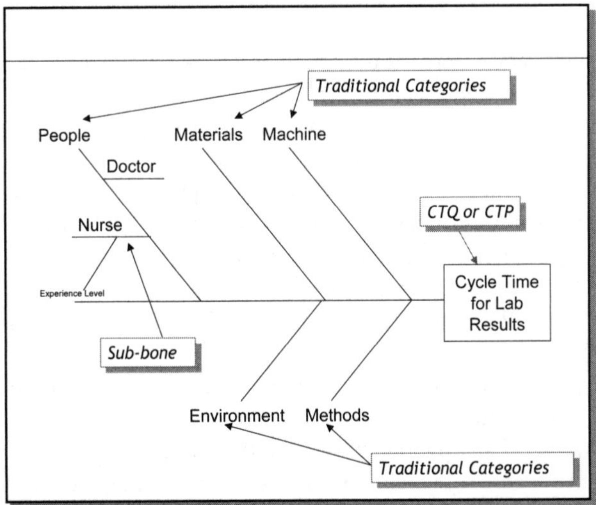

Figure 13-1 Example Cause and Effect Diagram

If appropriate, the team might create logical sub-categories if a category becomes too crowded. These become sub-bones on the diagram. During the session, the team should not argue over the placement of an item on the diagram. The categories are there for guidance and inspiration. The figure above shows the start of a Cause and Effect Diagram on cycle time for lab results in a hospital.

Cause and Effect Matrix

If the team only has one CTQ or CTP for the project, the next step is to decide which x data will be collected based on the brainstorming session. Two options for picking the x's are to use multi-voting (described in the Define chapter) or a Cause and Effect Matrix.

The Cause and Effect Matrix is a tool that is even more appropriate when the project has multiple CTQ's and CTP's. A Cause and Effect Matrix gives weights to each y indicating the importance of that y. Then, each x is rated in terms of its correlation to each y. Calculations are made based on the importance and correlation, and higher scoring x's are the best candidates for data collection.

The steps for developing a Cause and Effect Matrix are:

1. List the y's across the top. Each should be taken from the head of an associated Cause and Effect Diagram.
2. Clarify how the team will decide both the weightings for the y's and the ratings for the x's. (Some ideas are to use the highest suggested rating, to use the Champion to break any ties, take the average, etc.)
3. Assign a weight for each y. Use a scale from 1 to 10 where 1 indicates least important in terms of relative importance and a 10 indicates the most important in terms of relative importance. If all the y's are equally important, rate them all the same with any arbitrary number.
4. List in the first column all potential x's that may affect any of the y's. These should be taken from the multiple Cause and Effect Diagrams. If any overlap exists, list the x only once.
5. Rate the degree to which the x affects or is correlated to each y. Use the following scale:

   0 - no effect or correlation
   1 - small effect or weak correlation
   3 - medium effect or medium correlation
   9 - strong effect or strong correlation.

   This scale ensures that the x data that the team thinks has the strongest effect on the y will stand out.

6. Multiply each rating by the weight and sum across the row, putting the result in last column.

	Cycle Time	Cost	Sigma Level		<------ CTQ or CTP
	10	8	5		<------ Weight
----X Data----	--- Correlation of Input to Output ---				-------- Total --------
Application Type	9	0	3		105
Set Up Method	9	3	9		159
Customer Type	0	3	9		69
Insurer	1	1	1		23
SCALE : 0=NONE 1=LOW 3=MODERATE 9=STRONG					

(10*9) + (8*0) + (5*3) = 105

Figure 13-2 Example Cause and Effect Matrix

The x's with the highest totals are the ones that the team should try to collect. The figure above shows a partial example of a Cause and Effect Matrix from a process that is registering customers for service. The team's focus is to reduce cycle time.

Before using the Cause and Effect matrix, the team should consider a few things. If a particular x is already available, consider not rating it. This tool is for determining what data to expend effort on collecting. The team should plan on using x data that are already available. However, do not just collect a sample of convenience. The x that might be the strong driver of y may not be collected normally.

Data Types and Sampling

At this point in the Measure phase, the team will have selected the y and x variables to be measured. Before doing actual data collection, the team should consider some statistical concepts including data types and sampling because these will affect how the team goes about collecting the data.

Defining the Data Type

The x or y data can be one of two data types: continuous or discrete. Continuous data are data that can take on any value on a continuous scale. For example, weight, distance, amount of time to complete a process, and cost are all continuous data types. Discrete data can only take on a unique set of values. For example, the number of people who pass a test, the count of errors on a bill, the day of the week (Monday, Tuesday, etc), and the name of the supplier are all discrete data types.

A team should try to find a way to collect the y data so they are continuous. Continuous data require a smaller sample size, and there are more statistical tools available for continuous y's. If the team has only discrete data, fewer statistical tools are available.

Often, on-time delivery is the little y for a Six Sigma project. Many teams' first instinct is to treat this as discrete data; i.e., the product or service was either on time or not. However, they're losing information with this approach. The amount of time that the delivery is late could be collected and used as continuous data. By collecting the actual time early or late, the team gets a better understanding of the problem and will have more tools to analyze the data.

Discrete Data Examples	Continuous Data Examples
Billing errors Software errors Pass / Fail Customer Type Gender	Length Weight Dollars Voltage Cycle Time

Figure 13-3 Examples of Data Types

Sampling

When collecting data, the team will probably use a sample. A sample is a subset of a population, where the population is the set of all items of interest in a problem. It would not only be costly and time-consuming, but, in most situations, it is impossible to gather data from an entire population. Populations are usually too large. Also, if the test requires damaging the product, it would be unfeasible to test the entire population, no matter what the size. Therefore, a sample of data is collected from a population of interest and a statistic is computed and used to estimate a population characteristic or parameter.

For instance, if the average height of all men was to be measured, it would be impossible to measure all men in the world (the population). It would take too long and cost too much, and the population would be changing while collecting the data. Instead, a sample of men from around the world could be collected. The average height of the men in the sample could be computed (the sample statistic) to estimate the average height of all men in the world (the population parameter).

The important aspect is to get a good sample so that the statistic is a good estimate. A good sample is based on sample size, when the data are collected, and how representative the sample is.

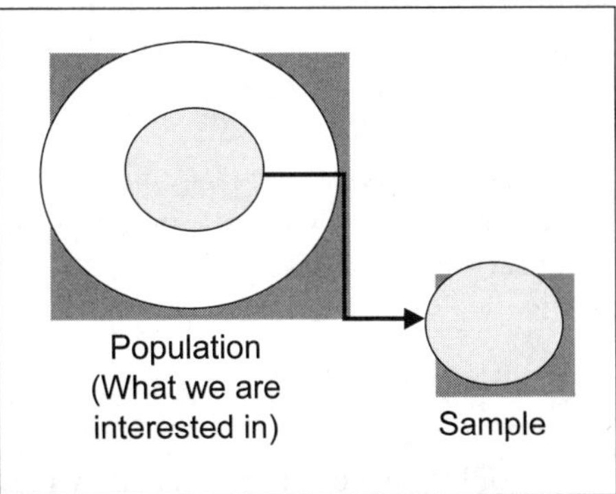

Figure 13-4 Population vs. Sample

There are two main issues with selecting the correct sample size: statistical appropriateness and logistics. Enough data must be collected to make a statistical determination but the team should also consider what is practical, timely, or doable. Many times these considerations are in conflict. Recognize the risks that occur if there are not enough data for appropriate statistical estimates.

The emphasis in a team should be to collect new data as opposed to using historical data. If new data are used, the team will know how the data were collected, if there are any special circumstances, etc. This may not be true if historical data are used. In addition, the process may have changed since the historical data were collected.

Even more important is the issue of obtaining a representative sample of the population. This is the key to feeling confident in the decisions based on the data. Suppose a team is studying wait times for customers at the bank. If they collect all their data from noon to one p.m. on Friday (a notoriously busy time for the bank), they will have a biased sample. Therefore, any estimates made from this sample will be biased. The team should use their process knowledge to determine the sampling strategy.

Understanding the Data

There are three important characteristics of continuous data: the location or center of the data, the spread or variability of the data, and the shape of the data distribution. Numerical analyses are used to understand location and spread. Graphical analyses are used to understand the location, spread, and shape. Various graphs are presented in the Analyze phase.

Numerical and graphical analyses are also used with discrete data. Discrete data can be summarized by counts, proportions, or time graphs of the data. Care must be taken to ensure that correct numerical tools are used for discrete data.

Numerical Analysis

The location or center of the data can be measured by the mean (average) or the median. The mean is the sum of all the data divided by the count of the data items. The symbol for the population mean is μ and the symbol for the sample mean is . The median is the fiftieth percentile; in other words, the data point at which 50% of the data fall below it and 50% of the data fall above it. See the figure below for calculations of the mean and the median.

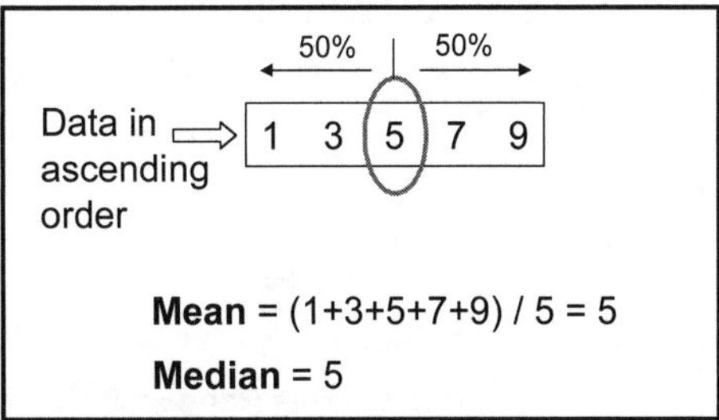

Figure 13-5 Sample Calculation of Mean and Median

The spread of the data can be measured by the variance, the standard deviation, the range, or the interquartile range. The sample variance, indicated by s2, is the average squared distance from the mean, calculated by:

$$s^2 = \frac{\sum_{i=1}^{n}(x_i - \bar{x})}{\ } \quad .$$

In this formula, *n* indicates the sample size, or the number of data points used, and xi represents each individual data point. This formula gives an estimate of the population variance, . The standard deviation is the square root of the variance σ^2. It is used more often than the variance because it is in the natural units of the data (versus the units squared). The symbol for the population standard deviation is σ and the symbol for the sample standard deviation is *s*.

	Population	Sample
Mean	μ	$\bar{x}$
Standard Deviation	σ	s
Variance	σ^2	s^2

Figure 13-6 Symbols Used for the Population and the Sample

The range is the maximum data value minus the minimum data value. The interquartile range is the seventy-fifth percentile minus the twenty-fifth percentile; in other words, the middle 50% of the data. See the figure below for calculations of the variance, standard deviation, range and interquartile range.

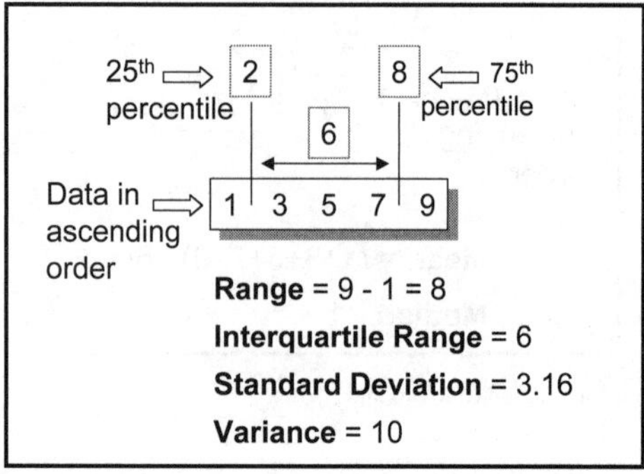

Figure 13-7 Calculations for the Variance, Standard Deviation, Range and Interquartile Range

Measurement System Analysis

One important step in the Measure phase, sometimes skipped by inexperienced Six Sigma teams, is conducting a Measurement System Analysis (MSA). An MSA is done to verify that the measurement system produces valid data, before the team makes data-based decisions.

A measurement system is defined as the collection of operations, procedures, gauges and other equipment, software, materials, facilities, and personnel used to assign a number to the characteristic being measured. The language and techniques of MSA are oriented more to the manufacturing environment. However, Measurement System Analysis is a critical part of any Six Sigma Project, regardless of the environment (e.g. transactional, service, etc.). The philosophy behind this kind of study is applicable to all project types.

Depending on the type of data, the statistical analysis will be different. For a continuous measurement, there are a variety of statistical properties that can be determined: stability, bias, precision (which can be broken down into repeatability and reproducibility), linearity, and discrimination. For a discrete measurement, estimates of the error rates can be determined for within appraiser, each appraiser versus standard, between appraisers, and all appraisers versus standard. The properties related to both continuous and discrete measures are discussed below.

Properties of Continuous Measurements

Stability

Stability is defined as the distribution of the measurement system remaining constant over time. Stability is often determined by measuring a standard or a golden unit on a periodic basis and plotting the results on a time-based chart, usually a Control Chart. (Control Charts are discussed in the Control chapter.) The purpose of this chart is to show that the variability and mean of the measurements remain the same over time. Assuming the standard or golden unit doesn't change, any changes in the variability or mean is due to the measurement system.

Bias

Bias is the difference between the observed average of the measurement data on a standard and the actual reference value. The purpose of doing a bias study is to determine if the measurement system is giving accurate values. To determine bias, a standard must be available that is traceable to the National Institute of Standards and Technology (NIST) or is an industry-agreed upon standard. To determine bias, a team would measure the standard several times (say 20 or 30) and compute the difference between the average of these readings and the reference value. This is the bias. The goal is to get this value close to zero.

Variability

Any measurements taken from a process will have variability. This variability can be broken into two main sources: process variability and measurement system variability. The goal is to have the variability due to the measurement system small in comparison to the process variability.

Precision is the measure of the measurement system variability, and is defined as the standard deviation due to the measurement system. A traditional way to determine precision is to take a sample of representative parts from the process and to have two or three people measure the parts two or three times. Usually, this is conducted as a blind study - the people aren't aware that they are part of the measurement system analysis. Precision can be split into repeatability and reproducibility. Repeatability is the variability of the gauge itself. Reproducibility is the variability associated with using different operators under different conditions. Variability from part-to-part is due to the process, while repeatability and reproducibility are due to the measurement system. The goal is to have the repeatability and reproducibility to be small.

To judge if the variability due to the measurement system is small enough, two metrics are commonly used. They are %R&R and % P/T. %R&R stands for percent repeatability and reproducibility. The formula is:

$$\%R\,\&\,R = \frac{6\hat{\sigma}_{Measurement\_System}}{6\hat{\sigma}_{Total}} \times 100$$

where the numerator contains an estimate (the Ù indicates an estimate) of the variability solely due to the measurement system and the denominator contains an estimate of the total variability. %P/ T stands for percent precision to tolerance. The formula is:

$$\%P/T = \frac{6\hat{\sigma}_{Measurement\_System}}{USL - LSL} \times 100$$

where USL is the upper specification limit and LSL is the lower specification limit. These give the allowable range of values for the process. For both metrics, usually the goal is to get these percentages to be under 10%.

Linearity

Often a measurement system is used to measure parts that have a range of sizes. Linearity is the determination of the bias and precision over the expected operating range of the gauge and determining if these are acceptable for all part sizes. For instance, if a measurement system measures length, it may work well (small bias and precision) for parts that are 12 inches long. However, it may not work at all for parts that are smaller than one-half inch. If the business makes parts of both sizes, the measurement system may not be appropriate for both types of parts.

Discrimination

Discrimination is the capability of a measurement tool to detect and adequately indicate small changes in a measured characteristic. For example, measuring the width of a strand of hair with a tape measure that indicates to the nearest sixteenth of an inch would not be adequate. The discrimination of the tape measure is not small enough. If the measurement tool is not adequate to detect small changes, determining how to find and fix errors in the process will be difficult.

Properties of Discrete Measurements

For discrete measurements, a blind study may also be done. An expert would usually determine whether the product is good or bad. Then, a variety of good and bad units is given to two or three appraisers. The appraisers each then determine if they think the product is good or bad. They are asked to look at the same unit more than once, without knowing that they had evaluated the unit previously. This is called the "within appraiser"error rate. It can then be determined how well all the appraisers are able to get the same result on the same product, the "between appraiser" error rate. In addition, it can be determined how well the appraisers agree with the expert, known as the "appraiser versus standard" error rate.

Analysis of Measurement System Data

Software packages can help with the analysis of data from a measurement study for both continuous and discrete data.

If a team is unable to get repeated measures, as described above, the calculations of these statistical properties may be difficult. The team may not be able to conduct a formal measurement system study. However, they should still review the measurement system and consider ways in which the data produced may have error. They should consider using a Cause and Effect Diagram with People, Machine, Method, Environment, and Material as possible categories and measurement system variability and inaccuracy in the head of the diagram. The team would then brainstorm possible reasons for inaccuracy and variability in the data due to the measurement system. The team would choose the most likely reasons to address. In addition, teams that don't do a formal measurement system study should place even more emphasis on clear operational definitions, discussed next.

Refer to the MSA Chapter in Part Six of this book for more information about MSA in non-manufacturing environments.

Preparing to Collect the Data

Data collection can be difficult. To help, the team should use operational definitions and data collection plans. They may also need to create data collection forms.

Operational Definition

An Operational Definition is a precise definition of the specific y to be measured. These will be used to baseline the performance. The purpose of the definition is to provide a single, agreed upon meaning for each specific y. This helps ensure reliability and consistency are built in up-front during the measurement process. Although the concept is simple, the task of creating a definition should not be underestimated.

Consider a team whose y was cycle time. They had the date and time of the start of the job in the computer; it was already being recorded. And, so was the date and time of the end of the job. Cycle time seemed to be a simple matter of calculation of the difference between these two numbers. The graph of their cycle time looked like the figure below.

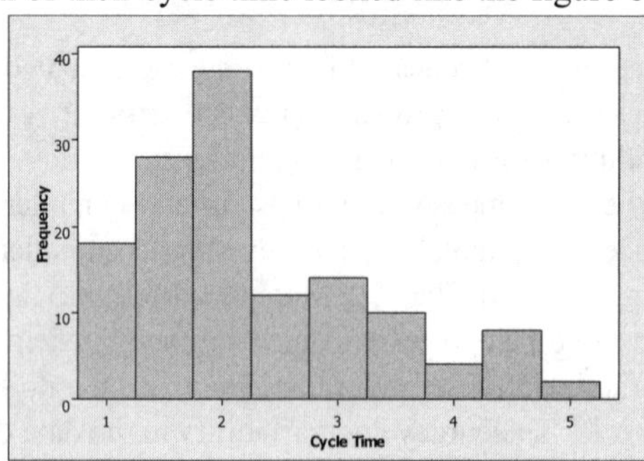

Figure 13-8 Cycle Time Graph

As cycle time is often skewed (has a longer tail to one side), they didn't worry too much about the look of their graph. However, as they progressed in their analysis, they found that almost all of the jobs that had long cycle times were started on Fridays. The company did not work on weekends so every job started on Friday had an extra two days automatically added to its cycle time - whereas most jobs required approximately one day to complete. An additional complication was that the company sometimes worked on Saturdays. They needed to figure out to account for this, as well as holidays, etc. They had to start their data collection and data analysis over.

A good Operational Definition will ensure that the first time the data are collected, they are collected correctly and the data will be useable.

Data Collection Plan

The data collection plan is an important deliverable from the team. It is a plan defining the precise data that will be collected, the amount of data that will be collected, a description of the logistical issues - who, where, when data will be collected - and what will be done with data collected. The purpose of the plan is to make sure that the data collected are meaningful and valid and that all relevant data are collected concurrently (the x's and the y's).

The steps for creating a data collection plan are:

1. List the data. Each row contains a separate y.
2. Enter the operational definition for each y.
3. List the sources from which the team will need to get the data.
4. Enter the sample size. Consider the cost and practicality of sampling, how representative the sample is, and the variability of the population.
5. Determine logistical issues - who will collect data, when will it be collected, and how it will be collected.
6. List all the x data to be collected at the same time. Take this information from the Cause and Effect Matrix or Cause and Effect Diagram.
7. List what will be done with the data. Include any analysis to be done as well as any graphs to generate.

Y Measure	Operationa l Definition	Data Source and Locatio n	Sample Size	Who Will Collect the Data	When Will Data be Collected	How Will Data be Collected	X Data that Should be Collected at the Same Time
Cycle Time	Difference between the barcode info on receipt and barcode on customer notification minus non-working hours (holidays, weekends)	IT system	300	Representatives in process	First three weeks of month	Data collection form; IT report	Customer type; application method; day of week; Representative initiating call; IT rep; wait times; process sub-step cycle times; rep sending notification

How will data be used?	How will data be displayed?
♦Identification of Big X;s ♦Identifying if Data are Normally Distributed ♦Identifying Sources of Variation	♦Histogram ♦Control Chart ♦Scatter Diagrams

Figure 13-9 Example Data Collection Plan

The figure above shows an example data collection plan for a project focused on reducing cycle time. Make sure to plan for data collection. Having to get data a second time because it wasn't done right the first time is demoralizing and painful for the team.

Data Collection Forms

Most teams need to do some level of manual data collection. Even in the most automated processes, some x data may not be available any other way. Changing an IT system to collect specific x data may not be a viable option.

For manual data collection, the team will have to develop a form to meet the data collection needs. The format should be straightforward and simple to minimize errors and the people gathering the information should be involved in its creation. If the team is collecting data about time, people in the process may be concerned about how the data will be used. They may worry that their performance is being judged. It is important to address these concerns to avoid possible bias in the results.

Try using the form on a trial basis before full implementation and examine the data for unexpected results or missing data. Adjust the collection form as required.

Another good idea is to have the team be as close to the data collection as possible to help with any problems. At a minimum, the collectors should have a contact number of a team member in case they have trouble. The figure below shows a sample data collection form used for a project focused on the cycle time for loan applications.

General Information

Loan ID #	
Application Type	
Amount	
Type of Loan	
Dealer	

Process Step	Start Time	End Time
1		
2		
3		
4		
5		
6		
7		
8		

Figure 13-10 Example Data Collection Form

Collecting the Data

One major determinant of the duration of a Six Sigma project is the time required to collect the data. The time required is dependent on how frequently the data are available in the process and the ease of collection.

The team needs to be involved in the data collection to make sure it is done right and that any anomalies or problems are recorded and understood. This will make the analysis of the data in the Analyze phase easier.

Baselining the y Data

Once the team collects the data, they must answer the question "How are we doing?" by baselining the y data. This is the final step in the Measure phase.

The team baselines the current process performance in terms of the CTQ's and CTP's. There are a variety of metrics that can be used to baseline these y's. These metrics include Sigma Level, C_p and C_{pk} or P_p and P_{pk}, yield, and Cost of Poor Quality. The team should select a metric that is appropriate for their business. For instance, if the business does not traditionally use Sigma Level, there is no requirement that they calculate it. It may have no meaning to the rest of the organization and therefore, it may not be an appropriate metric.

Sigma Level

Sigma Level is typically used with discrete data such as Pass/Fail data. Sigma Level is based on a calculation of defects per million opportunities (DMPO). The million in the calculation is a scaling factor. An opportunity is defined as a chance for a defect to occur per unit or delivery of service. An opportunity is a chance for a failure to meet customer requirements to occur per unit or delivery of service. In many cases, the number of opportunities should be defined to be one - the customer receives it right (no defects - therefore, no failure) or the customer doesn't receive it right (any number of defects - therefore, a failure). However, in processes where there is a need to differentiate between complexities in product, the team may define more opportunities.

The business would define each of the types of defects that commonly occur. For instance, a business that produces assembled printed circuit boards probably has a wide range of complexities in the types of boards they produce. In the process of assembling printed circuit boards, components are pasted on to a board. Some boards may have a few components spaced far apart. Some boards may have thousands of components tightly spaced. The difficulty in building boards with fewer components is very different from those with many components. The opportunities per board could be defined as the number of components placed on the board.

However, the opportunities shouldn't be arbitrarily defined. Increasing the number of opportunities will increase the Sigma Level, but this could be an unfair assessment.

The steps to finding the Sigma Level are:

1. Take the data collected based on the data collection plan, using an appropriate sample size.
2. Calculate the number of defects in the sample based on the Operational Definitions.
3. Define what an opportunity is. Be cautious in defining too many opportunities; this will artificially inflate the Sigma Level.
4. Calculate the Defects Per Million Opportunities using the following formula:

$$DPMO = \frac{Defects\ x\ 1\ Million}{Units\ x\ Opportunities}$$

5. Use the table to convert DPMO into the Sigma Level. A portion of the table follows:

DEFECTS PER MILLION	SIGMA LEVEL
5.4	5.9
8.5	5.8
13	5.7
21	5.6
32	5.5
48	5.4
72	5.3
108	5.2
159	5.1
233	5
337	4.9
483	4.8
687	4.7
968	4.6
1350	4.5
1866	4.4
2555	4.3
3467	4.2
4661	4.1
6210	4
8198	3.9
10724	3.8
13903	3.7
17864	3.6
22750	3.5
28716	3.4
35930	3.3
44565	3.2
54799	3.1
66807	3
80757	2.9
96801	2.8
115070	2.7
135666	2.6
158655	2.5
184060	2.4

241964	2.2
274253	2.1
308538	2
344578	1.9
382089	1.8
420740	1.7
460172	1.6
500000	1.5

Figure 13-11 DPMO to Sigma Level Conversion Table

This conversion table is based on the assumption that the data are taken over a short time period. Historically, Motorola found that processes over the long term exhibited a shift in the data of 1.5 σ standard deviations. This 1.5 s shift is incorporated into the calculations. Therefore, the Sigma Levels listed above are what processes would see in the long term. This incorporation of the shift into the calculations has become an accepted practice.

Suppose that a project is focused on a billing process. The team wants to have correct bills sent to the customer. They have defined one opportunity for this process - either the bill is correct or not. All of the bills produced are the same in terms of complexity. The team took a sample of 250 bills and found 60 defects. The DPMO was:

$$DPMO = \frac{(60)(1{,}000{,}000)}{(250)(1)} = 240{,}000$$

Using a conversion table, the team found the Sigma Level to be about 2.2. They used this information to baseline the current process performance.

Capability Indices

Process capability is the ability of the process to meet the requirements set for that process. One way to determine process capability is to calculate capability indices. Capability indices are used for continuous data and are unitless statistics or metrics. There are many capability indices but the two most commonly used are C_p and C_{pk} (or P_p and P_{pk}).

The formula for C_p is:

$$C_p = \frac{(USL - LSL)}{6s}$$

C_p is the potential capability indicating how well a process could be if it were centered on target. This is not necessarily its actual performance because it does not consider the location of the process, only the spread. It doesn't take into account the closeness of the estimated process mean to the specification limits.

Value	Interpretation of C_p
< 1.0	Poor capability
1.0-1.5	Marginal capability
> 1.5	Good capability
> 2.0	Motorola 6σ capability

Another metric used in conjunction with C_p is C_{pk}. The formula for C_{pk} is:

$$C_{pk} = \min\left[\frac{USL - \bar{x}}{3s}, \frac{\bar{x} - LSL}{3s}\right]$$

C_{pk} does take into account the location of the data by considering the closeness of the mean to the specification limits. For two-sided specifications, use the ratio of C_{pk} to C_p. The farther this ratio is from one, the more off-target the process is.

Value	Interpretation of C_{pk}
$C_{pk} > 1.5$	Motorola 6σ capability

To calculate P_p and P_{pk}, the same formulas are used except a different estimate for the standard deviation is used. Typically, C_p and C_{pk} use a pooled estimate of the standard deviation based on information from a control chart while P_p and P_{pk} use the long-term estimate of the standard deviation (from the formula presented earlier for standard deviation).

P_{pk} can be considered actual process performance and C_{pk} is what the process is capable of doing if there is no between subgroup variability. Often data are collected in subgroups, small samples collected over time, as described in the Control phase. Then variability from the process can be broken into two sources: within subgroup variability and between subgroup variability. Usually, C_{pk} will be smaller than P_{pk} since P_{pk} represents both within subgroup and between subgroup variability. C_{pk} only represents between subgroup variability. However, Pp and P_{pk} are interpreted the same as C_p and C_{pk}.

There is some controversy surrounding which set of these metrics to use. The important point is that their primary purpose is to compare the capability of the process before the team has made improvements to the capability of the process after the team has made improvements. Before calculating any of these indices, verify that the data are from a normal distribution and come from a stable process. Whichever metrics are used, be consistent - always use the same metrics for comparisons of before and after the process improvement. Motorola uses P_p and P_{pk} on a stable process and calls them C_p and C_{pk}.

Cost of Poor Quality

Cost of Poor Quality (COPQ) is another common metric used for Six Sigma projects. COPQ is usually defined as the costs associated with internal failures and external failures. Internal failure costs are costs resulting from nonconformance to quality standards and the effort associated with overcoming the impact of failures found before the product or service is formally released. Some examples of internal failures are scrap, excess and obsolete inventory, and rework. External failure costs are costs resulting from nonconformance to quality standards and the effort associated with overcoming the impact of failures found subsequent to the formal release of the product or service. Some examples of external failures are warranty costs, associated call center costs, etc.
If COPQ is the metric used by the team to baseline the current state, they should refer to a financial expert in the company for help. Each company may calculate COPQ differently.

Yield

Another metric used to baseline the process is yield. There are different ways to define yield. Two are presented here: first pass yield and final yield. First pass yield is the ratio of the units of a product that pass completely through the process the first time without rework over the number of units entered into the process. Final yield is the ratio of the total units delivered defect-free to the customer over the total number of units that entered the system. The figure below shows an example process with first pass yield and final yield calculations.

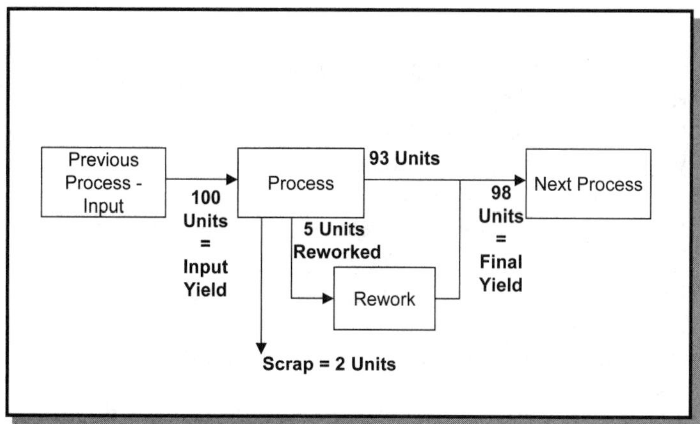

Figure 13-12 Example Yield Calculations

The first pass yield, also called first time yield, =

$\frac{\text{Units Passing Without Rework}}{\text{Input Yield}} = \frac{93}{100} = 0.93.$ In the example above, final yield =

$\frac{\text{Final Yield}}{\text{Input Yield}} = \frac{98}{100} = 0.98.$

Again, the important consideration is to be consistent in the way the metrics are calculated for before and after comparisons.

Reviewing the Measure Phase

After the team has baselined the current state, it is possible, though very unlikely, that the team may find that the process is doing better than originally thought. This may lead to some discussion regarding whether the project should continue.

Measure Phase Tollgate Questions

At the completion of the Measure phase, the team members, team leader, Master Black Belt, and Champion should feel comfortable with the answers to all of the following questions and any others that might be specific to the organization:

1. Has the charter been updated? If so, how?
2. Has the scope changed?
3. What are the x data collected?
4. What are the operational definitions?
5. How much data were collected?
6. What was done to assure the reliability and validity of the measurement process?
7. Has the data collection provided consistent information throughout the data collection period?
8. If the data collection were repeated, would the team get similar results?
9. Are the data collected representative of the population?
10. Do the data collected provide the information needed?
11. What is the baseline value for the data?

Case Study One

Sara and Team Scrap finished the charter and all the process maps. They had a Quick Win suggested by one team member. A minor change in the set-up procedure for the equipment would require less material. The change would impact the scrap amount but would only get them part of the way to their goal. They moved on to the Measure phase. They completed a Cause and Effect Diagram and Cause and Effect Matrix to determine the data to collect.

Before collecting the data, however, they needed to assess their measurement system. Scrap was determined by measuring the length and diameter on several characteristics of the base parts. The business used an optical comparator to measure both of these types of dimensions. The measurement device was heavily dependent on the operator. First, the team decided to determine the bias of the equipment. Standards, which could be traced back to the National Institute for Standards and Technology, were available for this measurement device. Using someone who normally operated this device, the following steps were taken:

1. For each standard, the operator measured the standard 20 times.
2. The average of these readings was computed.
3. The bias was calculated as the observed average minus the standard value.

The bias was unacceptable so the vendor was called in to calibrate the equipment.

In addition, they wanted to understand the variability of the equipment. The team decided to use a traditional approach to doing a measurement system study on the precision.

- They obtained 10 sample parts from the process. They used 10 parts that roughly covered the range of the dimensions found in the process.
- Three operators were selected from a group that normally used this device.
- Each operator measured each part two times.

From this study, the variability due to the parts was allocated as process variability. The variability due to operators was allocated to the measurement system. Their software package helped them calculate these. They were then able to assess %R and R. It was 50%, with most of the variability due to reproducibility. This was unacceptable and needed to be addressed. They interviewed the operators and found that the one that performed the best, Steve, had used some tooling to anchor the part in place when he did his measurements. The other operators didn't use the tooling, so their parts were moving as they did the measurements. The team revised the operating procedures for this equipment and had all the operators follow Steve's method. The study was repeated and the results for %R and R were lowered to 7%, in the acceptable range.

There was a variety of part sizes that were produced in the process. The team picked some other part numbers of different sizes and used these for bias and precision studies as previously described. They had already done a medium sized part, so they chose one of the smallest parts and one of the largest parts to repeat the study. This was their linearity study. All the study results were acceptable.

The team felt the discrimination of the measurement device was adequate. They used a rough rule of thumb that the data fall into at least eight distinct values to have adequate discrimination. Their data had many more distinct values so they deemed them okay.

The measurement device had not been properly managed in the past so the team spent some time training all operators on the new procedures. They also set up a new method to determine the stability of the measurement device. They had an operator measure a standard and plot the data on a chart (a control chart) every day so that they could verify if there were changes in the equipment between scheduled calibrations. The operator could determine if the bias or the precision was changing over time.

Once the team felt confident in their measurement system, they developed a data collection form to get all the x data in which they were interested. They worked with the people in the area to develop the form. In addition, they sat with the workers on the first day of using the form. They made some modification based on what they saw and then began the data collection in earnest. They made sure to leave cell phone numbers of the team in case any questions arose. The team collected data over a three-week period.

Case Study Two

Joe's team had a great start with the Define phase. The Charter was easily agreed upon and the maps were done in record time. However, things didn't go so well in the Measure phase. They did a Cause and Effect Diagram, but Joe forgot to explain the purpose of the tool was to define x data to collect. Instead, since everyone on the team had used this kind of tool before in its traditional sense, the team came up with a list of problems in the area. The most popular one had to do with getting a better copier in the area. Although getting a better copier might help the process, the team suspected this really wasn't the x that they were looking for.

They had to have a second session to re-look at the cause and effect diagram to clarify exactly what x data were needed.

Before collecting any data, the team needed to consider their measurement system. The first thing they concentrated on was their operational definition of cycle time. The cycle time was computed by their computer system and the way it was calculated was a match with the way the team had defined it. By verifying their operational definition, the team was trying to reduce the bias in the data. However, they weren't done. Just because the numbers looked good from the computer didn't mean that they didn't have measurement system variability. The team decided to investigate how the data were entered in the system.

On each application form there was a barcode that was supposed to be swiped at the first process step and at the end of the last process step. The team found inconsistencies from shift to shift on when this barcode was swiped. On first and third shift, the barcode was swiped as soon as it was received from the customer. On second shift, the barcode was swiped when someone decided to work on the application. Since this process area was often behind, this difference could change the cycle time by as much as a day. The team established a common procedure across all shifts to swipe as soon as the application arrived from the customer. By reducing variation in the way that the data were entered, the team was addressing the precision of their measurement system.

Another consideration of the team was discrimination. The way the IT system was designed, only the date was recorded, not the time. Therefore, they could not get values between one day and zero days for their cycle time. They were sure that cycle times might be a few hours on a slow day, so they asked the IT group to make a change to record the actual time. It was an easy change for the system and they had it completed in a week. Of course, it helped having an IT person on the team.

The team felt ready to collect data. Since the IT system was not designed to have the start and stop time of all intermediate steps and this would be a big change in the IT system, they decided to create a data collection form to indicate start and stop times of each step. Since the workers in essence would be timed, the team decided to spend more effort in the area to explain what they would do with the data. They had meetings on all shifts to explain the project and the goals. The workers on all shifts were trained on the data collection form and any questions that they had were addressed.

After collecting the data, the team established a baseline on the cycle time. They found the average cycle time to be 20 days with a standard deviation of 6 days.

The management had set a goal, based on industry standards, of getting the cycle time to be 10 days or less. They decided to use Cpk as their metric. However, their y (cycle time) was non-normal. To calculate the capability metric, they had to transform their data first by taking the natural log to make it look normal. Fortunately, their software package helped them with this. Their Cpk value was -0.62. The negative number indicated that their average was beyond their upper requirement of 10 days. They could not calculate a Cp value as they only had a one-sided specification. Clearly, this process needed some help in the Analyze phase.

Summary

The team's primary goal in the Measure phase is to baseline the current state of each CTQ or CTP, and measure x variables that affect the CTQs and CTPs. In order to do that, the team must make sure the measurement system is providing valid data. The team must also operationally define the data to be collected, and develop a concrete plan for data collection, along with tools for manual data collection. Then the data can be collected, and the process baseline can be established.

Chapter

14

Analyze Phase

In the Analyze phase, the question to be answered is "What is wrong?" In other words, in this phase the team determines the root causes of the problems of the process. The team identified the problems of the process in the Define and Measure phases of the project.

The key deliverable from this phase is validated root causes. Anything else done in the Analyze phase is a means to get to this end. The team will use a variety of tools and statistical techniques to find these root causes. This chapter discusses some of those tools and techniques.

Tools for Identifying Root Causes

The team must choose the right tools to identify these root causes. The tools chosen are based on the type of data that were collected and what the team is trying to determine from the data. The team should use a combination of graphical and numerical tools. The graphical tools are important to understand the data characteristics and to ensure that the statistical analyses are meaningful (e.g., not influenced by outliers). The numerical (or statistical) analyses ensure that any differences identified in the graphs are truly significant and not just a function of natural variation.

There are a variety of graphs that can be used to better understand the data in addition to the Pareto chart presented earlier: histograms, boxplots, dotplots, scatter plots, run charts and multi-vari charts. Each graph has its own purpose.

Histogram

The histogram is a bar graph that shows the frequency of values. The values are grouped together; these groups are usually called bins. Histograms are useful to understand the location, spread and shape of the data. In addition, potential outliers or missing data can been seen. The heights of the bars represent the frequency or count and the widths of the bars represent a range of values.

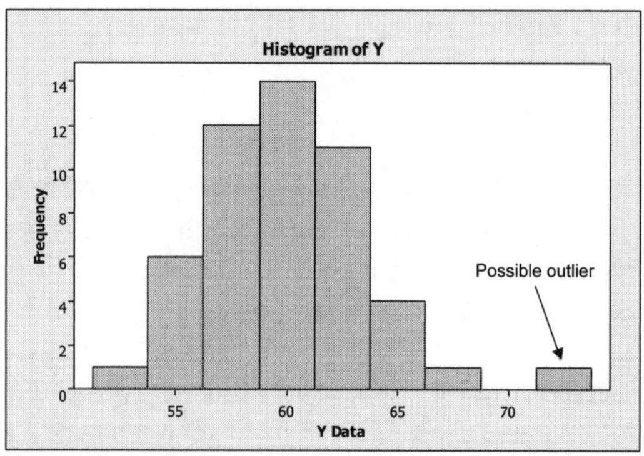

Figure 14-1 Example Histogram

In the figure above, the mean of the data can be estimated to be about 60, the minimum is about 50 and the maximum is about 72. The largest point on the graph looks like a possible outlier and should be investigated. Unusual data points should not be arbitrarily deleted. An outlier may provide some deeper insight into the process. However, if the data point represents a special situation that is not part of the population under consideration, then the data point can be excluded.

Boxplot

A boxplot, sometimes called a box-and-whisker plot, is an alternative to the histogram. It also shows location, spread, and shape of the data. Depending on the type of boxplot, it may also show outliers. Boxplots are useful for comparison of two or more groups

A boxplot consists of a box and two tails. The length of the box describes the middle 50% of the data - it is the interquartile range. The two tails extend out to the expected range of the measurements. Outliers, if they exist in the data, are indicated as individual points outside this expected range.

The figure below shows two boxplots comparing the cycle time of a process using two different methods or procedures. From the picture, it appears that method two takes longer than method one; the median for method two is higher than the median for method one. Method two has an outlier that should be investigated to ensure that it is a valid data point. The variability of the data appears to be about the same; the box lengths and the range of the data are about equal. The difference between the location (median or mean) of the two groups can be tested statistically.

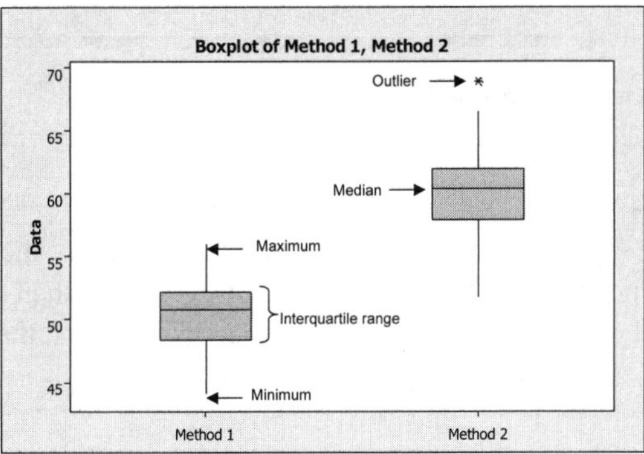

Figure 14-2 Example Boxplots

Histogram and Boxplot

The figure below shows a histogram and a boxplot on the same data. The data are skewed to one side. This is apparent in the histogram as well as the boxplot. The boxplot indicates skewness, since the median is not in the center of the box and the right tail is longer than the left tail. The labeled points at the end of the right tail are possible outliers that should be investigated.

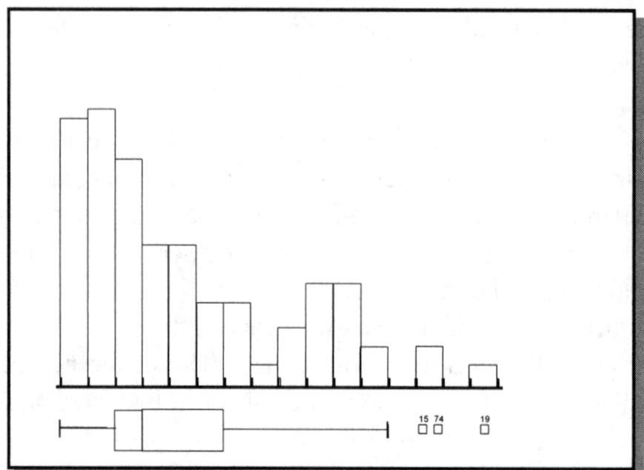

Figure 14-3 Example Histogram and Boxplot

Dotplot

A dotplot is a graphical display that shows the location, the spread, and each of the individual data points. Dotplots are used when there are relatively few data points. If there are too many data points, the plot will be too crowded to interpret easily.

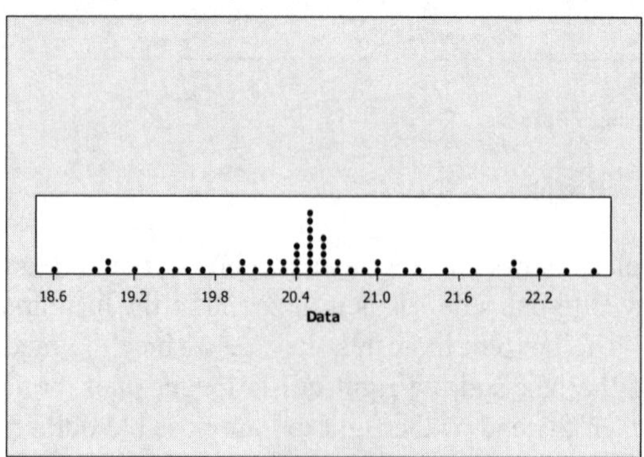

Figure 14-4 Example Dotplot

In the figure above, each dot indicates a unique data point. From this plot, one can estimate the mean to be approximately 20.5, the maximum to be about 22.6 and the minimum to be about 18.6.

Scatter Plot

A scatter plot is a plot of continuous x data versus continuous y data. Each point on the plot represents a pair of x and y data. The plot can be used to determine if there is a relationship between the x and y. However, the relationship does not necessarily constitute a cause and effect relationship. Both variables may increase together, leading one to hypothesize that x is causing y. There may be another variable influencing both. Or, it may be a spurious relationship - there really is not cause and effect, it just so happened that they both went up together. The figure below shows a positive relationship (as x goes up, y goes up) and a negative relationship (as x goes up, y goes down).

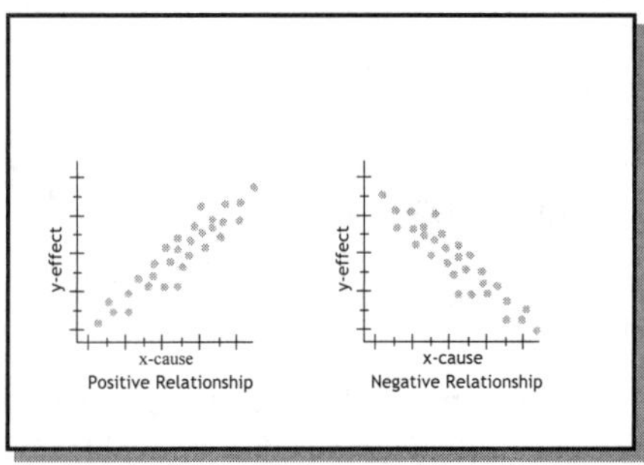

Figure 14-5 Example Scatter Plots

For example, the left graph in this figure may represent data from a software process where the time required to develop code (y) has a relationship with the number of requirements in the software requirements document (x). The right graph in the figure may represent data from a call center where call length (y) has a relationship with the experience level of the representative answering the phone (x).

Matrix Plot

Many software packages offer a matrix plot that allows the user to create many scatter diagrams at one time. Relationships between the X's may be just as interesting as the relationship between an X and a Y. Since the purpose of a project is to determine the Big X's to control, if two X's are related it may only be necessary to control one.

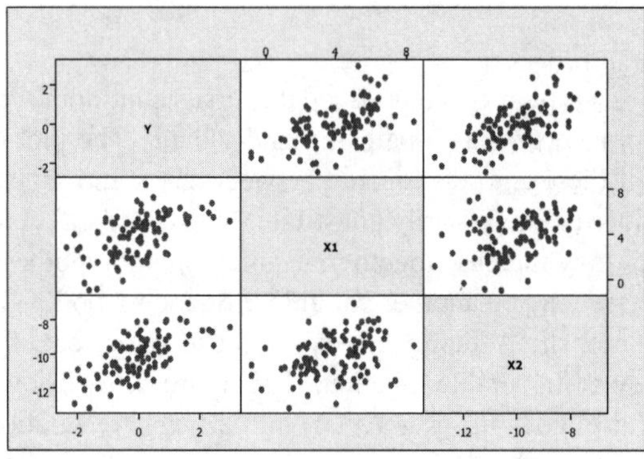

Figure 14-6 Example Matrix Plot

In the matrix plot in the figure above, there appears to be a relationship between x1 and y as well as x2 and y. However, there does not appear to be a relationship between x1 and x2.

Run Chart

Run charts, also known as trend charts or time series charts, are plots of the data over time. The vertical axis is the data value. The horizontal axis represents the time (e.g., days, weeks), lot or run number. The run chart helps determine shifts, drifts, patterns in the data over time or run order. A run chart is a powerful chart that can help solve problems that occur over time, lots or runs. In a run chart, the data are recorded in the order in which they occur. Run charts can tell a story that a histogram, or box-plot, cannot. In the figure below, the three run charts are all based on the histogram displayed above them.

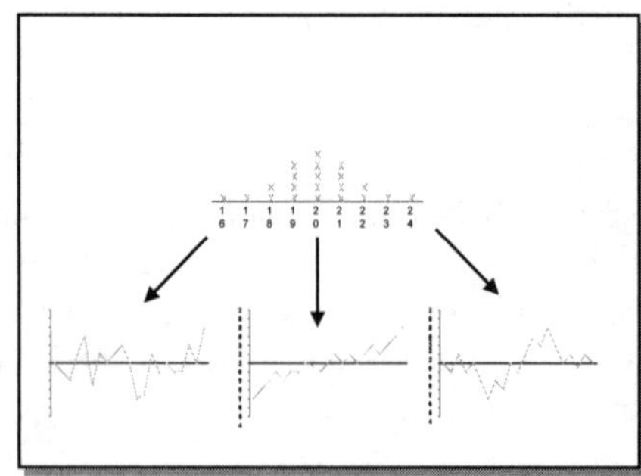

Figure 14-7 Example Run Charts Based on Histogram

Each run chart represents a different kind of process. The first run chart exhibits a stable process. The definition of a stable process is one where the location, spread, and shape of the data remain constant over time. The second run chart exhibits a process that has a change in the location of the process; there is a trend upward. The third run chart exhibits a process that appears to have a cyclical pattern to it. Although the histogram looks the same for each run chart, they each tell a different story. It is always a good idea to plot a run chart if the data were collected in time order.

Multi-Vari Chart

Another powerful graph is the multi-vari chart. This graph can be used to determine dominant sources of variation by plotting multiple x's and one y. This chart is used with discrete x's and a continuous y.

In the figure below, data were collected over three different time periods. In each time period, three parts were collected and nine readings were taken across each part. There is a dot representing the mean in each time period and a line connecting the three. Since this line is almost parallel to the X-axis, this is an indication that there is not much time-to-time variability. In each time period, there is a dot representing the mean of each part and a line connecting the three. Since these lines are not parallel to the X-axis, this is an indication that a large source of variation is part-to-part variability. For each part in each time period, there are nine dots indicating the nine readings on each part. Since the dots are relatively close together, this is an indication that there is not much within part variation. Numerical analysis can accompany this graph to help with the interpretation. This is discussed in the Sources of Variation section.

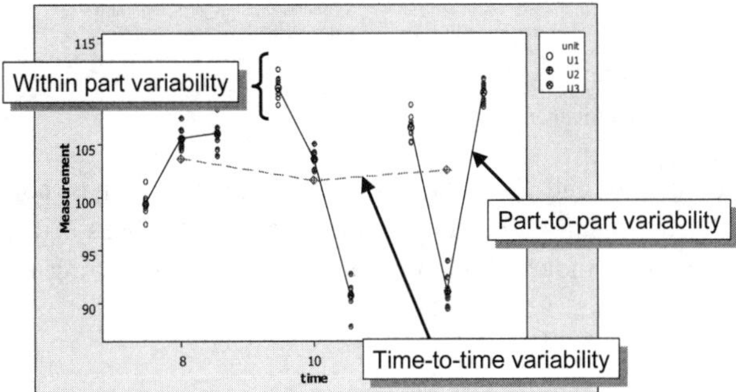

Figure 14-8 Example Multi-Vari Chart

Data Distribution

Why Normal Distribution Is Important

The data distribution can have many different shapes. The most common and useful distribution of the data is the normal distribution or sometimes called the Gaussian distribution after the developer Carl Gauss. Other commonly used distributions are Exponential, Chi-Square, F, Student's t, Poisson, and Binomial. However, there are many others. Distribution fitting techniques can help determine which one of these distributions should be used to describe a particular set of data.

Many statistical tools assume that the data come from a normal distribution. The normal distribution occurs quite frequently in the real world. For example, heights of all young women in the world are normally distributed. The key features of a normal distribution are that it has one peak, the two sides around the peak are symmetrical, the tails theoretically go from negative infinity to positive infinity, and the area under the curve equals one. The distribution is described by the formula in the figure below.

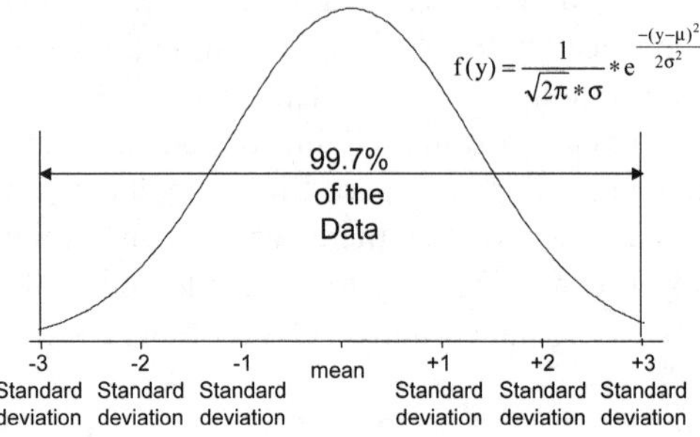

Figure 14-9 Normal Distribution

As can be seen in the formula, the normal distribution is completely described by the mean and standard deviation. Regardless of the values for the mean and standard deviation of the data, the following is true for a normal distribution:

Number of standard deviations on either side of the mean	Percent of data between these limits
1	68.26
2	95.46
3	99.73

Transforming Data

If the data are not normally distributed and it is desired to use a statistical tool that assumes normality, a common approach is to transform the data. The transformation will make the distribution look normal so that the tool can be used. Typical transformations are the log and the square root. These transformations are compressive in nature. Many times if the data are not normal, it is because there is a longer tail to the right, as shown in the figure below. Taking the log or the square root of the data will compress this tail and the data end up looking more normally distributed.

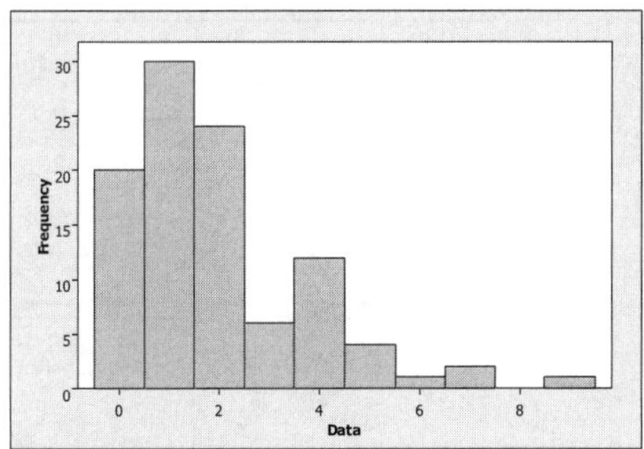

Figure 14-10 Example Skewed Histogram

If a transformation is required, the analysts must remember that they are working in the transformed world and transform everything to this scale. For instance, if the analysts calculated C_{pk} based on transformed data, the specifications for the process must be transformed as well. Many software packages provide tools, such as the Box-Cox transformation tool, to help find a proper transformation for the data. Another option for non-normal data is to conduct a non-parametric statistical analysis. A non-parametric analysis does not assume normally distributed data.

Tests for Normal Distribution

There are many tests to determine if the data are normally distributed. Two common tests are the Anderson-Darling test for normality and a normal probability plot. The Anderson-Darling test is a numerical test that results in a probability value. It tests the assumption that the data are normally distributed. If the probability value is less than 5%, the user may determine that the data are not normally distributed.

A normal probability plot is a graph of the ordered data using a log scale on the Y-axis. If the plotted data fall in a straight line, then it may be assumed that the data are normally distributed. Statistical software is typically used for these analyses. The figure below shows output from MINITAB® statistical software, made by the Minitab company, that includes both the Anderson-Darling test and a normal probability plot. The p-value (probability value) is greater than 0.05, and the data fall along a relatively straight line so the data are assumed to come from a normal distribution.

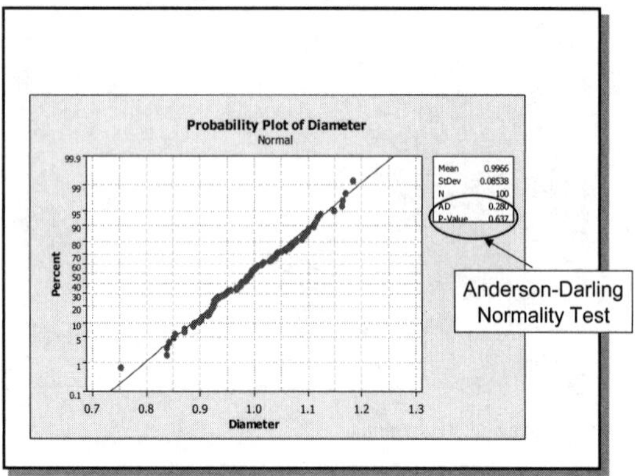

Figure 14-11 Example of Normality Test

The Central Limit Theorem

An important theorem used in statistical analysis is the Central Limit Theorem. The Central Limit Theorem states that if a population distribution is repeatedly sampled, using the same sample size and each time a sample mean is calculated, a distribution of sample means will be generated. The theorem states that this distribution of sample means will have the same mean as the original distribution, the variability will be smaller than the original distribution, and it will tend to be normally distributed. The standard deviation of the sampling distribution will be smaller than the standard deviation of the original distribution by a factor of $\sigma/\sqrt{n}$. As n gets larger, the distribution of sample means will become more normal. The Central Limit Theorem is important because many times the parameter of interest is the mean and the assumption of normality of a statistical tool may be met through the theorem. The figure below illustrates the concept of the Central Limit Theorem.

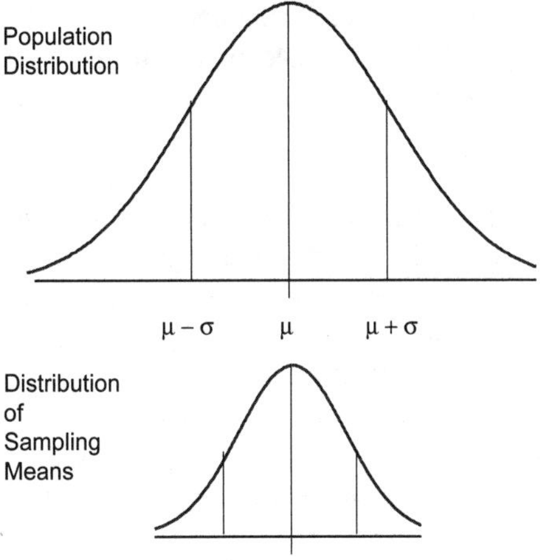

Figure 14-12 Central Limit Theorem

Variation

Variation is an extremely important concept in a Six Sigma project. Many companies report averages and make decisions on these values without the understanding of the associated variation. And, variation costs money. If the process always produced the same results, think how predictable and comfortable life would be. However, this never is the case. As variation increases, the process becomes less reliable. Inspection, rework, and scrap become a way of life. Material Review Boards take on a life of their own. Defects escape to the customer and time is required to investigate problems and create corrective action reports. Or, customer service is constantly handling customer complaints.

Although variation can never be completed removed, it can be reduced. The goal of Six Sigma and the team is to identify the sources of variation and eliminate these sources.

5 Why's

5 Why's is a tool that is often used in conjunction with a Cause and Effect Diagram. The idea of 5 Why's is to take potential causes of a problem and to dig deeper by asking "Why?" up to five times. The goal is get to something that is actionable.

Often, potential causes suggested by the team are really symptoms and not root causes. The figure below shows an example of using the 5 Why's in a hospital scenario. The project is focused on reducing cycle times for the lab results in an emergency room. One suggested problem was that the request was not being submitted immediately after the doctor orders the lab tests. The team dug deeper and found that the ward secretary wasn't submitting the order immediately because she was batching up orders to submit all at once. This was something against which they could take action.

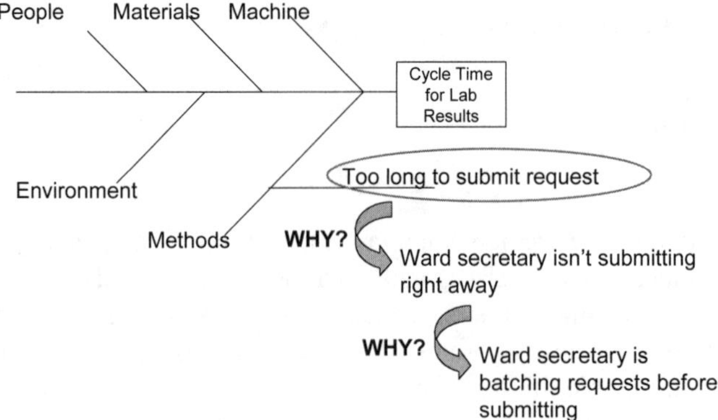

Figure 14-13 Example of 5 Why's

Any potential root causes identified using Cause and Effect Diagrams and 5 Why's should be validated through data. The root causes should be proven to affect the response using tools like correlation, regression, hypothesis testing, and designed experiments that are discussed in the following sections.

Comparative Methods

A commonly used set of techniques is comparative methods. The comparative methods discussed here are comprised of hypothesis testing and confidence intervals.

Hypothesis Testing

Hypothesis testing is a statistical analysis where a hypothesis is stated, sample data are collected, and a decision is made based on the sample data and related probability value. This testing can be used to detect differences such as differences between a process mean and the target for the process, differences between two suppliers, or differences between multiple employees.

To conduct a hypothesis test, the first step is to state the business question involving a comparison. For instance, the team may wonder if there is a difference in variability seen in thickness due to two different material types. Once the business question is posed, the next step is to convert the business language or question into statistical language or hypothesis statements. Two hypothesis statements are written. The first statement is the null hypothesis, H0. This is a statement of what is to be disproved. The second statement is the alternative hypothesis, Ha. This is a statement of what is to be proved. Between the two statements, 100% of all possibilities are covered. The hypothesis will be focused on a parameter of the population such as the mean, standard deviation, variance, proportion, or median. The figure below shows an example of a null hypothesis and alternative hypothesis.

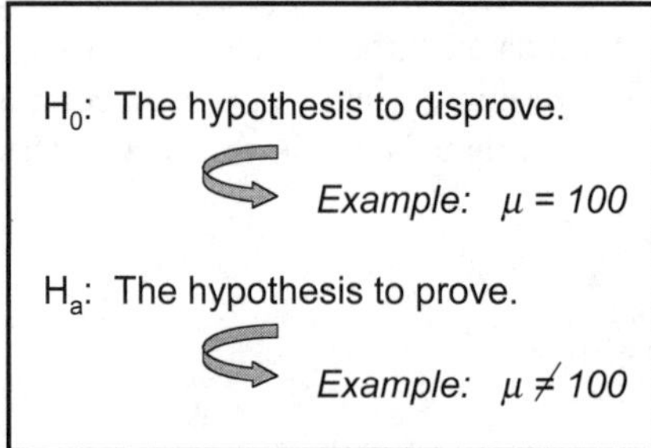

H_0: The hypothesis to disprove.
Example: $\mu = 100$
H_a: The hypothesis to prove.
Example: $\mu \neq 100$

Figure 14-14 Example of Hypothesis Testing

Assuming the null hypothesis is true, a probability value is determined for how likely it would be to get data like the sample value observed. If this probability value (p-value) is low enough, the risk of deciding the null hypothesis is false when it really is true will be low enough. The null hypothesis will be rejected and therefore, the alternative hypothesis (the hypothesis to be proved) will be true. If the null hypothesis is not true, the alternative hypothesis must be true since the two cover all possibilities.

The type of hypothesis test that could be conducted is based on the data type (discrete or continuous) of the y data. For instance, if the data are continuous, the analysts may want to conduct tests on the mean, median, or variance. If the data are discrete, the analysts may want to conduct a test on proportions.

There are two risks that are associated with hypothesis testing: the alpha risk and the beta risk. The alpha risk is the probability of rejecting the null hypothesis when it is true. The beta risk is the probability of failing to reject the null hypothesis when it is false.

The analyst sets the alpha risk desired. A typical alpha risk is 5%. Therefore, if the calculated p-value (probability value) is less than this set alpha risk, the analysis will reject the null hypothesis.

The beta risk is a function of sample size. One minus beta is called the power. Power is the ability of the test to detect a difference if one exists. An analyst must collect enough data to ensure that the test is powerful enough. A power value closer to one is the goal.

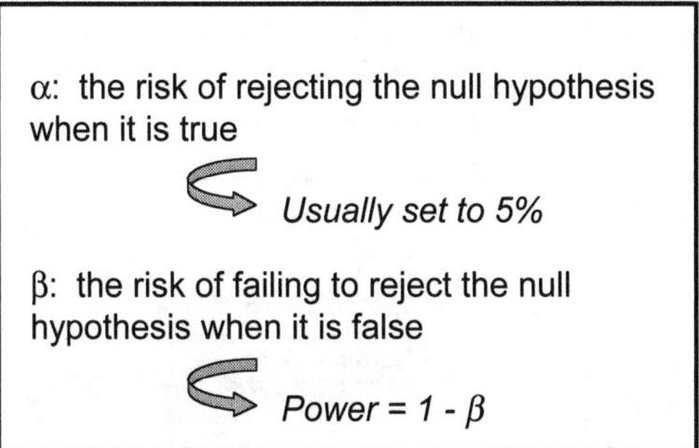

Figure 14-15 Risks Associated with Hypothesis Testing

Suppose that a team was wondering if there was difference in the material they were receiving from two different suppliers. The team gathered data from both suppliers. Their Y was the length of the material. The null hypothesis that they were testing is that the means for length of two suppliers are the same. In other words:

$$H_0 : \mu_{\text{Supplier A}} - \mu_{\text{Supplier B}} = 0$$

The p-value from the test was .04. This value was less than the alpha risk that the team was willing to take (.05) so the null hypothesis was rejected. The team concluded that there was a difference between the population averages of the two suppliers. The team might then pick the better supplier or determine if they could improve the other supplier. This analysis is shown in the figure below.

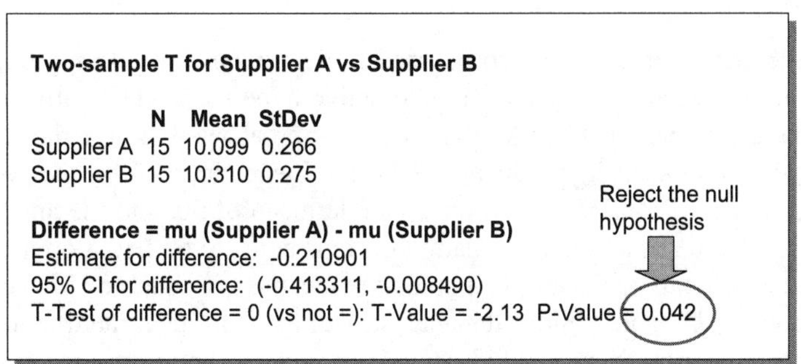
Two-sample T for Supplier A vs Supplier B

	N	Mean	StDev
Supplier A	15	10.099	0.266
Supplier B	15	10.310	0.275

Difference = mu (Supplier A) - mu (Supplier B)
Estimate for difference: -0.210901
95% CI for difference: (-0.413311, -0.008490)
T-Test of difference = 0 (vs not =): T-Value = -2.13 P-Value = 0.042

Figure 14-16 Example of Hypothesis Test

The boxplots of the sample data support the analysis. In the boxplots shown in the figure below, the two sample means are indicated by two circles. The circles are connected by a line. The larger the slope of the line, the larger the difference in the means.

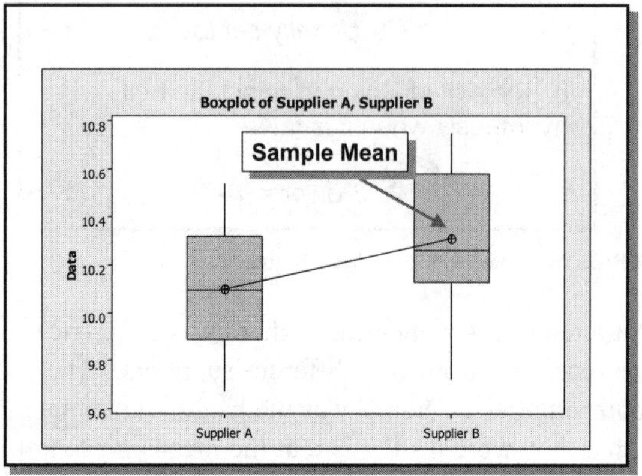

Figure 14-17 Boxplots Associated with Hypothesis Testing

Confidence Intervals

Confidence intervals are an alternative approach to hypothesis testing with p-values. Confidence intervals are upper and lower bounds about an estimate within which we are a certain percent "confident" that the true value is located. The methodology guarantees that if samples of size n were repeatedly drawn from the same population, 100(1 - a)% of the confidence intervals constructed would contain the true parameter.

Confidence intervals can also be used to determine if a null hypothesis should be rejected in favor of an alternative hypothesis. They are based on the same math as the p-values so both approaches will give the same answer. If an analysis produces a 95% confidence interval for the population mean of (45, 55) based on a sample mean of 50, and the analysts would like to know if the population mean is equal to 60 (H0: population mean is equal to 60 versus Ha: population mean is not equal to 60), a conclusion can be made. Since the analysts can feel confident the true mean is in the interval, they would conclude (based on the sample mean and standard deviation) that the population mean of 60 is not in the interval. Therefore, the null hypothesis would be rejected in favor of the alternative hypothesis (Ha: population mean is not equal to 60).

This is illustrated in the figure below.

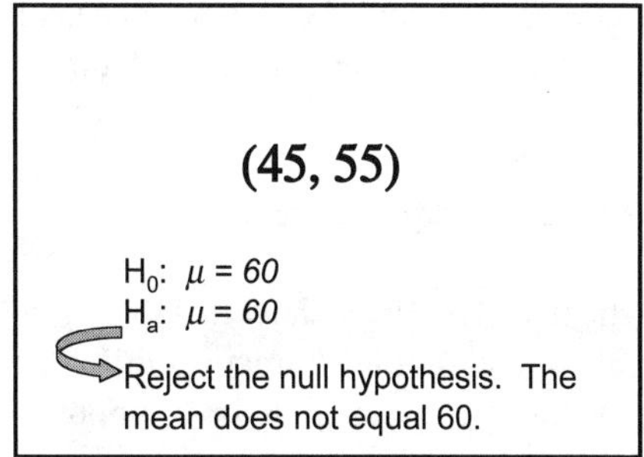

(45, 55)
H_0: μ = 60
H_a: μ = 60
Reject the null hypothesis. The mean does not equal 60.

Figure 14-18 Confidence Interval

In the hypothesis example on the two suppliers presented above, the MINITAB output also contained a confidence interval. Since the null hypothesis was that the means are equal, this can also be written as:

$$H_0 : \mu_{\text{Supplier A}} - \mu_{\text{Supplier B}} = 0$$

In other words, the difference between the two population means is zero. The confidence interval on the difference between the two population means is (-0.413311, -0.008490). Since this interval does not include zero, the team was confident that there was a difference between the two suppliers. This analysis is shown in the figure below.

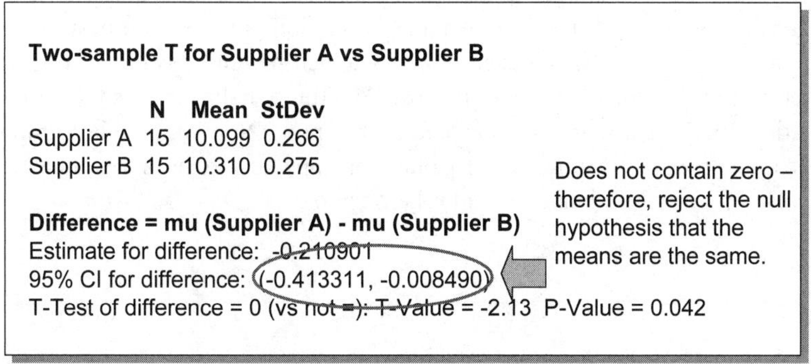

Two-sample T for Supplier A vs Supplier B

	N	Mean	StDev
Supplier A	15	10.099	0.266
Supplier B	15	10.310	0.275

Difference = mu (Supplier A) - mu (Supplier B)
Estimate for difference: -0.210901
95% CI for difference: (-0.413311, -0.008490)
T-Test of difference = 0 (vs not =): T-Value = -2.13 P-Value = 0.042

Does not contain zero – therefore, reject the null hypothesis that the means are the same.

Figure 14-19 Example of Confidence Interval on the Difference Between Two Means

Confidence intervals are important because they give an estimate of the variability of a reported sample statistic. People often treat the sample mean as the true population mean when in fact it is an estimate. And, the estimate has variability. The confidence interval calculations are based on the variability associated with the parameter estimated, the sample size and the alpha risk.

Statistical Versus Practical Significance

Although a statistical analysis may show that there is a statistically significant difference, there may not be a practical difference. In other words, the difference may not be big enough to be of importance to the business. The bigger the sample size used in an analysis, the smaller the deviation from the null hypothesis that may be detected. Always remember to check for outliers that may be influencing results. And finally, ask whether this difference means anything practically.

Correlation

Correlation is tool that can be used with a continuous x and a continuous y. The Pearson correlation coefficient is a statistic that measures the linear relationship between the x and y. The symbol used is r. The correlation value ranges from -1 to 1. The closer the value is to 1 in magnitude, the stronger the relationship between the two. A value of zero, or close to zero, indicates no relationship between the x and y. A positive value indicates that as x goes up, y goes up. A negative value indicates that as x goes up, y goes down. The Pearson correlation is a measure of a linear relationship, so scatter plots are used to depict the relationship. The scatter plot may show other relationships. The figure below shows the correlation coefficient with an associated scatter plot. Note that the last graph has a correlation value of zero but there appears to be a relationship between x and y, albeit a curved relationship.

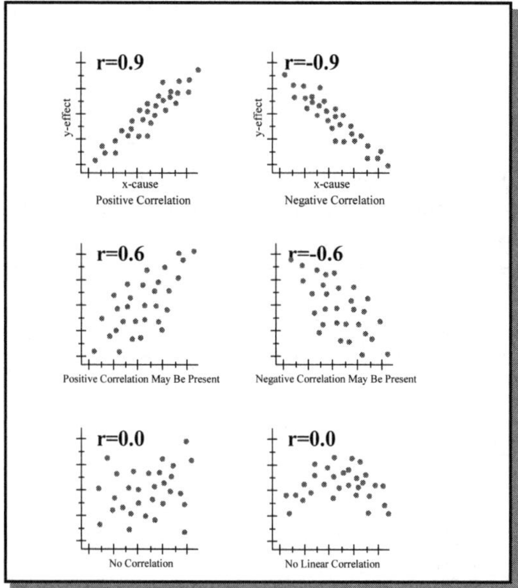

Figure 14-20 Example Correlation Coefficients with Associated Scatter Plots

Regression

A natural extension of correlation is regression. Regression is the technique of fitting a mathematical equation that relates the x's to the y. Regression is often used when Design of Experiments cannot be done because the x factor(s) cannot be controlled. Regression analysis is also used with historical data - data where the business already collects the y and associated x's. The regression equation can be used as a prediction model for making process improvements.

Simple Linear Regression

Simple linear regression is a statistical method used to fit a line for one x and one y. The formula of the line is $Y = \beta_0 + \beta_1 X$ where β_1 is the intercept term and β_1 is the slope associated with x. These beta terms are called the coefficients of the model. The regression model describes a response variable y as a function of an input factor x. The larger the b1 term, the more change in y given a change in x. Simple linear regression is highly beneficial in understanding the relationship between two selected variables.

The figure below shows the output from a simple linear regression analysis in MINITAB. The software fits a line of y=3.881 +1.241 x where 3.881 is the y intercept and 1.241 is the slope associated with the x. As evidenced by the graph, in this case the line doesn't do a great job of predicting y from x.

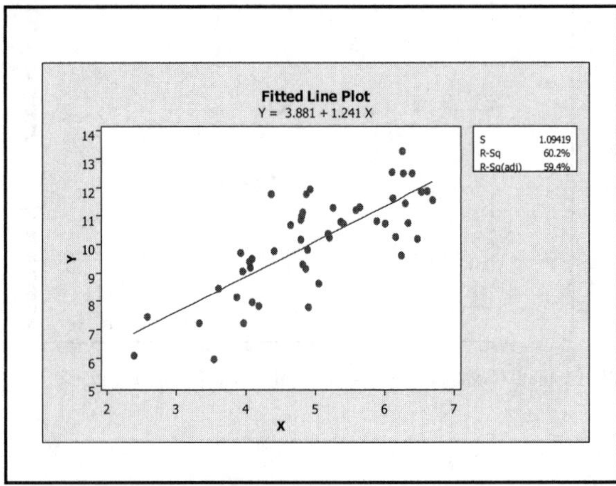

Figure 14-21 Example Output for Simple Linear Regression

Multiple Linear Regression

Multiple linear regression is a natural progression from simple linear regression. Rarely will using just one x do a good job of explaining the variability in the y. So, multiple linear regression is fitting an equation for a y with multiple x's in the form of $Y = \beta_0 + \beta_1 X_1 + \beta_2 X_2 + \beta_3 X_3 + ...$ etc.

Suppose that a team is studying the effect of various x's on the response of temperature in a chemical process. The x's are argon, oxygen, nitrogen, ion rate and the reaction time. All of the x's are continuous. The figure below shows the results from the analysis.

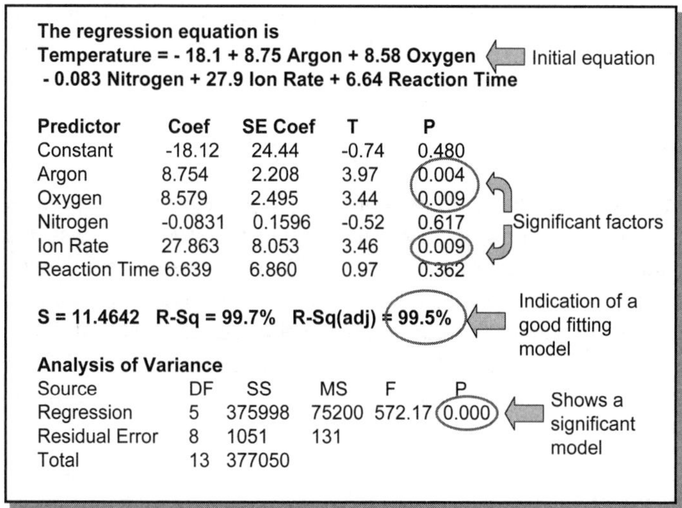

The regression equation is
Temperature = - 18.1 + 8.75 Argon + 8.58 Oxygen - 0.083 Nitrogen + 27.9 Ion Rate + 6.64 Reaction Time

Predictor	Coef	SE Coef	T	P
Constant	-18.12	24.44	-0.74	0.480
Argon	8.754	2.208	3.97	0.004
Oxygen	8.579	2.495	3.44	0.009
Nitrogen	-0.0831	0.1596	-0.52	0.617
Ion Rate	27.863	8.053	3.46	0.009
Reaction Time	6.639	6.860	0.97	0.362

S = 11.4642 R-Sq = 99.7% R-Sq(adj) = 99.5%

Analysis of Variance

Source	DF	SS	MS	F	P
Regression	5	375998	75200	572.17	0.000
Residual Error	8	1051	131		
Total	13	377050			

Figure 14-22 Example of Multiple Linear Regression Analysis

Listed first in the output is the fitted equation. In the ANOVA (Analysis of Variance) table, the null hypothesis is that the model is not significant. In other words, the null hypothesis is the coefficients in the model for each x are zero. Since the p-value is zero, the null hypothesis is rejected and the team concluded that the at least one x coefficient is not zero - therefore, indicating that at least one x is explaining some of the variability in the y. The output also exhibits the adjusted R-square value of 99.5%. The adjusted R-square value is an indicator of how much variability in the y the model is explaining. The closer this value is to 100%, the better fit of the model. Using the p-values for each predictor indicates that Argon, Oxygen, and Ion Rate are all important in explaining the y. The null hypothesis for each line in the "Predictor" table is that the b associated with that x variable is equal to zero. In other words, the null hypothesis is that the x is not significant in the model. After looking at this output, the team would then remove insignificant terms (nitrogen and reaction time) from the model and refit it. Of course, they would also check assumptions made when fitting this model.

Although the formula given above is the most commonly used form, there are methods other than linear regression. The "linear" refers to the parameters, the values of the betas, so polynomial models such as $Y = \beta_0 + \beta_1 X_1 + \beta_2 X_2 + \beta_{11} X_1^2 + \beta_{22} X_2^2$ are still considered linear models.

Non-Linear Regression

Examples of non-linear regression include growth models, such as predicting crop outcomes based on environmental conditions. In such cases, the model is probably very complex. Another example of non-linear regression is for a response (y) that is binary; it can only take on the values of zero or one. An example is whether a drug worked or it did not. In this case, the variance of the y is a function of the mean. In this particular case, the analyst would use logistic regression to solve this problem.

Sources of Variation

Sources of variation (SOV) is an analysis technique that identifies the dominant source(s) of variation. It includes graphical and numerical techniques. A multi-vari chart, as described previously, helps to see graphically the sources of variation. For each source of variation, a variance component can be calculated with the associated percentage of the total variation.

The steps for conducting a sources of variation study are:

1. Identify the major families of variation to be considered in the study. A traditional study includes time-to-time variability, part-to-part variability, and within part variability. If the sources of variation study is being done on a measurement system, common sources of variation considered are reproducibility and repeatability.
2. Design the sampling plan for the x data using a tree diagram. A tree diagram is a drawing that shows the structure of the sampling plan.
   a. Determine if each level of the tree is crossed or nested with the level above it. Two factors, A and B, are crossed if the same levels of A appear with each level of B, and vice versa. The upper diagram in the figure below shows a crossed tree diagram for two factors. At the bottom of the tree are repeated reading

for the combination of factors above it.

B is said to be nested within A if the levels of B are different for each level of A. This should be shown on the tree diagram. Typically in a SOV study, the factors are nested. The lower diagram in the figure below shows a nested tree diagram for two factors. At the bottom of the tree are repeated reading for the combination of factors above it.

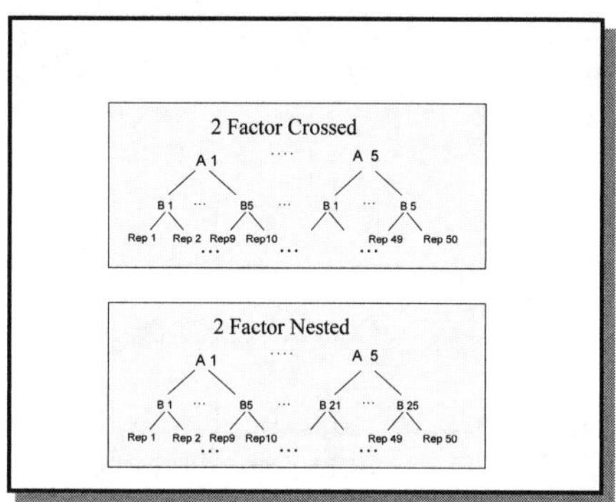

Figure 14-23 Crossed versus Nested Tree Diagrams

b. Determine if the factors are random or fixed. A random factor is one where the levels are a random sample of levels from a population of possible levels. Usually, with a random factor the analyst is trying to answer the question "How much does this factor contribute to overall variability?" A fixed factor is one where the levels are selected by a non-random process or if the values consist of the entire population. Usually, with a fixed factor the analyst is trying to answer the question "Which is better or is there a difference?" Typically in a SOV study, the factors are random. Variance components will only result with random factors.

3. Collect the data. Review the data to make sure there are no outliers that may influence the results.
4. Quantify the sources of variation by looking at a multi-vari chart and computing variance components.
5. Draw conclusions and make recommendations.

Suppose that a team is focusing on improving a drilling area. The area has many operators and many drills and many drill bits. The team collected data over four days, using different operators each day, with two drills selected at random and two random drill bits. The tree diagram might look like the one in the figure below.

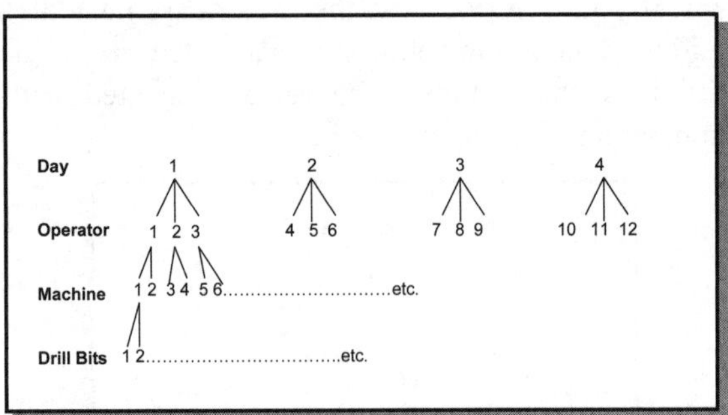

Figure 14-24 Example Tree Diagram for Sources of Variation Study

After analyzing the data in MINITAB, they had output like the table below. The results show that drill bits are the largest contributor to overall variability - drill bits contribute 70% to the total variability. Operator-to-operator variability contributed 20%.

Variance Components

Source	Var Comp.	% of Total	StDev
Day	0.026	2.0	0.163
Operator	0.264	20.0	0.514
Machine	0.106	8.0	0.325
Drill Bits	0.9247	70.0	0.962
Total	1.321		1.149

Largest Source of Variation

Figure 14-25 Example of Sources of Variation Study

The team took the results from the study and plotted them in the Pareto chart below. Their next step would be to identify why drill bits were the largest source of variation.

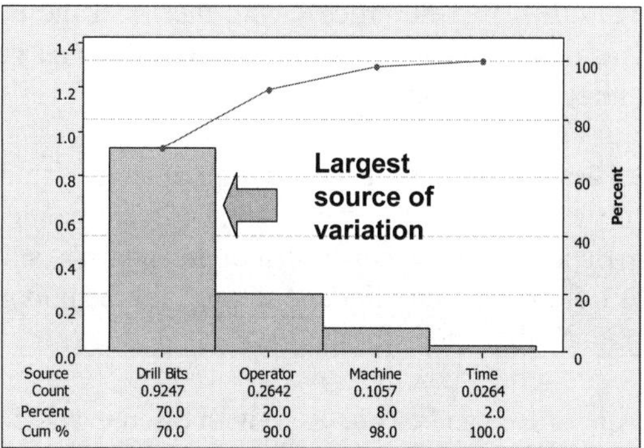

Figure 14-26 Example Pareto of Sources of Variation

Sources of variation is a tool that can be used in a variety of situations (manufacturing, service, and transaction) to determine the largest source of variation. After the largest source is determined, the team must then dig deeper to determine what might be causing that factor to be driving the variation in the y. A Cause and Effect Diagram can be a helpful tool for this.

Design of Experiments

One of the most powerful tools for a Black Belt is Statistical Design of Experiments. Design of Experiments (DOE) is used when the x factors can be changed to understand the impact that each factor has on the response, the y. DOE can be used to determine the important factors, to compare a variety of options, or to find the optimal setting for the important factors. Depending on the objective of the experiment, the experimental design will change. The experimental design is the list of all experimental runs. An experimental run is a defined combination of x levels to be conducted. Suppose there are three x's in an experiment (temperature, pressure and supplier) and a y, thickness. A run might look like:

Temperature	Pressure	Supplier
100 degrees	50 psi	Joe's Supplies

The team would then run the process with the x's at these settings and measure the resulting thickness. An experimental design is all the runs defined that the team will conduct.

It may be necessary to use a sequence of experiments to find the optimal values for the y's. For example, the first experiment would be to determine the important x's (narrow down a list of potential x's) and then another experiment would be used to find the settings for the important or valuable x's to optimize the values for the y's.

A designed experiment can be conducted with discrete and continuous x's and with continuous y's and, to a certain extent, discrete y's.

Two-Level Factorial Experiment

One of the most common and useful forms of an experiment is a two-level factorial experiment, known as a 2k design. Each x factor is varied over two levels. For a continuous factor, these are low and high values. For instance, temperature may be run at a low level (e.g. 100 degrees) and a high level (e.g. 200 degrees). For a discrete factor, it may be conditions of interest, such as Supplier A and Supplier B, where the levels are arbitrarily assigned a low and high level. The minimum number of runs in this kind of design, known as a full factorial, is 2k where k is the number of factors that are studied in the experiment. All possible combinations of the two levels of all factors are run. An example of a full factorial with three x's is shown in the table below.

X1	X2	X3
Low	Low	Low
High	Low	Low
Low	High	Low
High	High	Low
Low	Low	High
High	Low	High
Low	High	High
High	High	High

Figure 14-27 Example Experiment with Three x's

Fractional Factorial Experiment

If more than four x factors are considered in a design, a fractional factorial design is usually run. A fractional factorial design means that only a subset of the full factorial will be run. These designs are known as 2k-p where the p indicates the fraction run. For instance, if p is one, half of the full factorial is run. If p is two, a quarter of the full factorial is run, etc. The reason for using a fraction of the design is to minimize the runs made. The fractional factorial experiments can be just as informative as the full factorial experiments while greatly reducing the number of runs and the number of resources used. Software packages, such as MINITAB, will generate the appropriate fraction of runs to be made.

Designed Experiment Outcome

The outcome of a designed experiment is an understanding of the factors that are important to the y, or response. A hypothesis test is used for each factor to determine if that factor has a significant effect on the response. In addition, a mathematical model (based on the hypothesis testing) can be created and this model can be used to determine where the x's should be set to get the best results for y.

Suppose that a team wanted to gain a better understanding of their advertising strategy. They decided to study four factors:

1. A promotional idea with the levels being either they didn't do the promotion (low level) or they did do the promotion (high level)
2. Another promotional idea with the levels being either they didn't do the promotion (low level) or they did do the promotion (high level)
3. Two different product types with Product Type A assigned the low level and Product Type B assigned the high level.
4. Two different geographical regions in which they were considering offering the promotions with east assigned the low level and west assigned the high level.

They ran the experiment at several outlets where the products were sold. The experiment was run over a month and sales dollars were used as the response. Using p-values to determine the significant factors, they found that both promotional ideas seemed to have a positive effect on sales. In addition, an interaction between these two factors was found to be significant. Product type and geographical location were not found to be significant factors. The interaction had a negative effect on the sales. The team interpreted this to mean that using both promotions would not get an increase in sales equal to the addition of each promotion individually. Also, an adjusted R square value of 95.17% showed that the model fit well. The team used this information in the experiment to determine the best profit strategy.

Estimated Effects and Coefficients for Sales (coded units)

Term	Effect	Coef	SE Coef	T	P
Constant		810.2	384.5	46.32	0.000
Product Type	-1313.6	-656.8	429.9	-1.53	0.161
Promotion 1	11731.0	5865.5	429.9	13.64	0.000
Promotion 2	11145.7	5572.8	429.9	12.96	0.000
Location	1266.7	633.4	429.9	1.47	0.175
Product Type*Promotion 1	402.1	201.1	429.9	0.47	0.651
Product Type*Promotion 2	-1711.7	-855.9	429.9	-1.99	0.078
Production Type*Location	154.0	77.0	429.9	0.18	0.862
Promotion 1*Promotion 2	-3890.3	-1945.2	429.9	-4.52	0.001
Promotion 1*Location	-781.0	-390.5	429.9	-0.91	0.387
Promotion 2*Location	7.8	3.9	429.9	0.01	0.993

Significant factors

S = 1719.56 R-Sq = 97.71% R-Sq(adj) = 95.17%

Indication of a good fitting model

Figure 14-28 Example Output from a Designed Experiment

Experimental design techniques are extremely useful in a variety of situations to understand the cause and effect relationship between a set of x's and y's.

Response Surface Methodology

An advanced experimental technique is called Response Surface Methodology (RSM). This technique is used specifically with continuous x's and y's to find the optimal settings for each. In this experiment, a contour surface plot is generated based on a fitted equation. From this plot and the associated mathematical model, the best settings for the x's can be chosen to optimize the y.

In a RSM design, the x factors are run typically at 3 to 5 levels so that this surface can be better plotted and understood. RSM designs can be used to either to find the maximum value for the y's (e.g. maximize profit) or the minimum value for the y's (e.g. minimize cost). This would depend on the definition of the optimal y.

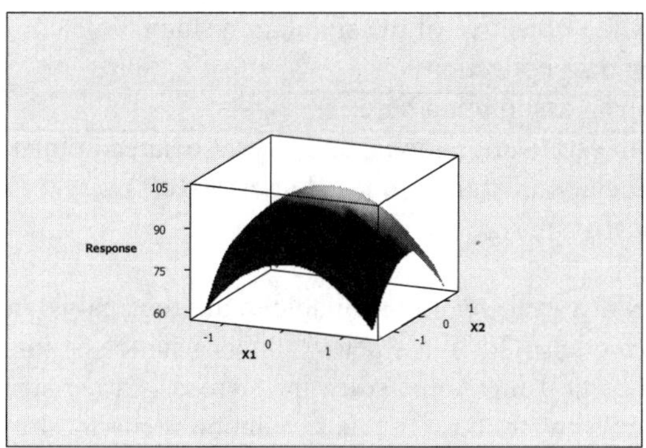

Figure 14-29 Example Response Plot

The figure above shows a contour plot based on the results of a response surface experiment with two x's. To interpret this graph, find the peak (to maximize the y's) or the valley (to minimize the y's) and then find the x's at this desired Y. This graph shows the peak in the center and this would be the optimal setting for maximizing y. Therefore, the optimal x values are somewhere toward the middle values. RSM is usually the last stage in experimenting, done after a full or fractional factorial.

Process Simulation

Process simulation is the method of creating a computer model of a process. The model allows the user to make changes to the simulated process to mimic what would happen if changes were made to the real process. Simulation is often accompanied by an animation of the process as well. The simulation allows the user to collect statistics about any output of interest like cycle time, utilization levels, defects generated, etc. The animation allows the user to visualize bottlenecks, work-in-process, inventory levels, etc. The animation also helps others to understand the proposed changes. Simulation is a powerful tool for complicated processes with a lot of variations or multiple paths. With a developed model, the analyst can ask "what if" to consider alternative ways to run a process.

The steps for conducting a process simulation are:

1. Identify the objective of the simulation study.
2. Perform data collection.
3. Develop an assumption base.
4. Create the simulation model and any associated animation.
5. Run simulations of the proposed alternative(s).
6. Statistically analyze the output of interest.

Simulation software is readily available to aid the analyst in the development of the model. The figure below shows a static version of a 3-D animation of a bank using Arena software. Arena is a registered trademark of Rockwell Software, Inc. In this simulation scenario, the analyst could ask questions such as "What would happen if we had one less teller?" or "What if the number of walk-in customers increased by 25%?" The software could show a dynamic view of what would happen in the process as well as provide statistics on such things as teller utilization, customer wait times, etc. A designed experiment is often used to study the proposed alternatives.

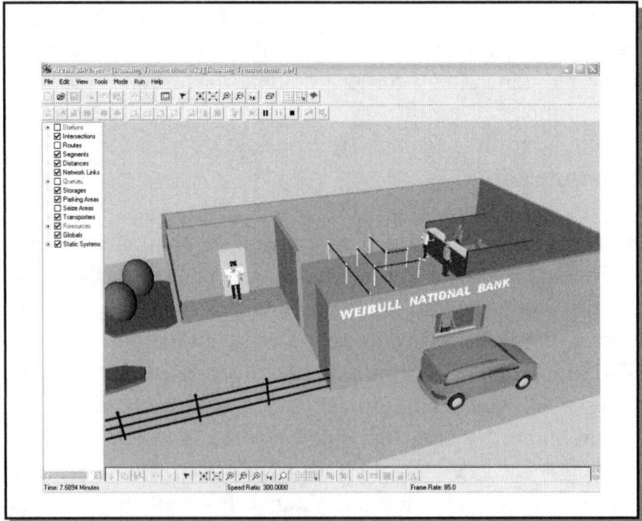

Figure 14-30 Example from Simulation Software

Nonparametric Analyses

Although there is not really an agreed upon definition of what is a "non-parametric" analysis in the statistics world, in this book it is used to encompasses those statistical analyses that do not require an assumption of normality of the random variable. A nonparametric test makes fewer and simpler assumptions about the distribution of the data. Nonparametric tests do not rely on the estimation of parameters such as the mean or the standard deviation. Most nonparametric tests are based on ranks (or the order) and not the actual data values. One advantage in using ranks instead of the actual values is that the hypothesis test is much less sensitive to the effect of outliers. Conversely, if the data are normal and free of outliers, nonparametric tests are generally less powerful than normal-based tests to detect a real difference when one exists.

As an example, a commonly used nonparametric test is the Spearman rank correlation coefficient. The correlation value is interpreted as described in the correlation section; however, the correlation is calculated using the ranks of the data instead of the actual data values. There are several other useful no-parametric tests, such as tests on the medians instead of the means. These tests are used for testing the statistical significance of an x value, for comparing two different x's, or for multiple comparisons.

Nonparametric tests should be used when all of the following conditions are true:

- Data are non-normal.
- Data cannot be readily transformed to normality.
- Sample size is small, say less than 30. If the sample sizes are large, the Central Limit Theorem says that parametric tests are robust to non-normality.
- Data are independent. This assumption also holds for parametric tests.

Statistical software can be used to perform these types of tests.

Time Series Analysis

If data are taken sequentially over time and that data exhibit the property of dependence among adjacent points, understanding the nature of the dependence falls in the world of time series analysis. As with regression and DOE, the result of the time series analysis is a model that represents the process. Time series analysis is a common technique in inventory control, economic data, forecasting, etc.

Time series techniques are generally used to understand and model a system well is usually to understand and model a system well enough to either predict or control the outputs based on inputs. Time series is an important tool that can be used as an alternative to statistical techniques that assume independence. If the data truly have dependence amongst them, it behooves the analyst to understand this data structure. The figure below exhibits data that show dependent data - there is an obvious pattern. Independent data should not exhibit any particular pattern.

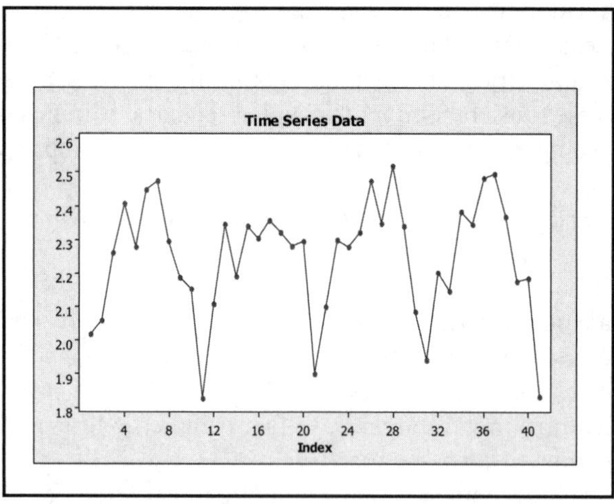

Figure 14-31 Example of Data with Time Dependence

For example, if a project involves sales data with a seasonal component, time series analysis may be an appropriate analysis tool. Any data collected in time order should be tested and understood to see if there is correlation that requires a time series approach. Statistical software can be used to perform these analyses.

Analyze Phase Tollgate Questions

As the team progresses through the Analyze phase, the team leader should schedule meetings with the Champion and Master Black Belt to review all that has been accomplished.

The following is a list of questions which the Sponsor/Champion and the Master Black Belt can use to prompt discussion in these meetings. Also, the team can use these questions as they progress through the Analyze phase to be sure that they have completed all the important items.

1. Have then been any revisions to the charter? Has the scope changed?
2. What was the approach to analyzing the data? Why were these tools chosen? What worked well/did not work well about these tools?
3. What are the root causes of the problems? How were these conclusions drawn?
4. How did the team analyze the data to identify the factors that account for variation in the process?
5. Once completing the analysis, what is the opportunity represented by addressing the problem? What is the impact on customer satis faction, retention, and loyalty?

Case Study One

After collecting three weeks of data, Team Scrap was ready for some analysis. They had completed a capability study on the drilled hole diameters of the bases. They had discovered that drilled holes that were out of specification were the primary reason for scrapping the bases. Therefore, this became the focus of their project. The customer required the diameter to be one inch with an allowable range of variable of + /- one-eighth of an inch. To answer the question, "How are we doing?" in the Measure phase, the team decided to use the Cp and Cpk metrics. The results from their study are in the figure below.

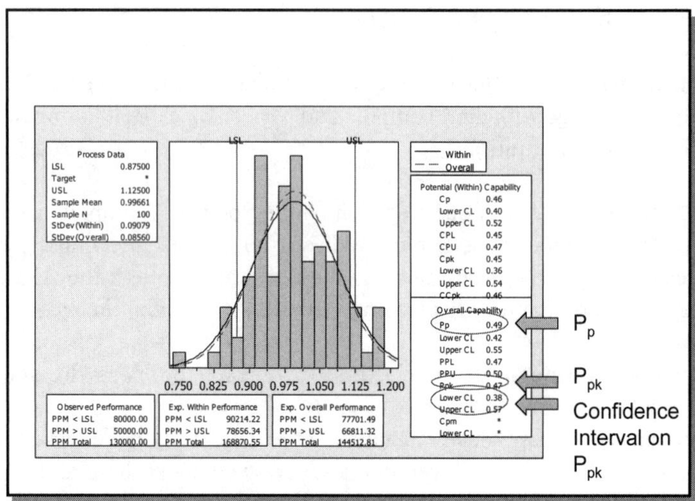

Figure 14-32 Example Capability Analysis

They checked to assure that they met the assumptions for normality and stability of the data before calculating the capability metrics. Because these metrics are statistics based on sample data, they had the MINITAB statistical software calculate confidence intervals around their estimates.

The team used Pp and Ppk as their estimates. This matches Motorola's use of capability indices. The two numbers were low (P_p=0.49 and P_{pk}=0.47) but close to each other indicating that the process was pretty well centered on target. They just had too much variation. The confidence interval for P_{pk} was .38 to .57 indicating a need for improvement.

The team decided to use a sources of variation approach to solving their problem. The team chose the following factors in the study: operator, machine, drill bit, and time. The team had two different operators, three different machines, three different drill bits, and ten different hours of production. The team found that drill bits were causing the largest source of variation. This surprised them since the team had suspected that operator-to-operator variability would be the largest source. The team decided to use a cause and effect diagram to delve deeper into the issue and to brainstorm what might be causing this variation.

Case Study Two

Since Joe's team had discovered that their C_{pk} value was negative, they realized that the first order of business would be to find a way to move the mean of their data. In this case, this just meant that the average of their cycle times was too long. They had collected a variety of x data and they started generating graphs to look for differences. Most of the x data were discrete and the team heavily used boxplots. The team saw a difference in the type of application used. They also saw a big difference by customer type. As they reviewed the data collection forms where the time for each step was recorded, they found that, although the average cycle time was 20 days, the actual time that people worked on an application was two hours! The rest of the time the application was sitting around waiting to be processed.

To verify that there was really a difference by type of application used and customer type, the team used hypothesis testing techniques. The team determined that there was a significant difference for both. The team could select the application type that decreased the cycle time but they couldn't select the customer type. They decided to determine why one customer type might take longer than another by looking closer at this issue.

Not surprisingly, the customer type that took the longest was a customer that really was a consortium of customers listed under one name. They found that this situation took longer to set up in the IT system. The software did not really support this situation and the customer representatives needed to put in a special request to the IT department to get the customer set up.

The team studied the times for each process step in further detail. As they studied the system, they found that many customer representatives were idle during the day. The representatives were not cross-trained and the applications often came in batches. So, the people in the first part of the process would be busy for a while but then they would have some long periods of downtime.

Summary

In the Analyze phase, the team determines the root causes of the problems of the process that it identified in the Define and Measure phases of the project. The team uses a variety of tools and statistical techniques to identify these root causes. This chapter has presented a number of those tools and techniques.

Chapter

15

Improve Phase

In the Improve phase, the team has validated the causes of the problems in the process and is ready to generate a list of solutions for consideration. They will answer the question "What needs to be done?" As the team moves into this phase, the emphasis goes from analytical to creative. To create a major difference in the outputs, a new way to handle the inputs must be considered. When the team has decided on a solution to present to management, the team must also consider the cost/benefit analysis of the solutions as well as the best way to sell their ideas to others in the business.

The deliverables from the Improve phase are:

- Proposed solution(s)
- Cost/benefit analysis
- Presentation to management
- Pilot plan

In this chapter, these deliverables are discussed as well as other tools and considerations to be successful in this phase.

The first task in the Improve phase is to develop ideas for improving the process. The traditional method for developing improvement ideas is to use conventional brainstorming. The next section presents alternative ways to generate creative ideas in a team.

Generating Solutions

If asked, "Where are you most creative or innovative?" most people would not answer "In the conference room where our team normally meets." And yet, when it comes to the point in the project where the team must generate new ideas and new ways to run the process, the team leader generally schedules a meeting in the same conference room with the same brainstorming technique with the hope that this will produce something spectacular. It might, but then it might not.

There are many books written on how to inspire creativity. In addition, there are techniques that the Black Belt can use to try to encourage creativity.

One of the proponents of improving thinking skills is Edward de Bono. De Bono introduced the idea of lateral thinking - a thinking style that is not linear, sequential or logical. Based on this concept of thinking, de Bono developed some thinking techniques. The techniques of de Bono's discussed here are Random Stimulation and Six Thinking Hats. In addition, Mind Mapping and Challenge Assumptions are presented.

Random Stimulation

De Bono emphasizes in his teaching that the brain is structured to look for patterns. And, when a recognized pattern emerges, the brain no longer is required to think. It follows the established pattern. He likens this to driving on unfamiliar roads - a person needs to think about getting from here to there. However, when someone finds a familiar road, thinking is no longer required - they drive on autopilot. In Random Stimulation, the purpose is to try to break this pattern of recognition and follow a new line of thought by introducing a random word selected from the dictionary. The team uses this random word to look for ways it could be related to the situation under consideration. This word is used to force the team to look at the problem in a new way.

The steps for Random Stimulation are:

1. State the problem under consideration.

2. Select a method to pick a random word from the dictionary such as a random page number and a random number of words from the top of the page. De Bono suggests continuing down the page until finding a noun.
3. Let the team free associate how the random word is like the problem at hand.

Six Thinking Hats

De Bono developed a methodology for trying to get have the whole team look at a problem from one point of view at a time until all point of views are covered. The technique, called Six Thinking Hats, involves the team using six hats to represent an attitude to adopt. The team wears the same color hat at the same time. The hats are as follows:

- The *White* Hat thinking requires team members to consider only the data and information at hand. With white hat thinking, participants put aside arguments and individual opinions and review only what information is available or required.
- The *Red* Hat gives team members the opportunity to present their feelings or intuition about the subject without explanation or need for justification. The red hat helps teams reveal conflict and air feelings openly without fear of retribution Use of this hat encourages risk-taking and right-brain thinking.
- The *Black* Hat thinking calls for caution and critical judgment. Using this hat helps teams avoid "groupthink" and proposing unrealistic solutions. This hat should be used with caution so that creativity is not stifled.
- The *Blue* Hat is used for control of the brainstorming process. The blue hat helps teams evaluate the thinking style and determine if it is appropriate. This hat allows members to ask for summaries and helps the team progress when it appears to be off track. It is useful for "thinking about thinking."
- The *Green* Hat makes time and space available for creative thinking. When in use, the team is encouraged to use divergent thinking and explore alternative ideas or options.
- The *Yellow* Hat is for optimism and a positive view of things. When this hat is in use, teams look at the logical benefits of the proposal. Every green hat idea deserves some yellow hat attention.

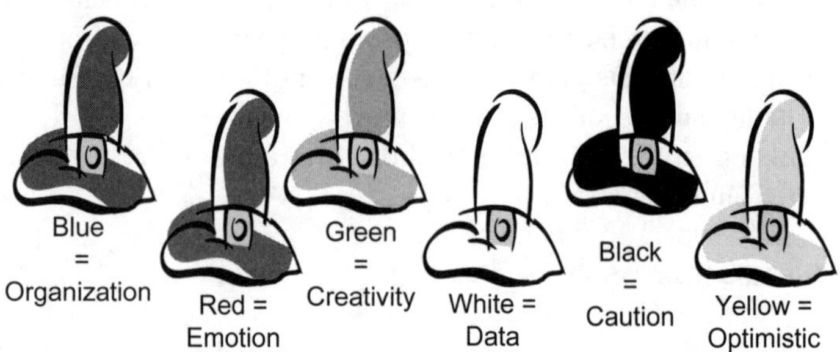

Figure 15-1 The Six Thinking Hats

This technique is especially useful when a team member refuses to consider a suggestion except from one angle. Many times, a team has someone who always wears the black hat; that is, that can never see the good in an idea. Six Thinking Hats can help reduce team conflict.

The steps for Six Thinking Hats are:

1. State the problem under consideration.
2. Start with the Blue Hat, the organization hat, to have the team decide the order of the wearing of the hats.
3. Have all team members wear the same color hat at the same time until all the hats to be used have been gone through.
4. Finish with the Blue Hat to conclude the session.

Mind Mapping

Mind Mapping is a technique developed by Tony Buzan that encourages team members to use right brain and left brain thinking to raise their levels of creativity. It focuses on finding associations and breaking the linear thinking usually used with traditional brainstorming. To create a mind map:

1. Start with a theme or a colored image representing the theme in the center. An image encourages creative thought while increasing memory significantly.
2. Brainstorm key words associated with the theme.
3. Words should be printed. Printed (not cursive) letters are best,

providing a more photographic, immediate, and casual appearance.

4. Words should be on lines and each line should be connected to other lines. This guarantees a basic structure to all mind maps for clarity and ease of comprehension.
5. Use one idea per line. This leaves ideas with more free hooks, allowing for flexibility and association with other concepts and visual separation of ideas. Secondary words branch off from the main or key words. Use images throughout the mind map.
6. To stimulate the whole brain, use plenty of words and pictures. Use color. Colors enhance memory, delight the eye, and establish connections among ideas.
7. The mind should be left as free as possible. Any thinking about where things should go, and even if they should be included, contaminates the process and slows progress down.
8. Use symbols for more advanced mind maps. Create a coding system using arrows, asterisks, exclamation marks, crosses, parenthesis, shapes, and any other creative images. Use them consistently with similar meaning for clarity and understanding.

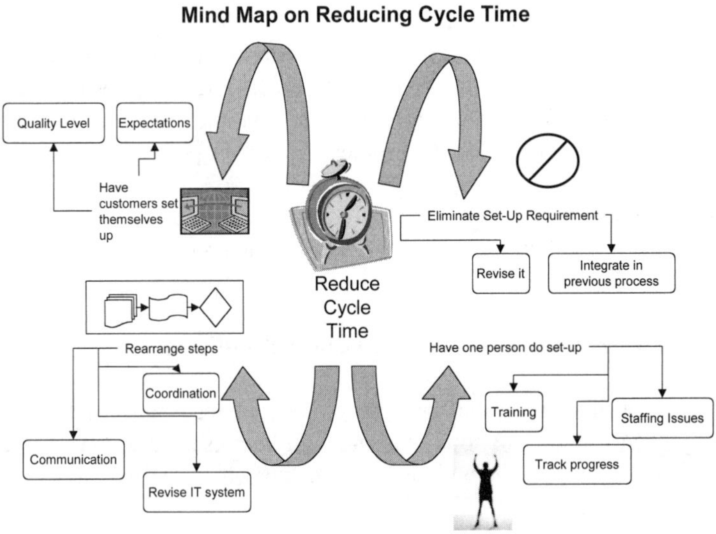

Mind Map on Reducing Cycle Time

Quality Level
Expectations
Have customers set themselves up
Eliminate Set-Up Requirement
Revise it
Integrate in previous process
Reduce Cycle Time
Rearrange steps
Coordination
Communication
Revise IT system
Have one person do set-up
Training
Staffing Issues
Track progress

Figure 15-2 Example Mind Map

Challenge Assumptions

Another technique to encourage generating new ideas is Challenge Assumptions. For every situation, people have assumptions whether consciously or subconsciously about what can or can't be done. In Challenge Assumptions, these assumptions are specifically stated and used to look at the situation in a new way.

The steps for this technique are:

1. State the situation under consideration.
2. Write down as many existing assumptions about the situation as the team can.
3. Try reversing the assumption. Transform it into the opposite of what it is now. For example, if the assumption is that approval from a department head is required for all transactions, assume that such approval will no longer be required.
4. Consider modifying the assumptions. Revise each assumption to make it better or easier to deal with. Change a name, time frame, location, etc. For example, assume that supervisors, rather than department heads, need to approve transactions.
5. Try varying the perspective. Try viewing the assumptions from the perspective of another person (boss, a customer, the president, etc.), work group, or organization. Imagine what the problem would look like to them. Describe the problem from their perspective.

The team should work on one to four assumptions that may be holding them back from thinking of new ways to do the process.

Decision-Making Tools for Selecting Solutions

After a list of solutions has been developed, the team will need to select the desired solution (or solutions) that will be presented to management. Three tools that can help with decision-making are pairwise ranking, solution matrix, and force field analysis.

Pairwise Ranking

The pairwise ranking technique is a team tool that can help prioritize a list of items. There are many different variations of this technique but all of them require the team to compare items by ranking each of them against the others. The combined results of these paired rankings help clarify the priorities. This tool can be used when the team's preferences cannot be determined informally.

The steps for this technique are:

1. Write the list of items to prioritize on a flipchart or board so that everyone can see them. Label them starting at "A."
2. Draw a pairwise ranking table on a flipchart or overhead, as shown in the figure below. This table is used to record the team's collective preferences, so it should be visible to everyone. Label the columns and rows starting with "A" as shown.
3. For each paired ranking, the team will indicate a preference for one item over the other. In order to further indicate the strength of each preference, agree on a point system that gives more weight to strong preferences than to moderate or weak ones. If, for example, the team had eight members, a point system like the one shown in Figure 15-3 could be used.
4. Write the point system on a flipchart or board so everyone can see it.
5. Compare each item to every other item, one at a time. Notice that there is a circle that represents the intersection between a column and row, which in turn represents the comparison of two different items on the list. Using a show of hands, the team should vote its preference for each possible comparison.
6. After each vote, the team leader should record the results by writing the letter (in the associated circle) that indicates which item the team preferred and the number that shows the strength of the team's preference.
7. After completing all comparisons, count the total number of points assigned to each letter. The point totals will illustrate the team's collective preference.

This technique can give structure to making decisions in the team.

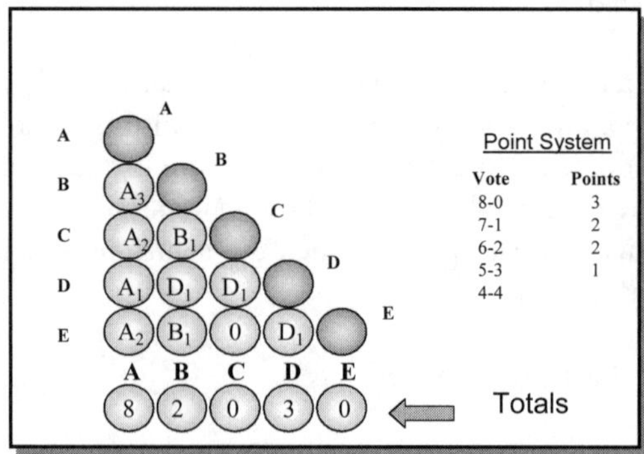

Figure 15-3 Example of Pairwise Ranking

Solution Matrix

Another tool for selecting desired solutions from the list of solution ideas is the solution matrix. The solution matrix is a modification of the Cause and Effect Matrix. See the figure below for an example of a solution matrix.

Example Solution Matrix

	Sigma	Time	Cost / Benefit	Other	<<Categories
	10	5	5		<<Importance
--Solutions --	---------- Ratings --------				------- Total -------
Solution 1	5	5	5		100
Solution 2	3	5	5		80
					0
					0
					0
					0
					0
					0
					0
					0
					0

Figure 15-4 Example Solution Matrix

The steps to creating a solution matrix are:

1. Remove any solutions that are impossible given the current business circumstances or that are outside the boundaries defined in the scope.
2. Consider business cultural issues and remove any solutions that may not match the culture.
3. Use teaming techniques (such as multi-voting or pairwise ranking)

to reduce the initial list to a few viable, supportable solutions.

4. Develop a list of criteria to evaluate the solutions. Common criteria to use are the Sigma Level, the time to implement the solution and the costs and benefits of the solution. Consider any other criteria that are important to the business. These criteria go across the top of the matrix
5. Weight each criterion on a scale of 1 to 10 where 10 is the most important criterion as it relates to other criteria and 1 would be the least important.
6. List all solutions to be rated in the first column.
7. Develop a rating scale for the Sigma Level. Each project team should develop its own scale since the scale will depend on the current state of the process. The figure below shows an example of a rating scale.
8. Develop a rating scale for the time to implement the solution.
9. Develop a rating scale for the costs and resulting benefits of solutions.
10. For each solution being evaluated, use the scale for each criterion and rate the solution.
11. For each solution, multiply the rating for each criterion times the weight for that criterion. Then, for each solution, total up the calculated values.
12. These total values are then ranked. The team should then decide how many of the top-ranked solutions will be presented to management.

Sigma Level	Rating
6.0	10
5.75	9
5.5	8
5.25	7
5.0	6
4.75	5
4.5	4
4.25	3
4.0	2
3.75	1

Figure 15-5 Example Rating Scale for Sigma Level

Force Field Analysis

Another tool to help understand the proposed solutions is force field analysis. In force field analysis, the team takes each solution one at a time. On a flipchart, the team would write two columns: Restraining Forces and Driving Forces. The team would then brainstorm what current conditions in the business would be obstacles to implementation of the solution. These are the restraining forces. In addition, the team would brainstorm what current conditions in the business would be beneficial to the implementation. These are the driving forces.

After completing the force field analysis, the team should consider strategies for eliminating or reducing the significance of each of the restraining forces identified and for reinforcing the driving forces. The table below shows an example force field analysis.

Restraining Forces	Driving Forces
• Need to change IT system • Need to re-define job descriptions	• Workers in area are frustrated with current process • Management of process is supportive of change

Figure 15-6 Example Force Field Analysis

By conducting a force field analysis on each solution to be presented to management, the team will have a clear understanding of what needs to be done to have a good implementation.

"Selling" the Solution

Finding a good solution is not enough. The team must convince others in the organization that the solution is implemented and implemented well.

Stakeholder analysis, the presentation to management, and the cost/benefit analysis are important to selling the solution to the business. The pilot plan is also important for a successful implementation.

Stakeholder Analysis

A tool to help with managing the proposed change is a stakeholder analysis. Stakeholder analysis is the review of the positions of the stakeholders of the process to be improved. A stakeholder is any one who has a stake or interest in the process. To conduct a stakeholder analysis, the team would list all the people or groups who are touched by the process. The next step would be to list the concerns that each stakeholder might have about the changes proposed. The team would also list the positive outcomes that the change might have on the stakeholder. Before presenting the solution, the team should work through the stakeholder concerns and try to emphasize the positive outcomes and mitigate the concerns. The table below shows a partial example of a stakeholder analysis.

STAKEHOLDER	CONCERNS ANTICIPATED	POSITIVE OUTCOMES
Quality Dept	Reduced headcount needed	Less parts on hold for inspection
Drill Dept	Need to learn new process	Will be able to produce more

Figure 15-7 Partial Example of a Stakeholder Analysis

Project Presentation

After the team has selected the solution(s), typically they will make a presentation to management for a decision on whether to proceed with implementation plans. The presentation should take on the following form:

1. Presentation of the team Charter.
2. Description of the method the team followed to arrive at the final recommended improvements. The point is to build credibility with data.
3. Presentation of the flowchart of the current process.
4. Review of the customer requirements and root causes of why the process is not meeting the customer requirements.

5. Discussion of the solution with an implementation plan, benefits of the solution and the control strategy.
6. Explanation of proposed process maps.
7. Closure with questions and discussions.

Often the presentation is in a storyboard format, a format that emphasizes graphs and pictures to tell the story of the Six Sigma project. The figure below shows an example storyboard. After the presentation, the team will usually need to schedule a time to get a decision on the solutions.

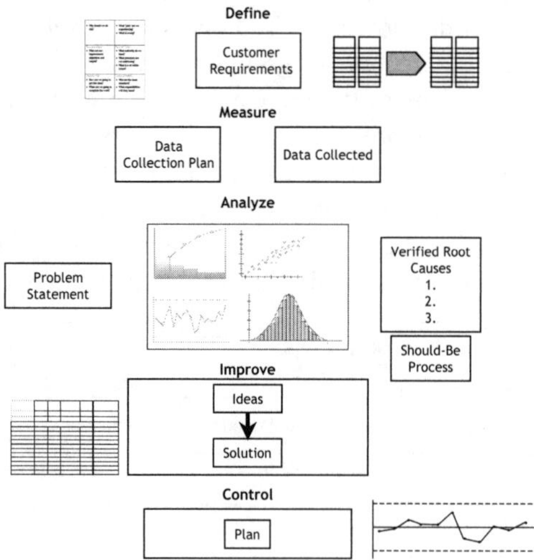

Figure 15-8 Example Storyboard

Costs and Benefits

Since Six Sigma projects are all about affecting the bottom line, conducting a cost/benefit analysis is an important step in any project. Typically, the cost/benefit analysis will be developed and reviewed in all phases of DMAIC. Initially in the Define phase, financial cost calculations are required to justify the project. As the team collects data in the Measure phase, the financial calculations are updated to reflect the new information. As the team proposes a solution with some estimated information about the benefits, the calculations again are re-visited in Improve phase. The Control phase also provides new information as the solution is fully implemented and then a final analysis of the costs and benefits are conducted.

There are two sources of financial benefits: quantitative and qualitative. To the extent that a quality impact can be quantified, it should be. The benefit of having an understanding of the estimated value creation is essential for easing the decision-making effort. Often the precise value of the financial impact is not known. In these cases a range of values can be used which attempt to bracket the magnitude of variation that one can reasonably expect. An example of a quantitative impact would be the removal of a process step that enables work to be performed without requiring overtime.

The financial organization of the business should define how to handle qualitative benefits. Ideally, all quality improvement solutions would yield quantifiable financial benefits. Unfortunately, this is not always the case. For example, at times it may be difficult to determine how an anticipated improvement in customer satisfaction of a given item will translate into financial benefits. The following are some of the more common difficult areas:

- Customer satisfaction improvements
- Cost avoidance
- Market share growth
- Brand enhancement
- Reduced management time

Attempts should be made to quantify the potential ranges of value using accepted internal estimation techniques. However, even if the benefit cannot be quantified it should still be visible in the decision-making process.

Also to be considered are the costs associated with the project. Sources of costs are direct, one-time, on-going, and indirect. Direct costs are costs that can be traced directly to producing a product or service. For example, costs associated with materials and direct labor charges are direct costs. One-time costs are costs that are incurred only once. Examples include new equipment or facilities or initial training. On-going costs are costs that will continuously be incurred. Examples include labor, supplies, and other operating costs. Indirect costs or support costs are costs incurred by a staff or service unit elsewhere in the organization. Frequently, these costs are charged at a standard rate by an internal billing system and are not always charged back to the initiative. An example of indirect costs might be cost associated with IT support.

It is often helpful to understand the degree of variation associated with cost and benefit estimates. While there are several methods that help provide understanding, the most common is the "best case, worst case" analysis. This analysis attempts to understand the total range of variation associated with the "most likely" situations. By displaying the total range of variation, the decision-maker is more equipped to understand the risk associated with this project and the sources of the risk.

Pilot Plan

Before conducting a full-blown implementation of the solution, a pilot study is recommended. A pilot is the trial implementation of the identified solution or the proposed design on a reduced scale. The pilot leads to developing an improved plan for solution implementation and its rollout. It also provides the opportunity to experience the solution without committing the entire organization and to better understand the impacts and obtain feedback from the process, customers, suppliers, staff, etc.

The purposes of conducting a pilot are to:

- Reduce the risk of failure by identifying potential failure points.
- Increase organizational buy-in.
- Validate and refine cost and benefit estimates.
- Perform adjustments and observe the solution implications.
- Evaluate the effectiveness of measures used to monitor the improvement.
- Test the validity of the solution.

The disadvantage of conducting a pilot is that it causes a delay to full-scale implementation and realization of benefits. However, the lessons realized can be quite comprehensive and rewarding.

A pilot plan should include the following components:

- Size
- Pilot success criteria
- Potential problem analysis
- Contingency plans
- Training
- Verification plan
- Communication
- Schedule

These components of the pilot plan are discussed below.

Size

The intention of the pilot is to test the validity of the solution and it should be focused on some representative portion of the process that ultimately needs to change. The complexity and size of pilot planning should be consistent with the magnitude of the solution.

Pilot Success Criteria

Two categories of success criteria established for a pilot are:

- Effectiveness of implementation - the degree which the organization understands and adopts the solution(s).
- Impact on target - the degree to which a solution helps move the overall process performance toward its improvement goals and targets.

Pilot success criteria should be established for both of these categories. One without the other will lead to either the failure of a good solution from bad planning or the good execution of a bad solution.

Potential Problem Analysis

Before conducting a pilot, a team must be prepared to address every potential barrier to pilot success. Many of these potential barriers can be identified prior to conducting the pilot. Good pilot planning takes into account the various challenges and barriers to success and addressees them before the failure occurs. Some approaches for reducing risks in implementation are:

- Mistake Proofing
- Risk Assessment Matrix and Controls Assessment Matrix
- FMEA

Mistake Proofing

Mistake proofing, also known as poke-yoke, was an idea developed by Shiego Shingo in the 1960's. Translated from Japanese, it means avoid mistakes. (Mistake proofing was also covered in the *Kaizen* teams chapters.) The basic idea of poka-yoke is to design or improve the process so that it is robust to humans. In other words, design the process so that mistakes are unlikely or at least easily detected and corrected. Occasionally, the only solution a team proposes is to train people. What the team is not recognizing is that mistakes are inevitable. There is not one person who can say they haven't made a mistake. Defects result from allowing these mistakes to reach the customer. Instead of focusing just on training, the team should consider incorporating poka-yoke techniques into the process.

Poka-yoke devices fall into two major categories: prevention and detection. A prevention device renders the process so that it is extremely unlikely to make a mistake at all. An example of a poka-yoke prevention device is that the microwave will not work if the door is open. A detection device signals the user when a mistake has been made, so that the user can quickly correct the problem. Detection devices typically warn the user of a problem, but they do not enforce the correction. An example of a poka-yoke detection device is the car beeps when the keys are left in the ignition.

Good poka-yoke devices share many common characteristics:

- They are simple and cheap. If they are too complicated or expensive, their use will not be cost-effective.
- They are part of the process.
- They are placed close to where the mistakes occur, providing quick feedback to the workers so that the mistakes can be corrected.

Poka-yoke is an important consideration for a process that involve humans; that is almost every process. Considering poka-yoke will only lead to a better process.

Risk Assessment Matrix and Controls Assessment Matrix

The risk assessment matrix is a tool for identifying potential problems before conducting a pilot study or implementing a solution. The controls assessment matrix is a tool to plan activities to protect the project from the risks identified as high priority. These tools are used together to have a problem-free implementation.

To create the risk assessment matrix, complete the following steps for each objective of the team's solution:

1. List the potential risks.
2. Rate the impact of the risk as High, Medium, or Low.
3. Rate the probability of occurrence as High, Medium, or Low.
4. Focus on the risks that have high impact and high probability.

The figure below shows a risk assessment matrix.

Project Objective:

Risks	(Hi, Med, Low) Business Impact	(Hi, Med, Low) Probability of Occurrence	Priority

Date: Prepared by: Reviewed by: Next Update:

Figure 15-9 Risk Assessment Matrix

To create the controls assessment matrix, do the following steps after completing the risk assessment matrix:

1. List the highest priority risk from a previously created risk assessment matrix.
2. List the controls that are in place to help mitigate or eliminate the risk.
3. Discuss the appropriateness and sufficiency of each control. Categorize the controls as excessive, adequate or inadequate.
4. Identify the actions that need to take place to address any control deficiencies. Remember, controls can be costly and cumbersome. Be sure that there is an appropriate balance of risks to controls and that other process objectives, such as efficiency and speed, are not sacrificed.
5. Repeat these steps for all the high priority risks.

The figure below shows a controls assessment matrix.

Project Objective: ______
Risk: ______

Controls	Existing Y/N	Is it sufficient? If not, why?	Next Steps What? Who? When?

Date: ______ Prepared by: ______ Reviewed by: ______ Next Update: ______

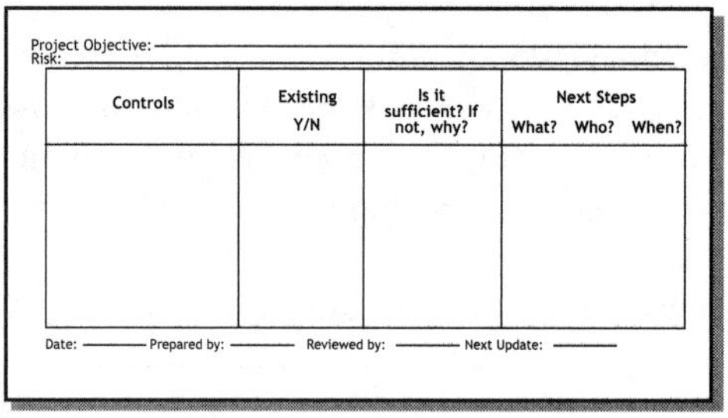

Figure 15-10 Controls Assessment Matrix

Failure Modes and Effects Analysis

Failure Modes and Effects Analysis (FMEA) is a spreadsheet-based team tool used to identify possible ways for the process to fail. The information is then prioritized based on severity of the failure, the likelihood of a cause creating the failure and the ability of any control to detect or prevent the failure or the cause. The FMEA can be used before the team implements any changes to the process.

The steps for conducting an FMEA are:

1. Review the process to be analyzed.
2. Brainstorm and group possible failure modes.
3. List one or more potential effects of each failure mode. Try to

answer the question: If this failure occurs, what are the consequences?

4. Assign severity rating for each effect. The severity is usually rated on a scale of 1 to 10, where 1 is the least severe and 10 is the most severe.
5. List potential causes for the failure. Answer the question, "If (potential failure mode), then (what?)."
6. Assign an occurrence rating for each failure cause. It should be based on the probability the cause will occur and it will produce the failure.
7. List current controls.
8. Assign a detection rating for each failure mode. It should be the ability to detect or prevent either the failure mode or the cause.
9. Calculate the risk priority number (RPN) for each effect by multiplying the severity rating times the occurrence rating times the detection rating.
10. Use the RPN's to help decide on the high priority failure modes to address. The higher the RPN, the more attention is warranted.
11. Plan to reduce or eliminate the risk associated with high priority failure modes.

The FMEA can make the team more comfortable with the proposed changes to the process by considering what might go wrong and taking steps to prevent any problems.

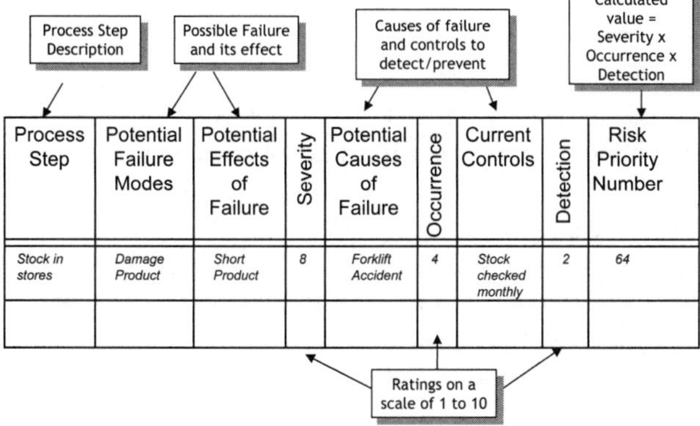

Process Step	Potential Failure Modes	Potential Effects of Failure	Severity	Potential Causes of Failure	Occurrence	Current Controls	Detection	Risk Priority Number
Stock in stores	*Damage Product*	*Short Product*	*8*	*Forklift Accident*	*4*	*Stock checked monthly*	*2*	*64*

Figure 15-11 Example FMEA

Contingency Plans

To build commitment to the solution and support for the final solution, it is important that the pilot implementation goes well. The team needs to assure they have contingency and recovery plans, in case something goes wrong, based on the FMEA or risk and control assessments conducted. The team may want to consider a SWAT-like team for emergencies. In addition, the team should be there as much as possible during the pilot process without creating an additional barrier or variable. In other words, they should not be influencing the behavior of the process. What the team learns and observes will be worth the time invested.

Training

The amount of training the people in the process need depends on whether the changes recommended are small or large. Training for a small change might involve something as simple as sitting around a table and reviewing new written procedures with everyone before trying them out together. Larger or more complex changes and those requiring new skills will benefit from planned training curriculums, training aids and structured practice time. At this point in the project, the team may consider involving a training expert to help define the skills and knowledge required and the best way to develop these.

Verification Plan

The team needs to be clear of what makes a successful pilot. A verification plan might be helpful. A verification plan includes:

- The identified pilot area.
- The objectives of the pilot.
- The steps to be conducted in the pilot, including how long the pilot will be run.
- The success criteria and associated goals.

And, to be filled in after the pilot completion:

- Pilot observations.

- Gaps that need to be addressed before further implementation.
- Next steps for the team.

See the figure below for an example verification plan.

Product/Service: ________

Function/Processes Tested: ________

Implementation Observations: ________

Implementation Plan Objectives: ________

GAP Analysis/Root Causes: ________

Steps/Activities: ________

Follow-up Actions: ________

Measurement Plan:

Measure	Plan Target	$\bar{X}$	s

Figure 15-12 Verification Plan

At the end of the pilot, the team needs to be able to answer questions like:

- Did the pilot have the anticipated results?
- Was the plan for conducting the pilot effective?
- What improvements should be made to the solution?
- Can the solution be implemented "as-is"? Should it be?
- Can the solution remain in place at the pilot location?
- What lessons learned and best practices can be applied during the solution implementation?

Depending on the results from the pilot, there are different paths the team may take. Six different paths are available:

1. Make appropriate changes to the solution and retest in the initial pilot area(s).
   - The gaps in the verification plan are significant.
   - Enhancements to the pilot are likely to close the gap to the level

that the initial area will not be piloted again. It is essential that the re-testing be kept to a minimum to:
(a) Keep the pilot team from losing confidence.
(b) Reap the benefits as early as possible.
(c) Increase the chances of organizational buy-in.

2. Pilot the solution in other area(s).
    - The gaps between expectations and the pilot results are minimal.
    - The test site is not representative of all the conditions likely to be encountered.
3. Abandon the solution and return to the Analyze phase.
    - Root causes were not validated correctly.
    - All the root causes were not identified and validated.
4. Abandon the solution and return to generating solutions.
    - The gaps are still significant after successfully conducting the pilot.
    - The assumptions used to select the solutions have been proven to be invalid.
5. Expand the pilot to include additional variables.
    - The gaps between expectations and the pilot results are minimal.
    - Additional variables can be considered - the initial pilot may have controlled or removed certain variables to prevent the pilot from being affected by too many variables. However, before the pilot is rolled out it is important to consider the impact that these additional variables will have on the validity of the solution.
6. Incorporate lessons learned and best practices, then implement the solution.
    - Minimal gaps exist in the verification plan.
    - The test site is representative of the population.
    - Organizational buy-in will not materially improve with additional pilots.

Communication

One of the most significant variables in a successful implementation is the quality of communications supporting a change. As the team progresses throughout the project, they should consider who they need to communicate with and what others need to know.

There are many methods of communication and the appropriate one should be chosen for the situation. Some methods include:

- town hall gatherings
- one-on-one meetings
- electronic mail
- posters
- bulletin boards
- small discussion groups or workshops
- videos

The timing of the communication is determined by the nature of the message and the method through which it will be communicated.

Seven key elements help to create a successful communications program:

1. Take the time to thoroughly plan the communication strategy.
2. Know the organizational culture and understand the impact on communication strategies.
3. Understand the target audience.
4. Allocate enough resources to the program.
5. Involve employees as much as possible.
6. Conduct a communications process check to verify that the message was understood.
7. Provide ways to obtain feedback about how message was "heard."

Schedule of Activities

Of course, the pilot plan cannot be complete without a schedule of activities that are required to implement the pilot. Traditional project planning is appropriate here with the development of milestones and then a list of tasks to support the completion of the milestones.

The team may also use a responsibility chart to list who is responsible, accountable, informed and consulted for each of the tasks defined for the milestones. See the figure below for an example responsibility chart.

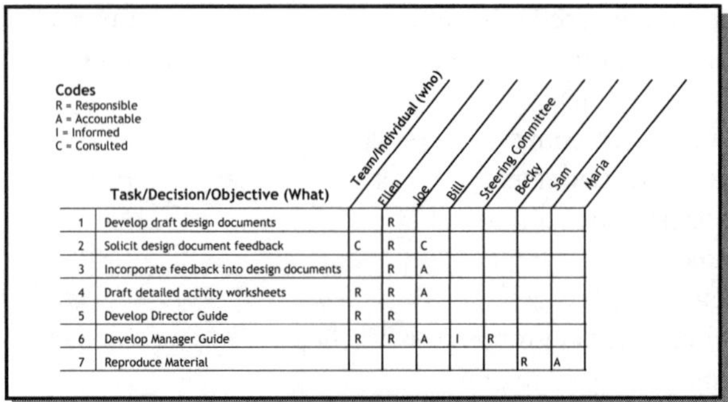

Codes
R = Responsible
A = Accountable
I = Informed
C = Consulted

	Task/Decision/Objective (What)	Ellen	Joe	Bill	Steering Committee	Becky	Sam	Maria
1	Develop draft design documents		R					
2	Solicit design document feedback	C	R	C				
3	Incorporate feedback into design documents		R	A				
4	Draft detailed activity worksheets	R	R	A				
5	Develop Director Guide	R	R					
6	Develop Manager Guide	R	R	A	I	R		
7	Reproduce Material						R	A

(The column headings after "Task/Decision/Objective (What)" are grouped under "Team/Individual (who)".)

Figure 15-13 Example Responsibility Chart

The more thought put into the pilot plan, the easier the full solution implementation will be. All components of the pilot plan will need to be considered with the solution plan.

Improve Phase Tollgate Questions

As the team progresses through the Improve phase, the team leader should schedule meetings with the Sponsor/Champion and Master Black Belt to review all that has been accomplished.

The following is a list of questions, called Tollgate Questions, which the Sponsor/Champion and the Master Black Belt can use to prompt discussion in these meetings. Also, the team can use these questions as they progress through the Improve phase to be sure that they have completed all the important items. By the time the team has completed the Improve phase, the team members, team leader, Master Black Belt, and Champion should feel comfortable with the answers to all of the following questions and any others that might be specific to the organization.

1. How did the team generate ideas? How was the list of ideas narrowed down to the best few? What methods were used to encourage creative thinking?
2. What criteria were used to evaluate the potential solutions? How does the preferred solution address the root cause of the problem?
3. How did the team develop the revised process design? What are the critical elements of the design?

4. Did the team conduct a cost/benefit analysis? What assumptions were made? Did a financial subject matter expert validate the cost/benefit analysis?
5. How was the compelling need for change explained? How will the team communicate this explanation to stakeholders? How is this reflected in the communication plan for implementation?
6. How will the team answer the stakeholders' "What's In It For Me" question? What can be done to mobilize their support? How is this reflected in the communication plan for implementation?
7. What training is required to ensure the people affected will be able to support the new process design with minimal frustration and maximum preparedness?
8. What were the lessons learned from the pilot?

Case Study One

Team Scrap had found in their Sources of Variation (SOV) study that drill bit variability was the largest source of variation. They needed to study this in more detail. They generated a cause and effect diagram on drill bit variability. Based on this diagram, they selected some x factors to include in an experiment. They normally used three different suppliers for the drill bits. They included supplier as an x. They also studied different lubricant types for the drilling operation. Their response in the experiment was hole diameter. After analyzing the results, the team found that one of the suppliers and two types of lubricants produced the best results; that is the lowest variation.

Operator-to-operator variability had also shown up as a source of variation in the SOV study. One operator in particular appeared to have less variability than the rest. The team listed very detailed process steps of each operator's use of the drill. The team discovered a difference in the way the operators set up the machine before drilling.

Armed with this information, the team got together with all the drill operators to brainstorm how to solve the identified problems. They decided that they needed to use just one supplier for the drill bits until the other suppliers could show the equivalent quality. The team needed to work out a strategy with the supplier management group to reduce to one supplier.

The team was reluctant to reduce to the two lubricants since the third lubricant type actually provided an easier clean up of the equipment. The team used the random stimulation creative thinking technique to provide a way to use the two lubricants that gave the best results. After a session, the team found a way to make using the two lubricants easier.

The team also studied the set up of the machine and used traditional brainstorming and affinity diagramming techniques to develop an agreed upon set of steps for set up.

The team was going to present all these pieces as a final solution to management. The Black Belt worked with a finance representative to develop a cost/benefit analysis for the solution to include in their presentation.

Case Study Two

Joe's team, trying to reduce cycle time, wanted to try a pull system with their process. In a pull system, work is processed at a step only when a signal is made by the process step next in line that more work is required. In other words, work is always triggered by the receiving step. The benefits from a pull system generally are reduced cycle times, better communication and therefore better defect identification between process steps, easier scheduling and more flexibility in responding to changes in demand volumes. However, the team wasn't sure that the pull system would really give their process these benefits. And, they weren't sure they could sell it to management or the people working in the area. It would be a big departure from the current way of doing things, not only in this process but also in all processes at the business.

The team decided to employ a simulation model to evaluate a few alternatives to managing the process. They involved a person from the Industrial Engineering department to help develop the simulation model. They validated the computer model by first designing it to match the current system and verifying that the results from the model were similar to the results obtained with the real process. When they felt confident about the model, they then designed an experiment to test a variety of different ways to staff and manage the process. The industrial engineer ran the experimental runs in the simulation model and the team analyzed the results.

Using an animated version of their selected staffing and management model, they were able to present to the executive group their solution. They were also able to use the animation of the current system to show the executive group what was wrong with the current process.

The animation was also used to convince the people in the process of the results that could be obtained with the new process. It would require that the people get cross-trained. The animation helped convince the people in the process the benefit of the training.

Summary

In the Improve phase, the team gets to put on its creative hats to generate solution ideas. The team then evaluates these ideas, and uses various decision-making tools to select the one or more ideas they deem the best. In order for the change to be successful, the team must convince the rest of the organization of the value of their ideas. They must win over key stakeholders, management, and any one affected by the change. A pilot can be used to test any solution idea, demonstrate its effectiveness, and highlight required adjustme

Chapter

16

Control Phase

In the Control phase, the emphasis is on a successful implementation and maintaining the gains achieved. The question the team is trying to answer is, "How can we guarantee performance?" From the Improve phase, the team has a successful pilot and has had a chance to tweak the solution. They need to take this information to plan the solution implementation. They also need to ensure that, when they finish the project, the success that they have seen will continue. This involves transferring the responsibility to the process owner.

The deliverables in the Control phase are:

- Solution implementation plan
- Successful solution
- Process control plan

In this chapter, these deliverables are discussed as well as other tools and considerations to be successful in this phase.

Solution Planning

The first step in this phase is to create a solution implementation plan based on the results from the pilot plan and pilot implementation in the Improve phase. A big piece of the solution implementation plan is to leave tools to help the owner manage the process after the team has gone on to other projects.

A large-scale project may require dealing with multiple processes and sub-processes, multiple implementation locations, a large number of implementation teams and several different disciplines and methodologies. A good implementation strategy and workplan can minimize the predictable problems. The coordination and integration of all efforts must be planned and managed across the scope of the project on a daily basis.

A solution plan should include the following components:

1. Potential Problem Analysis
2. Solution Implementation Schedule
3. Training Plan
4. Communication Plan
5. Costs and Benefits
6. Transfer to Owner Plan

The first five on this list were discussed in the pilot implementation section of the Improve chapter and, in general, are just extensions of the plans made for the pilot.

Potential Problem Analysis

Based on the information the team has learned in the pilot, a review of potential problems should be conducted. As in the Improve phase, risk assessment and controls assessment matrices and FMEA's are appropriate tools.

Solution Implementation Schedule

Since the solution implementation will be on a larger scale than the pilot implementation, the schedule of activities for implementation will also need to be more detailed and may potentially require the involvement of more people.

Training Plan

As the team considers the solution implementation training plan, they will need to decide if the training for the pilot can be used or if it will need to be expanded to accommodate the larger scale. Again, a training professional may be necessary depending on the impact of the change to the process.

Communication Plan

The team has already developed a communication plan for the pilot implementation. This plan will need to be adjusted for the larger audience with the solution implementation. Lessons learned from the pilot should be incorporated into the new plan.

Costs and Benefits

After the implementation of the solution, the team will be able to collect new data to do a final assessment of the costs and benefits. The team should follow the business standards for financial calculations and should consider involving an expert from finance.

Transfer to Owner Plan

As the team concludes its project, steps need to be taken to ensure the new improved process will continue as intended. Without attention and monitoring, things tend to revert to the initial state. To make sure the gains are maintained, clear ownership with responsibilities needs to be established for the process. Tools to help monitor and control the process also need to be institutionalized.

The team should consider the following:

- Process control plan to document the new process.
- Review meetings to communicate the state of the process.
- Updated flowcharts.
- Updated procedures.
- Control charts to monitor the process.
- Out-of-Control Action Plans to define how irregularities in the process are handled.

Each of these items is discussed below.

Process Control Plan

The process control plan is a document that can take several forms. The purpose, no matter what format, is to have a single record that defines how the process is to be monitored. Typically, the plan will have:

- The customer of the process.
- Applicable Critical Customer Requirements.
- The flowchart of the process.
- The X and Y data that should be collected to monitor the process.
- The measurement device used to collect the data.
- How often the data are to be collected.
- What chart (usually a control chart) on which to plot the data.
- The Out-of-Control Action Plan for each type of data.
- Revision number of the document to allow for updates.

This document, usually a page or two, allows anyone to identify what is important to maintain the current performance levels based on the results from the Six Sigma project.

Review Meetings

The process management team should establish periodic reviews of the performance of the process using the process control plan. Target performance levels for key X's and Y's will be an outcome of the team's project. Therefore, the process management team will be able to report on the process performance, any performance gaps, and will be able to prioritize needed improvements that are most important to the customers and the business.

Updated Flowcharts

The team created the "as-is" flowchart in the Define phase. In the final phases of the project, they should update the flowchart to document the improvement changes to the process. The team might consider, for explanation purposes, indicating the changes in a different color on the "as-is" chart. The flowchart contained in the process control plan should incorporate the new changes.

Updated Procedures

Procedures are documentation of the correct operating methods of the process designed to ensure consistency in the process and to institutionalize the improvements. The goal of a procedure should be to help simplify the execution of the activities and reduce the possibility of miscommunication and mistakes. However, it should not limit the flexibility of the person in the process. When creating the procedure, the team should use the people in the process to help create the procedures. If using a written procedure is a new idea in the process, the team may need to develop a plan to update the procedures if the process changes in the future.

Control Charts

Variation can be classified as common cause variation or special cause variation. Common cause variation is due to the natural variation of the process; that is, variation due to the way the process was designed. An example of common cause variation is the variation that might be seen by having several people working in the process. Each person might do things slightly differently. Special cause variation is variation that is due to assignable causes. An example of special cause variation is the variation that might result if someone untrained is allowed to work in the process. Special cause variation is variation that can be assigned a reason.

The best tool to determine if the variation is common cause or special cause is the control chart. A control chart is a specialized run chart. The Y axis is the metric of interest and the X axis is time, or a factor that indicates time such as lot or run number. The difference between a run chart and a control chart is a control chart has three statistically calculated lines: a center line, an upper control limit, and a lower control limit. There are many types of control charts but generically these lines can be described as:

Center line = Mean of the metric of interest
Lower control limit = Mean of the metric - 3 * Standard Deviation of the metric
Upper control limit = Mean of the metric + 3 * Standard Deviation of the metric

Special cause variation is identified by points falling below the lower limit or above the upper limit, trends, runs or any unusual patterns. Any indication of a special cause should be investigated to see if the process has changed. Here is a control chart with all points in control.

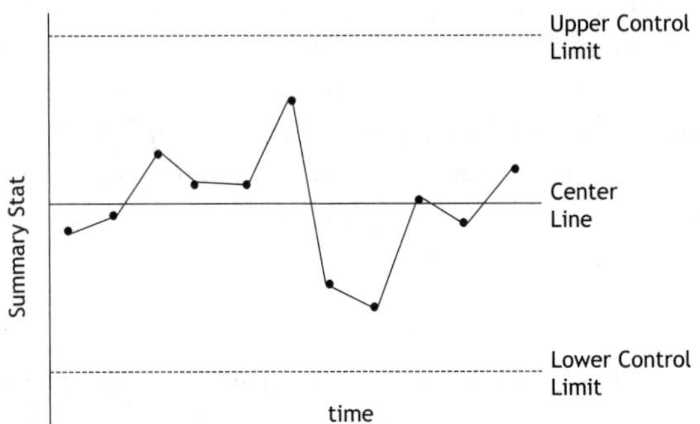

Figure 16-1 **Example Control Chart**

An important consideration when using a control chart is the subgroup size to use with the chart. In each time period, the data collected could be an individual value or a small sample, or subgroup. The determination of the size of the subgroup will be based on practical issues such as the cost of sampling, how quickly a change needs to be detected and the cost of investigating false alarms (false indications of instabilities).

To determine the subgroup size and frequency, consider the following:

- Determine the subgroup size so that the samples in the subgroup are produced under essentially the same conditions. The goal is for the effect of an assignable cause (x) to show up *between* subgroups, not *within* a subgroup. The variation within a subgroup should be due to non-assignable chance causes only.
- Ensure that the observations included in the subgroup are independent. Some processes may contain autocorrelation as a natural function of the process (e.g., chemical processes). Autocorrelation is the dependence of a current data point on previous data points within a given single stream of data.
- Determine the frequency of sampling so that the control chart can detect any changes in the process over time. As an initial starting point, sample twice as frequently as a change in the process could happen (e.g., if a change from shift-to-shift is possible, take two samples every shift.) Then, as the process proves to be stable over a period of time, consider reducing the frequency of sampling. Typically, for high-volume processes, taking small samples frequently is the best strategy.

Selecting the right control chart is based on whether the data to be plotted are continuous or discrete.

If the data are continuous, two charts are suggested - one to monitor the location of the data and one to monitor the spread of the data. The typical charts for data collected in subgroups are an X-bar and an R chart. In other words, the two charts are a plot of the subgroup means and a plot of the ranges of the subgroups. If the data are collected one value at a time, typical charts are an Individuals chart and a Moving Range chart; in other words, a plot of the individual values and a plot of the moving ranges. For the moving range chart, the moving range is defined as the range between two adjacent data points.

If the data are discrete, one chart is usually plotted. The chart type is dependent on the nature of the data - whether the data are counts of defects or defective units. There are four traditional charts for discrete data. C and U charts are used for defects, where c is the count of defects and u is the average number of defects per unit. P charts and NP charts are used for defective units where p is the proportion of defective units and np is the count of defective units.

To generate a control chart, these steps should be followed:

1. Identify the process quality characteristic to be charted and the potential sources of variation of that characteristic.
2. Ensure that the data collected are appropriate for the construction of the control chart. Two common assumptions about the data are that they are independent and normally distributed. If these assumptions are not true, then a non-traditional control chart should be used.
3. Calculate the control limits.
   a. Initially, calculate the control limits using 25-30 data points. If that much data are not available, establish the limits with what is available. However, re-evaluate the limits as more data are obtained.
   b. When creating the control chart with these 25-30 data points, if any point is below the lower control limit or above the upper control limit, investigate to see if there was a special cause:
      - If a special cause is found, eliminate that data point and recalculate the limits.

- If no special cause is found, keep the data point in the calculation of the limits.

4. Document the details of the complete control plan for the process.
5. Document the specific tests used to indicate out-of-control conditions and the associated corrective actions in the Out-of-Control Action Plan. This plan is described below.
6. Decide when the control limits should be revised. Revise the limits if there has been a change in the process. Also, consider re-evaluating the limits after some time has passed since they were calculated (e.g., 3-6 months).

Out-of-Control Action Plan

An important tool to use with a control chart is an Out-of-Control Action Plan (OCAP). An OCAP documents how instabilities will be detected and resolved.

To create an OCAP:

1. Define what tests to use to indicate out-of-control conditions. The one test for out-of-control applicable to any control chart is a data point falling above the upper control limit or a data point falling below the lower control limit. Other commonly used tests are:
   a. Test for a trend - several points in a row trending up or trending down
   b. Test for a run - several points in a row on one side of the center line.
2. Establish how to react to each out-of-control condition:
   a. Set up specific and clear guidelines to follow. These guidelines may include what corrective actions to take, procedures for documenting the out-of-control situation, procedures for notifying those who are responsible for maintaining and improving the process, etc.
   b. A flowchart format for the plan is recommended. Indicate which path was followed to stabilize the process. Data can be collected on this information to determine if there is a need for a more permanent corrective action. The figure below shows a section of an example OCAP.
3. Establish how to determine whether the corrective action has been effective. In other words, has process been re-stabilized? This may include extra data collection.

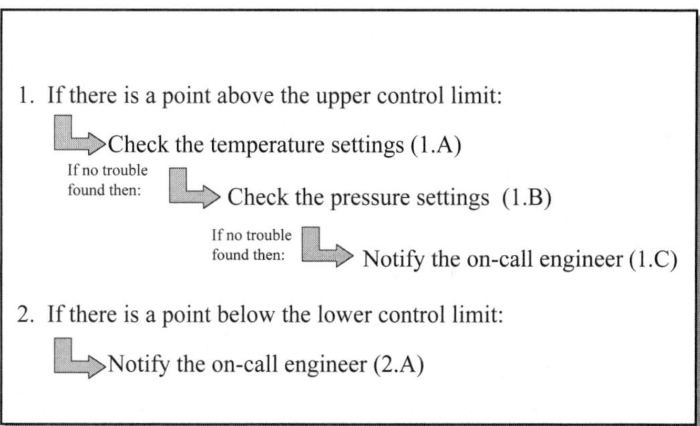

1. If there is a point above the upper control limit:
Check the temperature settings (1.A)
If no trouble found then:
Check the pressure settings (1.B)
If no trouble found then:
Notify the on-call engineer (1.C)
2. If there is a point below the lower control limit:
Notify the on-call engineer (2.A)

Figure 16-2 Example Section of a Flowchart OCSP

Standardization and Solution Replication

One of the powerful aspects of Six Sigma is to take successful implementations and expand them across the organization. This is accomplished with replication and standardization.

Replication is taking the solution from the team and applying it to the same type or a similar type of process. Standardization is taking the lessons/solutions from the team and applying those good ideas to processes that may be dissimilar to the original process improved.

The team should consider standardization and replication opportunities to significantly increase the impact on the sigma performance of processes to far exceed the anticipated results by the pilot and solution implementation.

As the implementation expands to other areas, four implementation approaches can be combined or used independently. The appropriate approach will depend on the resources available, the culture of the organization and the requirements for a fast implementation. The four approaches are:

- A sequenced approach is when a solution is fully implemented in one process or location, implementation begins at a second location.

- A parallel approach is when the solution is implemented at two or more locations or processes simultaneously.
- A phased approach is when a pre-determined milestone is achieved at one location, the implementation at a second location begins.
- A flat approach is when implementation is done at all target locations, companywide.

Project Conclusion Activities

There are some larger issues beyond just the extent of the project that it is appropriate for the team and management to address at the conclusion of the project. The team might consider evaluating how the team worked together, management may devise rewards to recognize the work of the team, and the team and Champion may share the knowledge gained with others. Each of these issues should be addressed as part of a larger business strategy.

Team Evaluation

When the project is officially over, a team peer evaluation may be done to assess how each individual did as a team member. There are many benefits to conducting a team peer evaluation, including:

- Team members become more aware of performance standards and behavior requirements because they are accountable for maintaining them.
- Peer pressure is a powerful motivator.
- Members who recognize that other people on the team will be evaluating their work should increase commitment and productivity.

However, there are some drawbacks to conducting the evaluations:

- They may be time-consuming.
- It is sometimes difficult to distinguish between the contributions of the team and those of individual members.
- Some members feel uncomfortable judging or evaluating other team members.

The first consideration should be the purpose of the evaluation. The evaluation may be used as part of performance evaluation for the employee. If the evaluation is used for performance evaluation, the Black Belt should follow any protocol set by the human resources department and should consider having an outside person conduct the evaluation.

On the other hand, it may just be used as a feedback tool for the individual team members to improve teaming skills. If used for feedback, the Black Belt might want to assess total team performance before having members conduct peer reviews. The team performance assessment may conducted as an open discussion forum with the team to discuss such issues as:

- How well did the team meets its goal defined in the charter?
- Was each meeting run with a clear, logical agenda?
- Was there balanced contribution and involvement from all team members?
- Were self-oriented behaviors, such as side conversations and individual interruptions, kept to a minimum?
- Was a profound understanding of the process reached?
- Was the team using data-based decision making?
- Was the DMAIC methodology followed through out the project?

After finishing the discussion on overall team performance, the Black Belt can follow this with the peer evaluation. The Black Belt will ask each person on the team to evaluate everyone else on the team, including the Black Belt. The figure below shows an example of an evaluation form. The form might also include a section for comments. The Black Belt would collect all the completed evaluation forms, average out the scores and share this information in on-on-one sessions with each team member. The team peer evaluation follows the spirit of Six Sigma - things can only get better if they get measured.

Example Team Peer Evaluation Form			
Initiates ideas	10 9 8 Frequently offers ideas and solutions.	7 6 5 4 Initiates only moderately, but supports initiating by others.	3 2 1 Tends to let others take most of the initiative and often reserves support.
Facilitates the introduction of new ideas	10 9 8 Actively encourages others to contribute without worrying about agreement.	7 6 5 4 Provides support for ideas with which he or she agrees.	3 2 1 Often resists the introduction of new ideas; looks for flaws.
Is directed toward group goals	10 9 8 Often helps to identify and clarify goals for the group.	7 6 5 4 Sometimes helps the group define its goals; sometimes confuses it with side issues.	3 2 1 Tends to place priority on own goals at the expense of groups.
Manages conflict	10 9 8 Regards conflict as helpful in promoting different perspectives and in sharpening the differences in views.	7 6 5 4 Generally disengages from conflict.	3 2 1 Tries to smooth over points of disagreement; plays a pacifying role.
Demonstrates support for others	10 9 8 Actively encourages the participation of others and asserts their right to be heard.	7 6 5 4 Encourages certain members part of the time, but does not encourage all members.	3 2 1 Does not offer support or encouragement for other members.
Reveals feelings	10 9 8 Openly expresses feelings about issues; ensures that feelings parallel views.	7 6 5 4 Sometimes disguises feelings or tries to keep them to self.	3 2 1 Denies both the existence of own feelings and the importance of expressing them in the group.
Displays openness	10 9 8 Freely and clearly expresses self on issues so that others know where he or she stands.	7 6 5 4 Sometimes employs tact and speaks circumspectly to camouflage real views.	3 2 1 Is vague about views on issues, even contradictory when pressed.
Confronts issues and behavior	10 9 8 Freely expresses views on difficult issues and on team members' nonproductive behavior.	7 6 5 4 Is cautious about taking advisable position on issues and on others' actions without first ensuring widespread approval.	3 2 1 Actively avoids issues and any conflict by talking about "safe" issues that are irrelevant to current group work.
Shares leadership	10 9 8 Assumes responsibility for guiding the group when own resources are needed or when problems lend themselves to his or her solving.	7 6 5 4 Competes with other members for visibility and influence.	3 2 1 Dominates group discussions and exerts disproportionate influence that subverts group progress.
Exhibits proper demeanor in decision-making process	10 9 8 Actively seeks a full exploration of all feasible options.	7 6 5 4 Becomes impatient with a deliberate pace in generating and evaluating all options when he or she does not	3 2 1 Moves strongly toward early closure of discussion to vote on a preferred option.

Figure 16-3 Example Team Peer Evaluation Form

Reward and Recognition

After the team proves that the change has made the intended improvements, the business may want to consider rewarding or recognizing the team. Rewards and recognitions should fit with the business culture and should be done in conjunction with the human resource department. Here is a list of some examples of rewards and recognition:

- Praise from management in a public forum
- Salary increases
- Days off
- Lunch or dinner
- Certificates or plaques
- Party
- Coaching opportunities

The reward or recognition given should be given as soon as possible after the desired behavior or achievement. It should always be stated why the recognition/reward is being given. In addition, the reward or recognition should be customized to take into account the significance of the achievement.

Knowledge Sharing

The team should capture their experiences throughout the improvement process. This knowledge capture can take many forms and should match any business requirements. Key suggested topics include:

- Survey or interview data gathered with a copy of the questions used
- Meeting minutes
- Analysis methods used
- Key X's determined
- Feedback from process participants
- Pilot and solution implementation results

Control Phase Tollgate Questions

As the team progresses through the Control phase, the team leader should schedule meetings with the Champion and Master Black Belt to review all that has been accomplished.

The following is a list of questions which the Champion and the Master Black Belt can use to prompt discussion in these meetings. Also, the team can use these questions as they progress through the Control phase to be sure that they have completed all the important items. By the time the team has completed the Control phase, the team members, team leader, Master Black Belt, and Champion should feel comfortable with the answers to all of the following questions and any others that might be specific to the organization.

1. Describe the implementation plan. How will the plan be monitored to ensure its success? Who is accountable?
2. What are the potential problems with the plan? What are the contingency plans?
3. What controls are in place to assure that the problem does not reoccur?
4. Who is the process owner? How will the responsibility for continued review be transferred from the improvement team to the process owner? How frequent are the reviews?
5. What is being measured? What evidence does the team have that

would indicate the process is "in-control"? How well and consistently is the process performing? Is a response plan in place for when the process experiences "out-of-control" occurrences?

6. How has the process been standardized? Have the process changes been documented?
7. How has the training plan been revised from the "Improve" phase? How has training been conducted to assure understanding of the process changes? How effective was this training? What continuing issues need to be addressed in the area of training?
8. What is the communication plan for implementation? How will the team use communications to manage this change, minimize resistance and mobilize stakeholders?
9. Based on the implementation and communications with key stakeholders, what are the barriers to successful change? What actions are planned to overcome these barriers?
10. Was the solution tested on a small scale? How representative was the test? How are the learnings from the pilot integrated into the implementation plan?
11. What gains or benefits have been realized from the implementation? How can the improvements be replicated elsewhere in the organization?
12. What did the team learn from the project? What are the best methods to share the lessons of the project?

Case Study One

Sara's scrap team received approval from management for their list of solutions. They conducted a pilot study for two weeks. The pilot showed using the two lubricant types and selected drill bits yielded the results they expected. However, the new method for machine set up, while helping to reduce scrap, increased cycle time. This was an undesired result. The team worked with the operators of the process to develop a way to use the new set up procedure but not influence cycle time.

The team decided to re-run the pilot to ensure that the new method would have the desired results. The second pilot also lasted two weeks. The solution was a success and the team felt that they were ready for a full-blown implementation.

The team developed a process control plan with the supervisor in charge of the area that defined which x's would be checked on a regular basis and who would be responsible for the checking. They also established Xbar-R control charts on the diameter of the drilled holes. The operators would plot the average and range of three holes drilled on every lot of product. An Out-of-Control Action Plan was developed to help the operators respond to any unusual patterns on the chart.

One of the team members worked on updating the flowcharts and procedures with the new changes. After the documentation was done, the team reviewed it with the operators to make sure that they understood the new procedures and felt comfortable with them.

The team estimated cost savings of $115,000 in reduced scrap over the next year.

Case Study Two

The cycle time team's simulation and animation results were convincing. The team gained support for their proposed changes. The team worked extensively with the human resource group and the training group since their solution depended heavily on changing the job definitions of the representatives working in the process. They also worked on a recognition and reward strategy for the representatives to provide incentive to them for learning new skills.

The team created a detailed implementation plan. Since the team ran the simulation as their pilot and did not do an actual realization, they wanted to make sure the solution implementation went well. They worked with the people in the area to create a schedule of activities to transition to the new process design and contingency plans in case of problems. In addition, they created a process control plan with a control chart to monitor the Y, cycle time, of the process. They also implemented a control chart on number of applications received each day to make sure volumes weren't changing over time.

After the new process had been up and running for two weeks and the team felt all issues due to the changes had been resolved, they started collecting data. Four weeks later the team used the data to reassess the costs and benefits with the project. The finance group approved the numbers and the business started to reap the estimated benefits of $215,000 per year due to reduced overtime from better scheduling and increased revenues by adding new clients faster. This number did not include any estimates on keeping customers who might have left the application process out of frustration for the process taking too long.

Summary

In the Control phase, the team first develops a plan to implement the one or more solutions selected in the Improve phase. That plan should include:

1. Potential Problem Analysis
2. Solution Implementation Schedule
3. Training Plan
4. Communication Plan
5. Costs and Benefits

The team also needs to ensure that, when they finish the project, the success that they have seen in implementation will continue. This involves transferring the responsibility to the process owner. This may require:

- Process control plan to document the new process.
- Review meetings to communicate the state of the process.
- Updated flowcharts.
- Updated procedures.
- Control charts to monitor the process.
- Out-of-Control Action Plans to define how irregularities in the process are handled

As the project wraps up, a couple of additional activities may be appropriate:

- The team might consider evaluating how the team worked together
- Management may devise rewards to recognize the work and success of the team

A Six Sigma project does not really "end" at the conclusion of the Control phase. There should be opportunities to extend the success of this project team in other areas. The team and Champion may share the knowledge gained with others, replicate the solution in other processes, and develop standards for other processes based on what they learned from their solution implementation. And the team may continue to examine the process to look for opportunities for continuous process improvement.

Chapter

17

DMAIC Summary

The DMAIC methodology is a powerful five-phase approach to addressing a process that needs improvement. Using DMAIC, a team does not have to worry about "what comes next" in the project. They know because it has been outlined for them. And although a team will not do every item listed in each phase, they can pick those that work for them. This approach allows flexibility in the structure.

As a Black Belt reviews a project, there are some key signs that indicate a well-executed project. For each phase, these signs are listed in the table below.

Unfortunately, a project can hit some roadblocks as well. The team and the Black Belt need to recognize the problems and take action to get the project back on track. The common pitfalls and the associated steps to rectify the situation are listed below.

Phase	Good Signs
Define	• Quick Wins were identified.Resources were adequately allocated to the project. • The Champion is responsible and involved. • The charter has been signed off and is supported by the team, the Black Belt and the Champion.
Measure	• The team developed a data collection plan. • Clear operational definitions were developed and validated. • The team showed conflict resolution skills.
Analyze	• The team used multiple tools to find root causes and validated the causes with data. • The team recognized the risks and power associated with the analyses.
Improve	• The team used non-traditional techniques to develop potential solutions. • The team conducted a pilot and included the lessons learned in the solution implementation. • Financial benefits were reviewed and revised.
Control	• Control charts have been incorporated as part of the process.

Phase	Common Pitfalls	To Get Back on Track
Define		
	There is too much or not enough detail in the process maps.	Strive to cover about 80% of all possibilities.
	The team assumes they know the Voice of the Customer.	Use data to verify the VOC. Conduct surveys, use listening posts, etc.
	The project is not connected to a Big Y.	The Black Belt needs to work with the Champion to clarify how the project can be re-defined to influence a Big Y.
	The scope is too big.	As the project progresses, use data to re-scope. For instance, use a Pareto chart to identify and select the biggest problem.
Measure		
	The team is using a sample of convenience.	Use the cause and effect diagram and matrix to find the data that should be analyzed.
	The sample is not representative of the process.	Study the process to better understand it and to design a sampling plan that will be representative. Then, use a data collection plan to develop a strategy for getting good data.
	The process is not stable.	Identify special causes in the process and work to remove them.
	The measurement system is assumed valid.	Work with the team to develop operational definitions and a methodology to study the measurement system.

Analyze		
	The assumptions weren't checked before doing a statistical analysis.	The analyst should refer to training materials for the steps to conducting any analysis. If necessary, the team should consult with someone who has more experience for help.
	The team only used simple graphical tools.	Simple graphical tools can be very powerful. However, the team should work with someone who has more experience to recognize opportunities to use statistical tools.
Improve		
	The team is relying on training to solve all process problems.	Use non-traditional creativity techniques to develop new solutions. Consider opportunities to use poka-yoke to prevent mistakes from happening.
	A dominant person on the team is pushing a particular solution and the team is intimidated into supporting it.	Use a structured decision-making technique, like a solution matrix, to select the best solution that all team members can support.
	The team is underestimating the impact of the changes on the people in the process.	Use a stakeholder analysis to understand any concerns and to help minimize them. Create a strong communication strategy to gain support for the changes.
Control		
	The team hasn't defined any tools for the process to use to monitor the important X's and Y's.	Consider opportunities to use a control chart to monitor the process.
	The team has not documented the changes to the process that they implemented.	Select the appropriate documentation techniques for the business such as a process control plan or written procedures.

The keys to having a successful project are:

- Establishing that the project is a business priority
- Understanding the true requirements for the process
- Using data to tell the story
- Picking the right tool for the right situation
- Communicating the project goals, accomplishments and successes
- Building credibility and support for the project

While easier said than done, excelling in these key areas will almost surely produce another Six Sigma success story.

Chapter 18

DMADV

Another methodology used in the Six Sigma world is called DMADV - Define, Measure, Analyze, Design and Verify. DMADV was developed out of the recognition that DMAIC was not powerful enough when faced with product or process design.

Studies have shown that a change required in the design phase of a product life cycle costs a company a fraction of what it would cost if the change is needed after the product is in production. By designing the product right in the first place, problems in manufacturing, assembly, service and support are diminished. By using a structured approach to design, the team may achieve:

- Streamlined development processes
- Shortened time to market cycle
- Designs that can be implemented

There are many similarities between the two approaches, DMAIC and DMADV. Note that the first three phases are the same though the emphasis of what is done may be different between the two. Much of the tool set for DMAIC is the same for DMADV. However, the questions to be answered in each phase for DMADV are different:

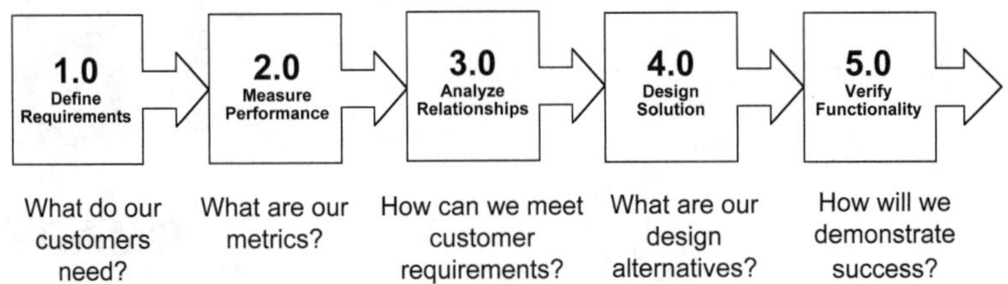

Figure 18-1 DMADV Flow and Associated Questions

In the Define phase, the emphasis is on understanding the customers and the customers' needs and wants. The Voice of the Customer is critically important in this phase.

In the Measure phase, the emphasis is on establishing metrics for the project and developing the X's that are important.

In the Analyze phase, initial design alternatives are developed.

In the Design phase, an approach is selected from among high level design alternatives identified during the Analyze phase and an initial implementation plan is developed.

In the Verify phase, the team verifies that the design will meet the requirements.

In the next sections, the five phases are covered with emphasis placed on additional tools used in DMADV. DMADV and its associated tools can be a very large topic. This chapter is intended to be an introduction to the tools and concepts of DMADV.

Define

In the Define phase of DMADV, the team needs to identify the business opportunity, identify the internal and external customers as sources of key VOB and VOC criteria, obtain the VOC and prioritize customer wants and needs, and translate the VOC's into measurable CCR's.

In the Define phase, some of the key deliverables are:

- Team Charter
- Project Plan
- Project Team
- Critical Customer Requirements
- Design Goals

As is evidenced by the list of deliverables, many of the activities for DMADV are the same that might be done in DMAIC and were discussed in the previous Define chapter. The team will still need to develop a charter, plan the project, and assemble the right people. In addition, the team will need to listen to the VOC to determine CCR measures and establish the goals associated with the product or process design. However, one of the major differences is the amount of time required to understand the VOC data to get to CCR's.

In the DMADV world, the main focus in the Define phase is usually on understanding the requirements of *external* customers where in DMAIC the main focus may be on *either* internal or external customers.

Additional tools may be used to understand who the customers are and what they want. In DMAIC, the Six Sigma team is often analyzing data to confirm what the customers want. In DMADV, the team may need to start from scratch. An additional topic that a team may need to understand is Marketing Research Strategies.

In a design project, the Six Sigma team needs to understand the marketing information that is driving their project. Although the Black Belt probably won't be leading marketing research efforts, the Black Belt needs to be savvy about how VOC data were developed and how the information is pertinent to the project. The marketing strategies are generally pre-Define activities, so the outcome of these activities is a key input into the Define phase. This information will also be a driver for the Quality Function Deployment (QFD) activities that will be discussed in the Measure phase of this section.

Two of many techniques that can be used to translate VOC data into useful information are Kano Analysis and Kansei Engineering.

Kano Analysis is a technique discussed in the Define phase of DMAIC and it might also be used in the DMADV world. In addition to the three categories presented earlier for Kano Analysis (must-be's, primary satisfiers, and delighters), there might be two more categories appropriate for design:

- *Indifferent*: The customers are indifferent to an added feature of the product or process. For example, a car company adds an indicator light on the dashboard to show that the radio is on. The customers may find no value in this feature.
- *Reverse*: The customer is unhappy with an added feature of the product or process. For example, a car company has a voice-activated system that will continue to tell the owner that the car door is open. The customer may become annoyed with the voice reminder.

Another technique that the team might benefit from is called Kansei Engineering. Kansei is a Japanese term that implies any or all of the following:

- Emotional
- Feeling
- Intuitive
- Subjective

This term has no direct translation in English.

Kansei engineering is a technique that can be used to sift out what the customer wants - subjectively by communicating feelings or emotions of the VOC. Kansei engineering studies subjective preferences (versus objective preferences) in the VOC. When a customer goes into a store and has five cell phones to look at, for instance, why will that customer go to a particular phone first? What drew that customer to that phone? This is what Kansei Engineering techniques strive to quantify.

Kansei is a family of product design techniques that relies on correlating the feelings a product invokes with its physical properties, such as shape and texture. For instance, a team might start with building a list of keywords that describe consumers' feelings towards a type of product (e.g.

automobile, trousers, cell phone, radio). These keywords are used to create rating scales, and representative consumers use the rating scales to evaluate product examples. For instance, suppose a company makes carriers for pets and is considering designing a new cat carrier. The marketing group develops a list of keywords, three of which are:

- Exciting
- Easy to use
- Cozy

They may then develop a rating scale using these words and their direct opposites. This rating scale would be given to the consumers with product examples. The scales with a customer's ratings might look like:

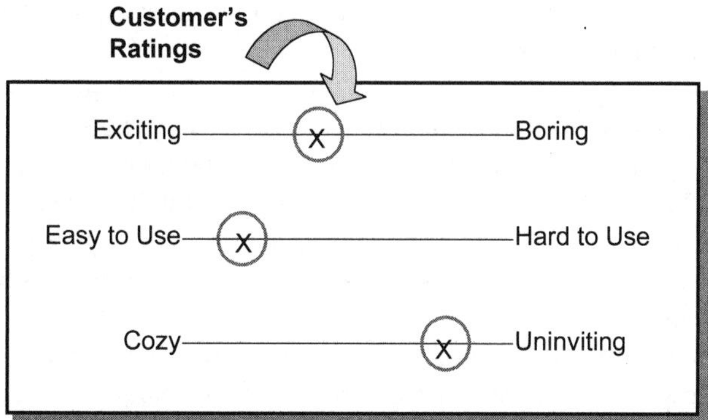

Figure 18-2 Example Rating Scales

The data from the consumers would then be statistically analyzed to determine how product design features correlate with consumers' ratings.

Figure 18-3 shows that marketing research, Kano analysis and Kansei engineering are techniques used pre-DMADV and this information is the basis for understanding CCR's in a Six Sigma project.

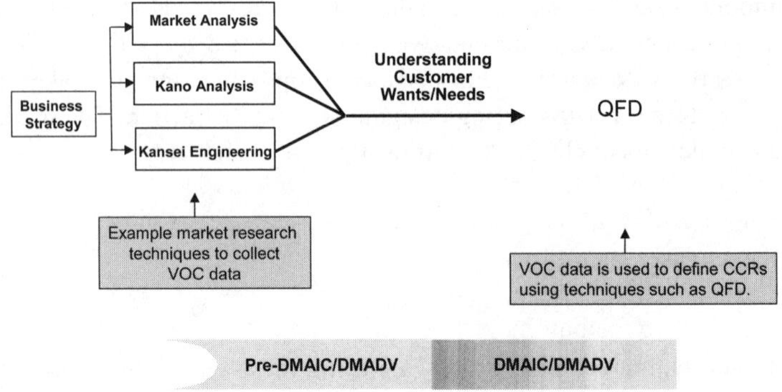

Figure 18-3 Listening to the VOC from Market Analysis

When using the data from these marketing techniques, the team might ask the following questions:

- What's the market segment for our project and who should we target?
- Why should we target them? What's the benefit to the customer?
- What are the features or service levels that will deliver the benefits?
- Have Kano concepts been applied to customer requirements?
- Is it appropriate to consider subjective Kansei satisfiers not just hard features?
- How are we differentiating our product or service from competitors?
- How much data analysis has been done? What are the results?
- How were the data collected? Were proper Six Sigma methods used to characterize the data (e.g., MSA on customer satisfaction)? How was the sample drawn?
- How do we know the market data are good?

As the team exits the Define phase, they should have confidence that they understand what is important to the customer.

Define Phase Tollgate Questions

As with DMAIC, there are some tollgate questions for DMADV that could be used to ensure that all important things have been considered before the team says they are done with the Define phase. The Tollgate Questions are:

- Why are we working on this project, rather than others?
- How does this project relate to the Big Y's?
- What is the result we expect from this project?
- Have we done everything necessary to capture and understand VOC from the customer's perspective?
- What are the CCR's? How were they determined?
- What specifically is within or outside of the project scope?
- Do we have the right team, and have the team members committed sufficient time to the project?
- Do we have the buy-in of key stakeholders?
- What are the risks on this project? What have we done to minimize them?
- What do we need from our Champion to ensure success?

Measure

In the Measure phase, the team needs to identify key customer metrics responsive to the VOC, assess measurement capabilities pertinent to VOB/VOC/CCR's, define data sources relevant for VOC analyses, and translate these CCR's into functional requirements.

In the Measure phase for DMADV, the key deliverables are:

- Qualified Measurement Systems
- Data Collection Plan
- Capability Analysis
- Refined Metrics
- Functional Requirements

Again, many of the activities are the same for Measure in both DMAIC and DMADV. But there are some differences. In DMADV, the team will need to look at how to measure X's that are new X's in a product or process development whereas in DMAIC the team is considering what current X data to collect. The team may need to develop and validate measurement systems instead of just validating a measurement system. Also in this phase, the team will need to consider measures on the project itself, including financial metrics. In DMADV, these financial metrics become increasingly important.

An additional tool used in the Measure phase for DMADV beyond those for DMAIC, and actually used in many of the phases, is Quality Function Deployment (QFD). QFD is a method for translating customer requirements into technical requirements at each phase of the product realization cycle. It is an integrated set of tools for recording user requirements, engineering characteristics and the trade-offs necessary between competing requirements. It provides a basis for deriving and communicating the "CTQ Flow-Down" (discussed in the Analyze phase of this section) throughout the product/process design initiative. QFD is also known as the "House of Quality" since the outline of the tool looks like a house.

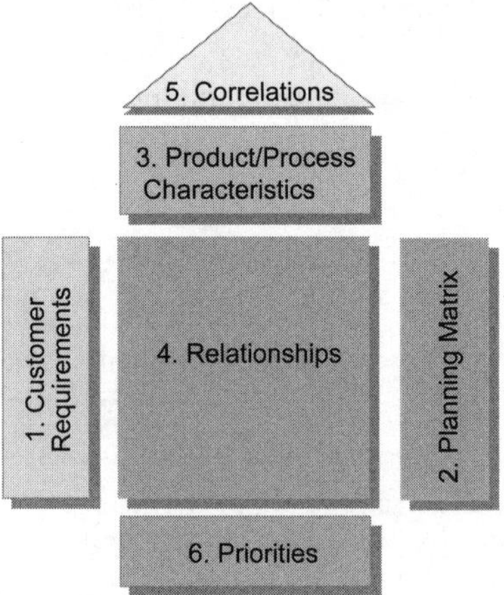

Figure 18-4 Quality Function Deployment (QFD)

There are six steps that can be followed to conduct a QFD analysis:

Step 1: The first step is the articulation of customer requirements. Techniques used could be interviewing, observation, prototyping, conceptual modeling, etc. The data from marketing research are also used. These requirements are also known as the "What's".

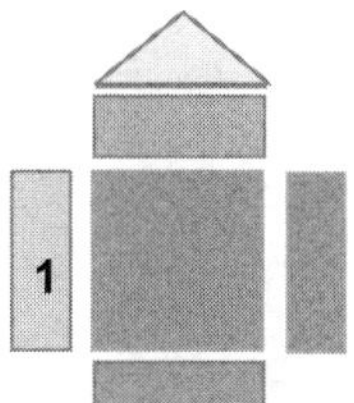

Figure 18-5 Customer Requirements

Step 2: In the second step, the company's current product is ranked against the competitors.

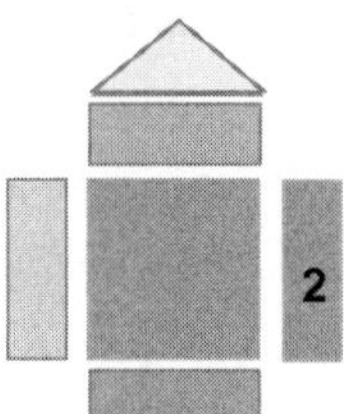

Figure 18-6 Planning Matrix

Step 3: In the third step, the team looks at Product/Process Characteristics, in other words, the "How's" of meeting the customer requirements. Candidate CCR's are listed across the top and for each their relevance is considered and ranked as to which will address customer needs.

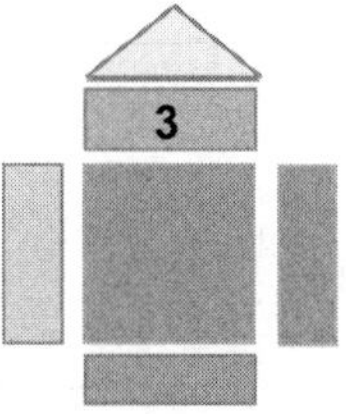

Figure 18-7 Product/Process Characteristics

Step 4: In the fourth step, the team relates customer and technical requirements with ratings such as "high", "moderate", "low", and "no" correlation. The team evaluates the degree to which customer wants and needs are addressed by the product/process characteristics.

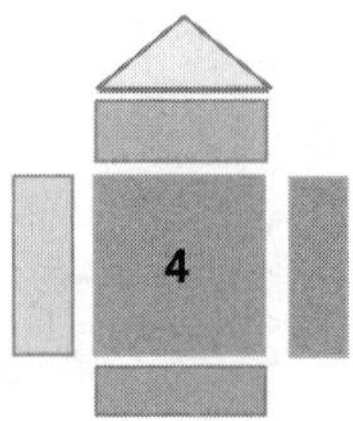

Figure 18-8 Relationship Matrix

Step 5: In the fifth step, the roof of the "House" focuses on relationships among product/process characteristics. It shows whether the "Hows" reinforce or conflict with one another.

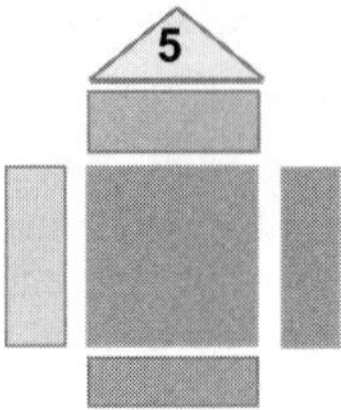

Figure 18-9 Correlation Matrix

Step 6: In the final section of the QFD matrix, the team summarizes the key conclusions. It ranks the relevance of product or process characteristics to the attainment of customers' wants or needs.

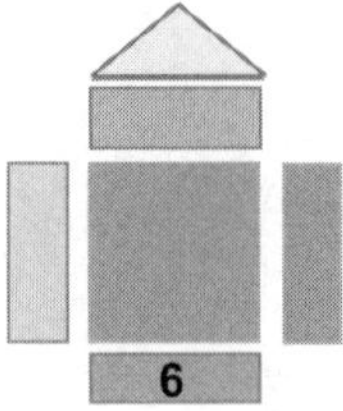

Figure 18-10 Priorities

Cat Carrier Example

For example, consider the cat carrier project. Suppose that through marketing techniques, including Kansei Engineering, the VOC data told the Six Sigma team that a few of the important customer requirements were that they desired the carrier to be:

- Lightweight (a must-be)
- Easy to open/close (a primary satisfier)
- Has a cozy feel (determined from Kansei Engineering techniques)

This information went in the Customer Requirements section of the QFD, Step 1. The team also prioritized these requirements on a scale of 1 to 5, with 5 being the most important to the customer. They gave "Lightweight", "Easy to open/close" and "Cozy feel" ratings of 5, 4, and 3 respectively.

In Step 2, the team looked at their current product offering compared to Company A's product in terms of Customer Requirements. They found that their product fared better than Company A's product in everything but "Lightweight".

In Step 3, the team defined the product characteristics that would meet these customer wants. For example, a partial list from the cat carrier team included:

- Material Choice
- Door Design
- Shape
- Size

In Step 4, the team evaluated how each of the product characteristics in Step 3 would influence the customer wants listed in Step 1. As an example, the team decided that:

- Material Choice has a HIGH correlation with "Lightweight".
- Material Choice has a HIGH correlation with "Has a cozy feel."

In the software the team used, a symbol was used to indicate high, moderate and low. A blank indicated no correlation.

For Step 5, the team did not find many relationships between the items listed in the Product Characteristic section. However, they thought there was a weak positive relationship between shape and size.

In the last step, the team summarized the findings of the QFD. The most important product characteristic was material choice.

Cat Carrier		Customer Importance	Material Choice	Door Design	Shape	Size	Product Characteristic	Product Characteristic	Planning Matrix ($ Our Company, 🚶 Company A) 0 1 2 3 4 5
Customer Requirements	Lightweight	5.0	◉	△	△	◉			$ 🚶
	Easy to open/close	4.0	△	◉	○				🚶 $
	Has a cozy feel	3.0	◉	○	○				🚶 $
	Customer Want								
	Customer Want								
	Customer Want								
Weighted Importance			76.0	50.0	26.0	45.0	0.0	0.0	
Relative Importance									

Correlations; Product / Process Characteristics

Figure 18-11 QFD Example for Cat Carrier

QFD is a flexible tool and is used in many different ways. The example presented above is just one way to use this tool. Often, the QFD is used in stages throughout the project providing traceability through the whole design process. Figure 18-12 shows how QFD is a tool that can be used to cascade from product requirements, to design features of the product, to required manufacturing processes to the actual process operations. The "How's" from one QFD become the "What's" of the subsequent QFD. In a DMADV project, CTQ's will eventually be derived from a stage of this tool. In the cat carrier example, this QFD would be just the first stage.

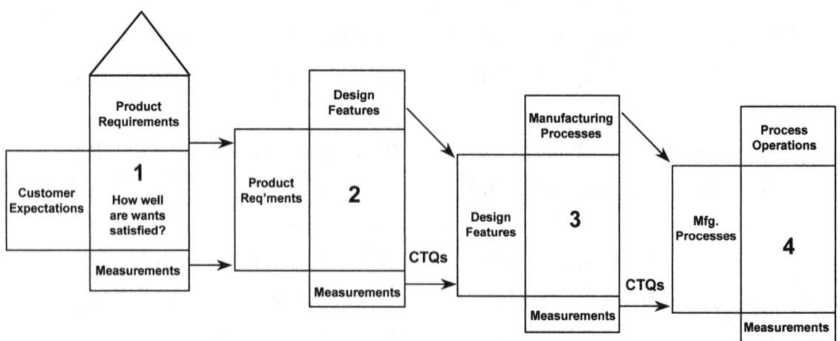

Figure 18-12 Multi-Stage QFD's

At the end of the Measure phase, the team should have a better understanding of what product or process characteristics are important to meet customer requirements, what measurement systems are to be used, what data will be collected, and if data are available, what the capability has been in the past.

Measure Phase Tollgate Questions

Some questions that the team and Champion may discuss at the end of the Measure phase are:

- What are the key process measures and project metrics?
- Were the measurements systems established/evaluated? What performance baselines have you established?
- What is the data collection plan? Have appropriate data sources, sampling plans, etc. been identified?
- Have we avoided jumping to solutions?
- How were functional requirements determined?

Analyze

In the Analyze phase, the team needs to identify key design factors influencing CCR's, quantify the impact of key design factors, analyze Sources of Variability (SOV), refine functional requirements and identify design alternatives.

In the Analyze phase, the deliverables are:

- Data Analysis
- Initial Models Developed
- Prioritized X's
- Variability Quantified
- CTQ Flow-Down
- Documented Design Alternatives

In many cases with DMADV, there are relevant data around to analyze. Usually there is relevant data on past products, which will help determine the X's that are important. So in the Analyze phase for DMADV, many tools are the same as with DMAIC. An additional topic presented is CTQ flow-down.

In Six Sigma, a key concept is to determine how the Y's are affected by the X's. This was presented at the beginning of Chapter 3 as:

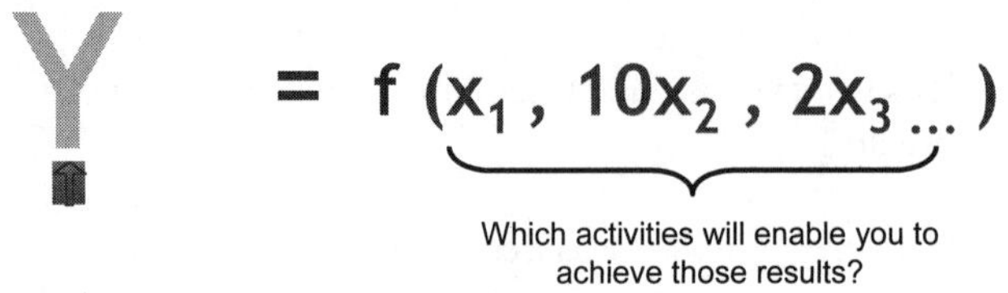

$$Y \uparrow = f(x_1, 10x_2, 2x_3 \ldots)$$

Which activities will enable you to achieve those results?

Figure 18-13 Relationship of Big Y's and vital X's

When a function converts the X's into a value of an output, the Y, this is known as a transfer function. Transfer functions can be developed from scientific principles, statistical analysis of empirical data, financial relationships, experiments, simulations, etc. Once it is known what X's drive the Y, then the next step is to find what drives those X's. These X's, therefore, then become the y's. This idea that the required capability at the top, the Y's, imposes the performance at the lower levels, the X's, is called CTQ flow-down. This concept is illustrated in Figure 18-14.

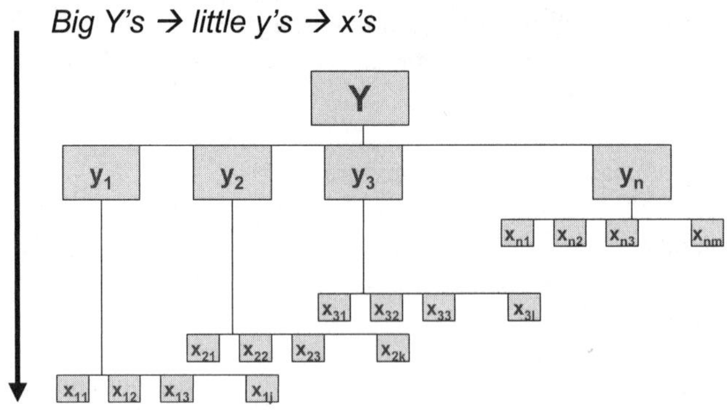

Figure 18-14 CTQ Flow-down

Cascading QFD's can help with this activity of translating requirements at the top to requirements at the bottom. For example, in the cat carrier project "Easy to open/close" may eventually be driven down to a CTQ on the distance from the edge of the top of the container to the placement of the hinge for the door (i.e. the distance between the edge of the top and the hinge placement should be 2 inches).

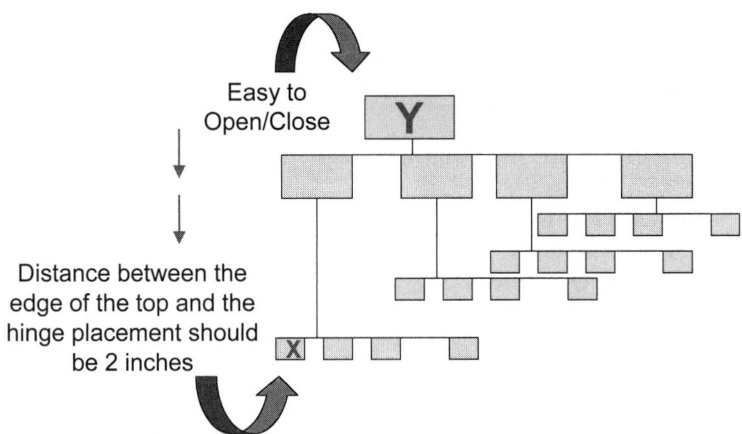

Figure 18-15 CTQ Flow-down for Cat Carrier Example

As the team looks to developing design alternatives in the Analyze phase, creative thinking techniques (as discussed in the Improve chapter) are important to help the team "think outside the box."

Analyze Phase Tollgate Questions

The team should leave the Analyze phase with some top-level design alternatives. The Tollgate Questions for this phase are:

- What are the key X's?
- What are the design options?
- Have you tracked the CTQ Flow-Down?
- What data did you collect and analyze? What were your conclusions?
- What assumptions (paradigms) did you uncover?
- What are the principle sources of variability?

Design

In the Design phase, the team needs to translate high-level design characteristics into optimized design parameters, identify and document the CTQ Flow-Down from CCR's through design parameters, assess sensitivities of CCR's to variability in design parameters, perform tolerance analysis, identify design gaps versus CCR's and review priorities, and approve trade-offs as needed.

In the Design phase, the deliverables are:

- Validated and Refined Models
- Feasible Solutions
- Trade-offs Quantified
- Tolerances Set
- Predicted Impact

From the Analyze phase, the team has a set of documented design alternatives and an understanding of the variability allowed in the Y's. The team will now refine the models developed between the X's and the Y's. In addition, the details need to be defined for the design alternatives to end up with feasible solutions, with trade-offs quantified for the solutions and the predicted impact in terms of the design goals.

The details required for the selected solutions will be done to the lowest levels. The designs the team develops will contain a number of subassembly parts or processes. They need to ensure that the quality requirements for the total system or product are met. Quantifying how the capability of Y drives limits on the allowable variability in the X's (and vice versa-how variability in the X's translates into variability in the Y) can be difficult. The team needs to quantify the CTQ Flow-Down so that attainable variability in the X's yields acceptable capability for the Y. Values for the X parameters should be chosen which not only produce the desired value for Y, but also minimize the variability of Y. The team will need to understand how variance can be allocated across the X's and conduct a sensitivity analysis associated with the design selected. Simulation, Design of Experiments and Response Surface Methodologies are tools that are useful for these activities.

A problem with CTQ Flow-Down is that there are many solutions so the team needs to find one that is tractable and one that makes sense. The VOB considerations (such as cost) may make some of the alternatives much less desirable than others. This decision process of distributing the technical slack is called Variance Allocation.

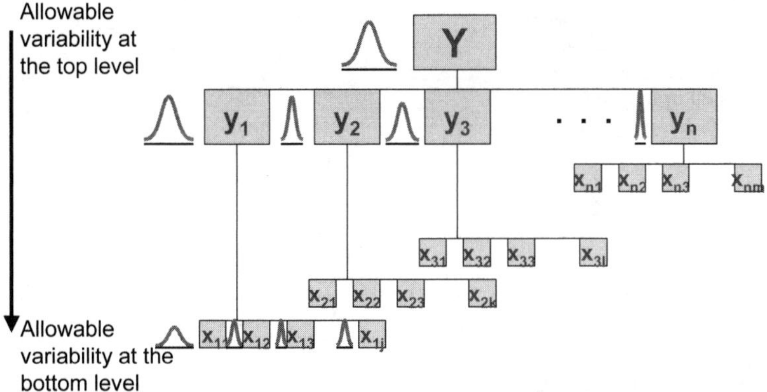

Figure 18-16 Variance Allocation

To help determine variance allocation, sensitivity analysis is used. Sensitivity analysis establishes the link between the variability in the X's and the variability in the Y's through the transfer function. Its principle use is to understand whether input variability will cause unacceptable changes in output performance. It is used:

- As a guide to understanding the need for and benefits of reducing variability in the X's
- As a means for defining and allocating what constitutes acceptable variability in the X's

At the end of this work, the team should have CTQ's with tolerances set, such as:

- The distance between the edge of the top of the carrier and the hinge placement should be 2 inches with allowable variability of ± 1/16th of an inch.

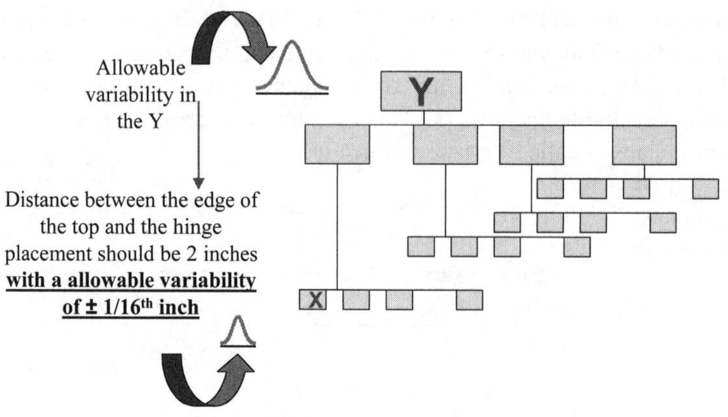

Figure 18-17 Variance Allocation for the Cat Carrier Example

Design Phase Tollgate Questions

At the end of the Design phase, the team should have a final design for the product/process selected with an understanding of the pro's and con's associated with the design. The team and Champion might also discuss the following Tollgate Questions:

- What design or improvement options did you consider? What ideas were discarded?
- How do other companies approach this? Is this an industry-leading solution? What is your ideal solution?
- How did you gather the data you used to make the best choice of design parameters?
- How did you optimize the details of the design?
- What trade-offs had to be made between competing Y's?
- Describe the CTQ Flow-down. What is the variability in the X's?

Verify

In the Verify phase, the team needs to identify potential design/process failure modes, use predictive tools to exhibit how the design satisfies the customer requirements, assess the capability of the design/process, verify functionality via pilots and prototypes, and document and train as required to assure successful hand-offs to the process or product owner(s).

In the Verify phase, the key deliverables are:

- Detailed Design
- Validated Predictions
- Pilot / Prototype
- FMEA's
- Capability Flow-Up
- Standards and Procedures

In DMADV, the team is building or creating a product or process which will often require a large investment. So as a practical matter, the team needs to determine a way of ensuring the design selected from the previous phase is going to work. There are a variety of ways to do that. Of course, one of the best ways is through experimentation. Another familiar tool used in this phase is Failure Modes and Effects Analysis (FMEA). In a product design project, the FMEA will focus on product failure modes and measures to prevent these failures before the product is accepted as final.

In addition, the team may conduct a Capability Flow-Up. This is the concept of:

- Setting the X's according the optimized values
- Using the Transfer Function to "flow-up" to the CCR's
- Determining whether the Y's hit the target values

By doing this, the team is using predictive tools, such as simulation, to verify mathematically the capability of the design. Figure 18-18 shows how the work done in the Design phase is checked in the Verify phase through Capability Flow-Up.

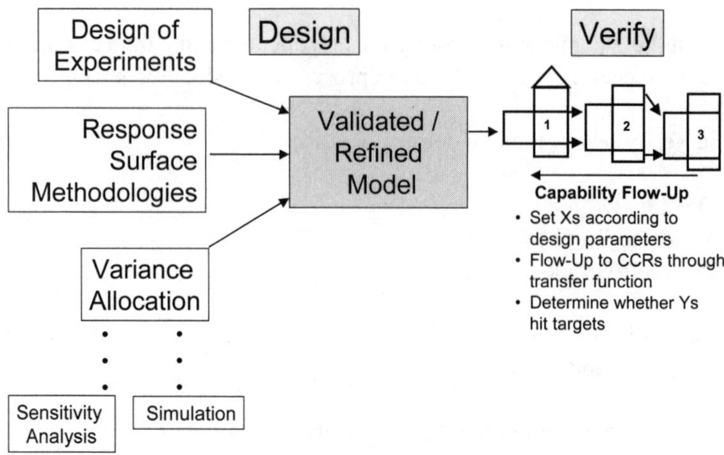

Figure 18-18 Capability Flow-Up

For final verification, the team needs to consider what pilots or prototypes are required for validation. Functionality can then be verified through these pilots or prototypes.

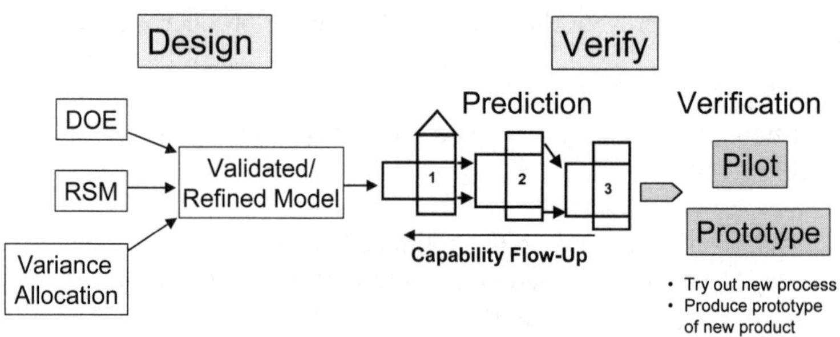

Figure 18-19 Pilot or Prototype to Verify Functionality

After the pilot or prototype, the team needs to deliver the detailed design. They will need to create a hand-off plan, including documentation, to transfer responsibility to the product or process owner.

Verify Phase Tollgate Questions

The Tollgate Questions in this phase are:

- Have you attained the expected capability?
- Are there gaps in the capability? What CCR's are at risk?
- What efforts were undertaken to validate predicted performance?
- What procedures do you plan to put in place to assure a successful hand-off to operations?

Summary

DMADV is an outgrowth of DMAIC that has a considerable overlap of toolsets. DMADV was motivated by the need for stronger VOC, disciplined CTQ Flow-Down, and more active investigative tools. DMADV is somewhat more amenable to quantum leaps in design and performance where DMAIC is somewhat more continuous improvement oriented.

A team may choose DMADV because it has some tools available that are not typically used in DMAIC such as stronger VOC and QFD. This doesn't mean the team has to use DMADV. DMAIC and DMADV are not all that different:

- With DMADV, there is a heavier emphasis on the VOC and CTQ Flow-Down.
- If the team needs to take a quantum leap, then DMADV may be appropriate.
- DMADV projects may be slightly more complex. But it's slight. It's worthwhile enough to have the additional toolset.

The important consideration is the scope of the project and understanding it well enough to identify the tools needed. The process used is more a matter of emphasis.

Sometimes a team may get to the Improve phase and switch into DMADV. But usually, teams will try to look ahead and make a subjective decision about which to use. Is this a design or redesign?

So, DMAIC or DMADV? That's not what's important. What is important is that the team accomplishes its goals.

Part Five:

Six Sigma Impact Measurement

The chapter in this section focuses on measuring the results of a Six Sigma campaign. It will first review the importance of metrics and measurement to all improvement activities. Then it will focus on the need to have a selected, few critical strategic metrics to drive the Six Sigma campaign. It will discuss some historical difficulties with metrics and improvement campaigns, and suggest guidelines for selecting meaningful metrics and goals for organizational performance improvement. The discussion will not be limited to financial goals and metrics. Business objectives that drive Six Sigma Campaigns must be multifaceted. They must balance among:

- customers and markets
- shareholders and financial results
- employee, associate, and personal development
- processes (the way things get done), and
- the environment.

Nevertheless, financial results are often considered the ultimate measure of organizational performance. The chapter will present two approaches to building useful financial measurement systems. The chapter will then end with a broader discussion of the Six Sigma Business Scorecard, a tool for comprehensive organizational performance measurement and tracking.

Chapter

19

Financial and Performance Measurement

The Importance of Metrics to Organizational Performance

"What gets measured gets done." This old management adage may not hold true 100 percent of the time, but by and large, it applies to individual and organizational performance.

That's why an effective, practical measurement system is a cornerstone of any successful Six Sigma Campaign. Effective measurement starts with the work done by the leadership team. It's critical to identify the priority metrics and goals for the organization. Without that guidance, it is difficult for the campaign to stay focused on issues that are strategic to the business.

In his book *Good to Great*, Jim Collins presents the findings of a his research team's study of "Fortune 500" companies that, for a fifteen-year period produced average performance, then began to out-produce the general stock market and their own sectors. [1] The team identified eleven companies that were able to produce cumulative financial returns at least triple those of the general stock market, and then sustain those cumulative returns over fifteen years. Collins describes the common attributes of those eleven companies. One of the attributes was measurement. Each of these companies identified one metric as the supreme metric that stood above all others. In Six Sigma language, that's an organization's "Big Y".

All eleven companies changed their top priority metric to something new, early in their leap to greatness. [2]

---

[1] Jim Collins, *Good to Great: Why Some Companies Make the Leap...and Others Don't* (New York: HarperColliins Publishers, Inc., 2001)

[2] ibid

- Kimberly Clark moved from Return on Net Assets (RONA) to profit per brand.
- Gillette switched from profit per division to profit per customer.
- Wells Fargo swapped profit per loan with profit per employee.
- Kroger changed profit per store to profit per local population.
- Walgreen's, similarly, changed profit per store to profit per customer.

The common change made by these organizations in their priority metrics was that they switched from an internally oriented metric to a more customer-focused metric. This focus is remarkably similar to Six Sigma's emphasis on the Voice of the Customer (VOC), and it ensures that the organization's strategic objectives align with customer expectations. And, just as an organization uses its strategic objectives to drive the Six Sigma Campaign, each of these organizations used its key metric for decision-making and driving and guiding the behaviors of the organization and individuals.

A Few Critical Metrics

Every organization should promote its number one priority metric. And, many successful organizations typically have up to three metrics, listed in priority order. Once the number of "critical" metrics exceeds five, leadership (and the rest of the organization) loses focus.

It is imperative that one of the "Big 3" metrics be a customer-focused metric. The other metrics should reflect a balanced approach to leading the organization and its people.

Be careful not to select metrics that are too broad Financial metrics like Economic Value Added (EVA) and Return on Investment (ROI) or RONA are good overall numbers for investor analysis. But they are the result of everything the organization does. Once you move past the executive level, it's hard for any individual to know how their actions impact those numbers. The examples from *Good to Great* shown above are also broad numbers, but they provide a little more focus on where people should direct their efforts. That's what a Six Sigma "Big Y" is a measure that is relevant, focused, and important to success.

Selecting a strategic metric is not the same as the traditional Management

By Objective (MBO) program. MBO drove a lot of compartmentalized behaviors when people tried to optimize small pieces of the process for which they had control with little regard for the big picture. The strategic objectives and metrics used will drive behaviors. So the organization needs to select meaningful Big Ys that will drive cooperation across functional boundaries.

Historical Issues with Metrics and Improvement Campaigns

Many improvement initiatives have come and gone over the last twenty years. Leadership in many organizations have adopted, and then abandoned, the initiatives for several reasons. One reason was that the improvement campaign's actual results did not meet leadership's expectations. Another more common reason was that the improvement campaign actually produced the expected results, but the measurement system failed to show those results to leadership. The organization simply did not have a way to keep track of and demonstrate the "real" impact of the improvement initiatives. The reported savings strained credibility, so over time, the initiatives simply drift away, to be replaced by the next wave. The balance of this chapter has some information about performance metrics, but most of it focuses on the problems with improvement initiative's visibility and makes a few suggestions on ways to address these issues.

This chapter will discuss some of the mismatches between accounting systems and improvement campaigns. But it will start by examining another common reason an organization abandons an improvement campaign - lack of clear goals and metrics for the improvement campaign that are aligned with the organization's business goals and objectives.

Lack of Clear Goals and Metrics Linked to Measurable Business Goals

In the early 1990s, one of the largest electrical utilities in the United States started a Total Quality Management (TQM) campaign. The organization did many things right. Leadership had a Vision directly linking customers and quality. The leadership team from the CEO on down was actively involved in the improvement effort and spent considerable time; company management partnered with the union for the first time ever, and both groups lined up support for the quality effort. People were trained, the methodology they followed was strong, and employees for the most part were enthusiastic, although cautious.

The voice of the customer and the voice of the regulator were clamoring for reduced rates, and the industry was evolving to a more open, competitive environment that would naturally drive rates down. "Reduced electrical rates" should have been the mantra of the improvement campaign. But the leadership team on both the management and union side refused to accept this or to commit to using the improvement campaign to achieve it. Instead, they committed to three soft business goals that really had little to do with business objectives and certainly had nothing to do with rates. That was a fatal flaw in the improvement campaign.

This fatal flaw ultimately marginalized the overall business performance improvement effort. Here are two examples of problems that arose because the improvement effort's metrics were unclear and not aligned with business objectives:

1. As the different power generation and distribution facilities looked for improvement opportunities, they would look to the goals and metrics set by leadership. Since they were not aligned with the business goals, the improvement opportunities worked were not always truly important to the business. Many of the projects were good ones, but precious resources and time were wasted on unimportant activities.

2. When "real work" improvement opportunities came up, there was a reluctance to address them. Consultants often call these situations the organization's "political problems." At one point, a union welder suggested that some of his peers were not welding in the right way. The cross-functional team of employees that reviewed and prioritized improvement opportunities argued over whether or not this was an "appropriate" opportunity. In another case, a young engineer submitted a technical project to the cross-functional team. In the latter case, the Vice President of Engineering heard about it and became upset that the young man had used another channel for an improvement idea.

The point is not whether each idea was, in fact, a "good" improvement idea. The point is, rather, that without hard, measurable business goals for guiding the improvement activities, the organization did not have a platform to have a reasonable discussion about the opportunities. Project selection criteria were reduced to opinions and beliefs, a situation all organizations must strive to avoid.

The failure of the improvement initiative was, in fact, preventable, especially since the organization had done so much right. But in the end, without clear, meaningful goals they lost focus. If the organization had started the TQM effort with the number one goal of "reduce rates while maintaining shareholder value," the outcome could have been much different.

Mismatches between Traditional Accounting and Improvement Campaigns

Accounting is a tradition-based profession. Much of business accounting, especially in manufacturing organizations, focuses upon getting the cost of inventory right. Traditionally, accounting also focuses on financial metrics. Because one of the metrics in most improvement campaigns is financial, leadership relies on its accounting system to show improvement opportunities and results. Unfortunately, more often than not, the accounting system hides the real benefits, making it difficult for leadership to make the right decisions.

Traditional Accounting Can Hide True Performance

This is a quick example of where traditional accounting-based decision-making can mislead or misdirect an improvement effort.

	A	B
Selling Price	$150	$175
Cost of Goods Sold:		
Materials	$30	$30
Labor @ std.	$20	$30
Overhead	$60	$90
Total Cost of Sales	$110	$150
Gross Profit	$40	$25
Cost of Special Services	$30	$0
Product Profitability	$10	$25

Notes:
Product A has a scrap factor of 50%
Customers for A return 30% of the product, demand special pricing....
Product A consumes 2x as much energy as Product B
Product B has no physical selling cost, people place orders electronically

Figure 19-1 Traditional Accounting View of Product/Service Profitability

Suppose as part of an improvement effort an organization is looking to expand its product or service line. The above statement of product/service profitability is being used to assess which product or service should be targeted. If the decision is made at the "Gross Profit" level, 'A' looks to be a better product. If other costs are taken into account, then 'B' starts looking better. Traditional accounting allocates overheads, usually based on labor or machine hours. Labor today probably constitutes less than 20% of a product's cost. Overhead is likely to be at least three times that number. It does not take much to see how this could begin to mislead an organization. While not shown in this example, overhead is also allocated to inventory. This can lead to the interesting financial result that sales and shipments go up for a month, while profits go down because of decreasing inventory levels.

A manufacturer of confectionery products in Australia had a similar problem. The organization was in the process of strategically repositioning their customer base. The company was dropping small "mom and pop" stores, where it was definitely losing money, along with mid-size distributors where the organization thought it was losing money, in order to focus on large retailers. When the company performed a simple customer profitability analysis, management was alarmed to learn that the most profitable business was the middle market and that the company was barely breaking even with the large retailers, given all of the specialized services and promotional discounts it was providing.

To avoid situations like this, many companies do profitability and customer analyses, but often get a misguided picture due to accounting allocations. One way to begin is to roll-up products into product families. Womack and Jones, in their *Lean Thinking* book, discuss Value Streams for certain products, which is a very similar tactic. And when an organization moves to cellular manufacturing, management is essentially moving toward the ability to assign direct (real) costs to a family of products or perhaps a group of customers. This enables the product profitability numbers to become much more real. The organization can almost begin to look at profit centers within the business. There are a host of people and management issues that arise with that approach, but at least the numbers are reliable.

Real Numbers - Management Accounting in Lean Organizations by Jean

Cunningham and Orest (Orie) Fiume with Emily Adams (Managing Times Press, 2003) treats this concept in more detail. The authors discuss materiality and other accounting requirements. They provide practical Lean accounting principles and examples. They also discuss the problems of valuing inventory, and show how lean performance improvements on the operational side can provide more flexibility on the accounting side.

Complex Roll-Up Metrics are Difficult to Comprehend

Many of the performance metrics provided by traditional accounting reports are roll-ups of many other metrics. Outside of accounting, most individuals in an organization would be hard pressed to articulate where these performance numbers come from and what affects them. For example, how many people would know what the impact would be on RONA if Accounts Receivable decreases by 25%? Very few!

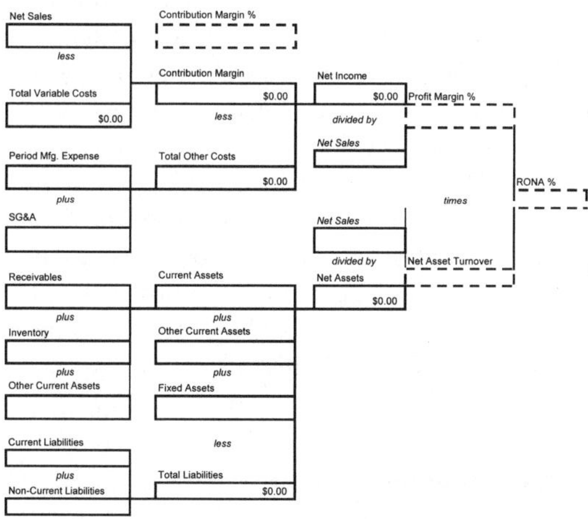

Figure 19-2 Relationships Between RONA and Accounts Receivable

An organization can take steps to educate itself about this type of performance number. For example, this diagram developed years ago by The DuPont Corporation shows the relationships between RONA and Accounts Receivable as well as other contributing metrics. This form can be easily put into an electronic spreadsheet to create a model showing the impact on RONA if any one of the individual boxes changes. The problem with RONA as the top performance metric is that it's the end result of everything. It's an important number, but it is not a good driver for performance improvement.

Accounting Information Lacks Timeliness

Traditionally, accounting information is also less than timely. Imagine you are bowling. You pick up a ball, take three steps, and roll the bowling ball down the lane toward the pins. Suddenly, a screen drops down between you and the pins. Shortly after that, you hear the pins crash, and you shout out, "How did I do?" After a minute a voice on the other side of the screen responds, "Don't worry about it. I'll send you a variance report." The lack of timeliness is typical of traditional after-the-fact accounting reports. Improvement initiatives require information in real-time or almost real-time. This is especially true of Six Sigma campaigns, where decisions are based on data.

Accounting Processes Need Improvement

There is a phenomenal amount of waste in most accounting departments that interferes with the timely reporting of information. People are busy working but many of their activities are not value-added activities. Tasks such as making files for accounts payable vendors, filing and storing copies of customer invoices, calculating journal entries down to nickels and dimes for accrued rents, and other miscellaneous transactions do not add value to the customer. A new mindset is needed. Given some of the ethical misdeeds over the last few years, no one is suggesting taking short cuts, but significant improvement opportunities exist in this function. Accounting can become a real value-adding peer to engineering and operational groups in improvement initiatives.

Meaningful Performance Metrics

Characteristics of Meaningful Metrics

Performance metrics and their related goals should be critical to success. Earlier in this book, Dr. W. Edwards Deming was quoted as saying, "85% of all productivity problems are process related." From the perspective of improvement then, it would make sense to focus on process metrics. People who operate processes can more easily relate to these types of metrics because they deal with processes on a daily basis.

Art Byrne, Wiremold's CEO, says if he were forced to use just two metrics for the entire company, he would choose customer service and inventory turns. Byrne explains, "A company cannot have high customer service - which might be measured in part by the percentage of on-time deliveries - and high inventory turns without doing a lot of things right. High inventory turns mean the business has achieved high velocity and eliminated significant waste through-out the organization." [3] These two metrics would also show change faster than a metric like RONA or EVA, thus allowing corrective actions to take place earlier.

There are five key requirements for meaningful performance metrics. This chapter has touched on them in many of the shared stories; they are summarized here.

1. Metrics need to support the organization's business strategy
2. There cannot be too many metrics or the organization loses focus.
3. Most metrics should be process-related (leading rather than lagging - refer to the Lean Teams chapter).
4. Metrics should measure actual results on a timely basis.
5. Metrics should reflect a balance among the four balanced dimensions of organizational performance.

Balanced Metrics

Strategically, metrics should maintain a clear focus on all the key dimensions of organizational performance:

1. Customers best captured from a customer perspective, not an internal perspective
   a. Loyalty factors (referrals, retention, new product buys, % of wallet)
   b. Delivery (on-time, complete, undamaged)
   c. Response (timeliness, complete, # of call backs)
2. Processes find the leading indicators that promote early action
   a. Quality (# of defects, returns, misquotes)
   b. Time (lead, process, cycle, on-time)
   c. Material (first pass yields, scrap, rework, inventory)
   d. Productivity (downtime, yields, rate of improvement)

[3] "Real Numbers - Management Accounting in Lean Organizations" by Jean Cunningham and Orest (Orie) Fiume with Emily Adams, Managing Times Press, 2003.

3. Financial
   a. Revenues (growth, new products % sales)
   b. Income (margins, certain cost as a % of sales)
   c. Cash flow (order to cash)
4. People
   a. Safety (lost time, injuries)
   b. Skills (learned, used, certified)
   c. People (turnover, days needed to hire replacements, referrals)
5. Environment
   a. Material (toxic waste, regulatory compliance)
   b. Energy (units used, cost per unit)

The metrics shown are not intended to be the prescribed metrics. They are simply representative of types of metrics that could be used. Different business depending on their life cycle, industry, and competitive position would look at different things. The metrics selected must be relevant, meaningful, fairly easy to capture, and easy to understand.

Metrics that Provide Continuous versus Discrete Performance Data

Consideration could also be given to continuous versus discrete data. For instance, if a Six Sigma team was interested in on-time delivery of products, the data could be collected as continuous or discrete. A continuous version of the data might be to collect exactly how many hours (or days) the product was early or late. A discrete version of the data would be to collect whether or not the product was on time (e.g. Yes, it was on time or No, it was early or late). Continuous data are preferable because more analytical tools are available and it usually requires a smaller sample size to make informed decisions.

Stretch Goals

One could also argue that there should be stretch in the goals set for every metric. In their book *Built to Last*, Jim Collins and Jerry Porras describe BHAGs - Big, Hairy, Audacious Goals. Similarly, Cunningham and Fiume in Real Numbers discuss the importance of selecting metrics that motivate the right behavior.

Setting "Good" Performance Goals

The goals should motivate, but if people think the numbers are unfair or meaningless, they frequently "work the system" to make the measure look good, even if it comes at the expense of the overall business. Motorola University consultants have come across numerous incidents where people hid maintenance costs, customer problems, etc. One call center had a metric of a maximum of 60 seconds per a phone call. The customer service representatives would actually cut off customers after 60 seconds, pretending the call was finished.

Track and Show Progress

Many of the performance improvement activities over the last twenty years fell by the wayside because it was difficult to track the actual financial and other benefits. Organizational performance did improve, but it was difficult to know how much benefit, if any, came from TQM, CQM, TOC, etc. A few performance improvement activities like Cost of Quality and Activity Based Costing actually resulted in second accounting systems. The complexity and non-value added work from keeping two sets of books ultimately doomed those initiatives as well to the long list of three letter improvement programs that were no longer embraced by leadership teams.

The *Six Sigma Management System* strives to prevent that from being the fate of Six Sigma Campaigns. T*he Six Sigma Management System* recognizes there are many drivers of improved "Big Y" performance. Leadership selects the objectives and metrics of organizational performance that will be tracked. This guides the selection of DMAIC projects. Each DMAIC project should then define the x-variables it aims to improve, the little y-variables and "Big Y"-variables that will be impacted. Rigorous governance of DMAIC projects ensures results are tracked, mid-course corrections are made, and results are published to the organization. That helps keep the Six Sigma campaign on track and mobilized.

Constructing Useful <u>Financial</u> Measurement Systems

Earlier in this chapter, the importance of balancing financial with non-financial metrics as part of the measurement of strategic organizational changes was discussed. Modern business leaders clearly recognize the need for non-financial metrics. In most cases, these non-financial metrics provide better information for corrective action than financial metrics. They are timelier and show a better picture of actual process performance. But at the end of the day, leadership, shareholders, and other key groups want to know about the financial results. For better or worse, they are the "ultimate" metric.

This section presents two tools organizations can use to track the financial gains from these multi-faceted performance improvements, in order to provide the credibility required to sustain a company-wide performance improvement effort over time. The basis for both of these tools is the "Financial Scorecard." A Scorecard is the conceptual framework that determines which dimensions of organizational performance will be measured.

A common type of scorecard is the scorecard that scorers and fans use to track what happens in a baseball game. A baseball scorecard allows for recording of each batter's performance (strikes out, reaches base on balls, singles, etc.), each runner's advancement (sacrifice fly, stolen base, error, etc.), and the involvement of each defensive player in getting outs (line out, double play, force out, etc.). Many scorecards also have space to track summary metrics such as number of strikeouts for the pitcher, number of errors a team makes, and the number of runs scored by each team. These are the views of the two baseball teams' performance that are measured.

A Financial Scorecard focuses the view of organizational performance on financial metrics. The chapter will look at two financial scorecards. Then it will cover a broader, multi-dimensional scorecard - The Six Sigma Business Scorecard - that takes into account many views of organizational performance.

Financial Scorecard #1 - Track the Year-to-Year Changes in Earnings

One way to show savings is to track improvements in earnings from one year to the next. It is helpful to develop a "scorecard" showing earnings changes in various categories. Some categories of changes would be related to improvement activities and some to general business conditions. The categories for any given organization should reflect its own environment. The following is one example.

Actual Y-T-D Changes to EBIT (000s) all in	Volume Related EBIT Margin Contribution	Product Mix Related EBIT Margin Contribution	Cost Savings	Productivity	CoPQ	Future Business Investment Dollars	EBIT Cumulative Impact (current year)	Cost Avoidance Benefits	Estimated Annualized Benefits	Targeted Annual Benefits	Comments
EBIT last year											
Six Sigma Project Savings Hard Dollars:							0				
Axial Vibration Project					50		50		100	125	Would explain why short?
Direct labor efficiency improvement				180			180		260	250	
Accel. New Product Dev	200						200		300	400	NPD proc. Chg resulted in $2 mil rev. inc. with !0% EBIT
Slip Ring Cost Reduction			75				75		150	200	Would explain why short?
Total Six Sigma Hard Dollar Savings	200	0	75	180	50		505	0	810	975	
Six Sigma Project Savings Soft Dollars:											
Axial Vib.Proj. saved business							0	30	30	30	Correction of Quality problems with Axial vibration allowed us to keep $2.7 mil in rev.
On time delivery Marine Bus. - increased revenues	25						25		50	75	On time delivery marine bus, kept $500K sales
Project x - reduction in Engineering Hours needed			30				30		50	40	
Total Six Sigma Potential Savings - soft dollars	25	0	30	0	0		55	30	130	145	
Specific General Business Changes											
Major customer pricing or raw material changes	(25)						(25)		(50)	(50)	Price reduction for marine business
Quality problem in gear mfg					(125)		(125)		(125)	(125)	Unexpected warranty cost
Other general business changes in EBIT						(35)	(35)		(35)	(35)	Six Sigma Training costs
EBIT Cumulative	200	0	105	180	(75)	(35)	375	30	730	910	

Figure 19-3 Financial Scorecard #1

EBIT in this example represents Earnings Before Income Taxes. Some of the changes from one year to the next are changes in product mix and changes in volumes. Product mix's impact on earnings changes could be done based on the difference in Gross Margins. The accounting system for most organizations could provide information on volume differences.

Last year's EBIT plus or minus the following items yields the current period (or YTD) EBIT:

- Major pricing changes (products sold or materials purchased
- Changes in unit volumes
- Changes in product mix (including margin improvements or give backs)
- Changes in productivity (especially if this improvement is due to a capital expenditure - in some instances this may be part of Six Sigma project savings)
- Cost of Poor Quality/Warranty

It's important not to get too granular with a worksheet like this. If the focus gets too detailed, the scorecard would suffer from the same problems as "activity based costing" and become a second accounting system. Organizations may only want to track projects with more than $50,000 in savings. Businesses with less than $10 million in revenues may wish to have a $10,000 threshold. An organization should really be looking for a reasonable picture of what is going on with the business.

In order to tie out to the financials, the last column would be "all other." An accounting technical term for this is "plug." What this report provides is a view of what is happening with earnings that leadership typically does not see. This information should be able to provide new and meaningful insights into the business. If it does not, try a different tool.

Metrics That Impact Earnings

Real gains and benefits come from very specific places that are usually measurable to some degree. They can also be tracked into three different categories: hard savings, soft savings, and cost avoidance. Improvement initiatives typically target one or more of these metrics.

1. Hard dollar savings are directly traceable to the bottom line and should be primarily attributable to one of the following:
   a. Revenue growth - growth in volume or margins from specific sources such as existing customers and existing products or new products and new customers. Projects should indicate which groups are targeted and how much revenue is expected. When tracking margin growth or increased volumes, be careful not to double count the growth and any related cost savings. For example, one organization tracked how freed sales time was used for more face-to-face customer time, and recognized the margins on increased revenue as a benefit. One would expect to see increased revenues in the following quarters as a result of freeing up those resources.
   b. Cost savings - savings related to a specific product/service or product/service family or specific customer or customer group. Cost savings come from a net reduction in resources used, such as materials, people, outside contractors, transportation cost, etc. Time is only a hard savings if the time is productively used for some new activity. Overhead or other indirect costs that are elim-

inated or reduced could also be direct savings, for example, eliminated warehouse space, reduced machine maintenance, reductions in energy usage, etc.

c. Productivity improvement - where the same resources are generating more output. The improvement in productivity can yield measurable direct savings improvements. However, if it does not decrease resources or increase throughput, then it is probably not a direct "business" cost savings. Productivity improvements upstream of a bottleneck operation or process step will usually not yield measurable bottom line (EBIT) improvement.
d. Reduced Cost of Poor Quality (CoPQ) - resulting from reducing defective product or service delivered. Many improvement projects focus on the Cost of Poor Quality. CoPQ. Improvements will often yield direct savings: scrap reduction and reduced rework or warranty costs, including costs material and labor, could all be direct savings.
e. Cash flow improvement - can yield hard savings through actions to reduce inventory, reduce cash receivables by cycle-time reduction, more accurate production scheduling, better supplier management, faster customer communication, reduction in errors, and better shipping management, etc.

2. Soft dollar savings are generally not directly traceable to the bottom line, but do represent a business benefit if the organization manages the freed time or resources in a prudent way. Soft dollar savings come from:
   a. Faster product/service cycle times - improved cycle times yield soft dollar savings in time or capacity savings when the time/capacity savings cannot be traced to a measurable resource reduction or revenue increase. An indirect, or soft cost savings, can be measured (e.g. set-up time reduction) based on the number of hours or minutes of saved time or capacity and the cost per hour of production or service delivery. Those soft savings could be turned into hard savings if the time or capacity savings ultimately produce, for example, increased revenue. A cycle time reduction project may have as its goal, "Reduce cycle time of 'n' automotive products in order to increase sales with 'z' customers by x%". The cycle time reduction and any related increase in sales would have to be tracked as metrics to see if the expected

hard savings were actually generated.

b. Freed up engineering and/or sales/service time - reducing the amount of sales time required for both sales and non-sales activities, or the amount of service representative time required for both customer service or support activities and non-service activities, can produce indirect or soft savings for an organization. The same is true with engineering time. The soft savings would be tracked as costs per hour, unless they could be traced to specific revenue increases or direct cost savings.

c. Freed up other indirect time - similarly, reducing any indirect time or overhead resources (such as less material handling, less time spent bill processing, less time expediting orders) can produce soft savings. The calculation is fairly straightforward - cost of the indirect time times hours saved. Again, there is only a "hard" dollar savings if resources are actually reduced.

3. Cost avoidance savings are tricky and organizations should not spend a great deal of time capturing them. Cost avoidance savings are categorized as indirect savings. They typically have no impact on gross margins or bottom line earnings because the expense was never incurred. If cost avoidance savings are pursued too aggressively, an organization could save their way into market or technological obsolescence and ultimately, into bankruptcy. So organizations need to employ common sense when looking for this type of savings. These are some examples of cost avoidance savings:
   a. Eliminating or reducing planned added labor and/or equipment and the associated costs through effectiveness or efficiency improvements.
   b. Avoiding a major capital expenditure through process or organizational improvements.
   c. Retaining customers who would otherwise be lost through product/service improvements or efficiency improvements that enable price decreases.

Financial Scorecard #2 - Goals versus Improvement Matrix

Another way to focus on savings, without tying directly into the financial statements, is to use a "Goals versus Improvement Projects" matrix.

Improvement Indicators, Goals and Projects				
		(Goals Examples for Illustration)		
Goals		Market Share	Revenue Growth	New Product Development
		Performance Indicators		
		Process Lead Time (Hours)	Total Revenues ($M)	NPD Cycle
Current FY Baseline		92.0	$850	12 months
Current FY Contributing Projects				
Lock in Specifications			$40	4 months
Quick mold changes		-15.3		
5S and Visual signals		-1.4		1 month
Materials staged in production sequence		-2.8		
Remove admin work from Salesperson			$100	
More face time with Customers			$50	2 months
Current FY Planned Performance	Curr. FY	72.5	$890	6 months
% Improvement		21%	5%	50%
Improvement Goals	FY+1	15%	10.0%	30%
From contributing projects	FY+2	25%	10.0%	25%
	FY+3	25%	15.0%	25%
Continual Improvement Target		25%	20.0%	25%

Figure 19-4 Financial Scorecard #2

This is a straightforward example with improvement goals stated in terms of percentage increases in market share, revenue, and new product development. The goals could also be stated as numbers. If an organization did not have too many improvement projects, it might list the major project deliverables in the left hand column under current year projects. If there are many projects, the matrix shows only the project name and the impact of that project in the relevant goals and metrics columns. Note that this example matrix takes a multi-year focus, with improvement goals stated for one, two, and three years out and beyond.

The Six Sigma Business Scorecard

The Six Sigma Business Scorecard provides the intellectual framework for measuring a comprehensive view of organizational performance. The Six Sigma Business Scorecard was developed by Motorola University consultant Praveen Gupta [4] in response to organizations' needs to measure and track the progress of all aspects of a business employing Six Sigma

---

[4] Praveen Gupta, Six Sigma Business Scorecard: *Creating a Comprehensive Corporate Performance Management System* (New York: McGraw-Hill, 2004)

process improvement. The Six Sigma Business Scorecard aligns departmental performance to the customer expectations of better, faster and cheaper products or services. Such business objectives can be achieved through employees' innovation, managers' improvement and leader's inspiration.

Measuring what is considered 'value' is a basic tenet of Six Sigma. Six Sigma is a major initiative that many corporations have adopted, but are not effectively measuring progress. Instead, organizations fall back to the financial measurements which are contrary to a balanced approach. Motorola Consultants have found that after the initial success of a Six Sigma initiative, it is difficult to sustain the initiative without compelling measures of success. There must be measures that track progress towards goals, that are accessible to everyone, and that can be regularly discussed or reviewed for trends in accomplishments. The Six Sigma Business Scorecard can be a critical management tool for organizations to measure the impact of Six Sigma projects and drive sustained improvement through Six Sigma.

Seven Elements of the Six Sigma Scorecard

The Scorecard focuses on process measurements and organizes those measurements into seven elements:

1. Leadership and Profitability (LNP)
2. Management and Improvement (MAI)
3. Employees and Innovation (EAI)
4. Purchasing and Supplier Management (PSM)
5. Operational Execution ((OPE)
6. Sales and Distribution (SND)
7. Service and Growth (SAG)

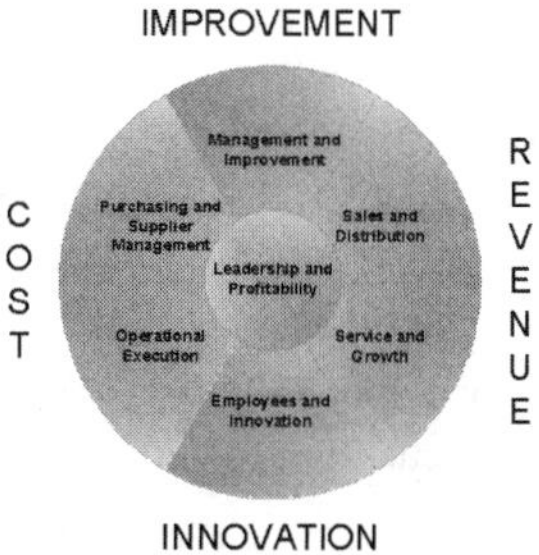

Figure 19-5 Six Sigma Business Scorecard Framework

The Six Sigma Business Scorecard framework balances sales and purchases, management and employee roles, customer service and innovation for growth, and execution for profitability.

Measurements

Gupta suggests some example areas of measurement and actual metrics for each category. [5] The actual metrics an organization selects for any given category should align with the objectives of collecting categorical data.

Category	**Objective**	**Areas of Measurement/ Metrics**
Leadership and Profitability (LNP)	Lead company to wellness and profitability	Communication Inspiration Planning Accuracy Community Perception Employee Perception Employee Recognition Compensation Asset Utilization Return on Investment Debt to Equity Ratio Profitability Shareholders' Value Growth

[5] ibid, page 70

Management and Improvement (MAI)	Drive dramatic improvement	Goal Setting Rate of Improvement Planning for Improvement
Employees and Innovation (EAI)	Involve employees intellectually	Innovative Recommendations per Employee Investment per Employee Number of Patents or Publications per Employee
Purchasing and Supplier Management (PSM)	Reduce cost of goods or services	Material Acceptance Total Spent/Sale Suppliers' Defect Rates (Sigma) Cost of Goods/Services Sold
Operational Execution (OPE)	Achieve performance excellence	Operational Cycle Time Process Defect Rate (Sigma) Customer Defects, Total Defects
Sales and Distribution (SND)	Manage customer relationships and generate revenue	Number of Inquiries New Business Dollars/Total Sales Dollars Profit Margins ($)/ Sales ($)
Service and Growth (SAG)	Gain competitive advantage and grow	Customer Satisfaction Customer Retention Repeat Business ($)/ Total Sales ($) New Products or ServicesPatents or Trademarks

The challenge to leadership is to select the measurements for the Scorecard. Gupta counsels, "The main purposes of the measurements are to challenge the existing system and identify opportunities for improvement and profitability. There is no absolute system that can be used from company to company. Each company's culture and measurement system

[6] ibid, page 85

are different. The system is not prescriptive; local adaptation is almost mandatory. This is so because the final measurement system must reflect true indicators of variances in a company's performance." [6]

Business Performance Index

Part of the Six Sigma Business Scorecard is a Business Performance Index (BPIn). The BPIn provides leadership with a snapshot of organizational performance. It enables leadership to monitor the organization for its growth and profitability in a straightforward and manageable form.

The BPIn is an index based on the ten measurements that are most critical to the organization's health and success. To build a BPIn, leadership must select those measurements, and assess how well the organization is performing versus its plan and projections for each. Leadership must also experientially weight the importance of each category of measurement. These data roll up into the BPIn.

Here is an example Business Performance Index.

Measurements	**Category Abbreviation**	**Importance**	**Initial Performance Guidelines**	**% Score**	**Weighted Score**
1. Employees Recognition (% of employees)	LNP	15	.2%- 25 .5% - 50 2% - 75 >5% - 100	100	15
2. Profitability	LNP	15	2% – 50 4% - 60 8% - 80 >12% - 100	65	9.75
3. Rate of improvement in process performance	MAI	20	<20% - 50 30% - 60 40% - 80 >50% - 100	20	4
4. Recommendations per employee	EAI	10	.5 – 50 1 - 60 2 - 70 >5 – 100	64	6.4
5. Total Spent/Sales	PSM	5	>60 - 30 45 – 50 35 - 75 <25 - 100	57	2.85
6. Suppliers' Defect Rates	PSM	5	3σ – 25 4σ – 50 5σ – 75 6σ – 100	48	2.4
7. Operational Cycle Time Variance	OPE	5	>50% – 25 40% – 50 25% – 75 <10% – 100	87	4.35
8. Process Defect Rate	OPE	5	$<3\sigma$ - 25 4σ - 50 5σ - 75 $>6\sigma$ – 100	31	1.55

9. New Business/ Total Sales	SND	10	20% - 25 30% - 50 40% - 75 50 %– 100	50	5
10. Customer Satisfaction	SAG	10	80% – 60 85% - 70 90% - 80 100% – 90	60	6
Total (BPIn)					57.3
DPU (-ln(BPIn/100)					.55687
DPMO (DPU/Exe)*1M					61974
Sigma					3.06

Figure 19-6 Example BPI

This BPIn is determined using ten measurements from the seven categories. The measurements encompass measurable processes, as well as soft processes, since total organizational performance is a function of all processes. In the example, the measurements include employee recognition by the chief executive for exceptional performance towards growth or profitability, and intellectual participation of employees through their recommendations. In a Six Sigma organization, idea management must become a well-defined process and implemented to utilize human capital. The example also includes Rate of Improvement in Process Performance as a critical measurement. This measurement is difficult to implement. A small percentage of employees know how much they, or their company, have improved during the last 12 months. Considering global competitive environment, one must focus on accelerating the rate of improvement. Continual improvement may not be sufficient; instead continual dramatic improvement would be necessary to sustain growth and profitability.

The ten measurements are listed in the left-most column. The category for each measurement is identified by abbreviation. The category significance, or Importance, assigned by leadership is allocated among the measurements in that category. For example, the total significance assigned to the Leadership and Profitability (LNP) category is 30; that is divided between the two LNP measurements. The significance score of 20 assigned to the Management and Improvement (MAI) is given to the single MAI measurement.

Initial Performance Guidelines are the standards that leadership has set for acceptable performance on the measurement, at different levels of improvement. For example, an achievement of 3s in Process Defect Rate is a 25% improvement, while greater than 6s represents 100% achievement of goal.

The % Score factors reflect leadership's assessment of organizational performance against plan on each measurement. A higher ranking means leadership believes it has a sound plan in place that is being executed effectively and producing desired results. This factor is multiplied with the Importance rating to produce the Weighted Score for each measurement.

The Weighted Scores additively produce the resulting BPIn as an overall indicator of corporate performance. It brings together all of the information into a single index number that can be tracked over time. The simplicity of one number is attractive to many organizations. However, organizations need to be cautioned against relying solely or too heavily on this single number.

Corporate Sigma

Once the BPIn is calculated, it is used to calculate corporate sigma. Organizations must have an indicator of overall sigma level of performance on those measures that are most critical. This is frequently overlooked in the Six Sigma journey.

To calculate organizational sigma level, the BPIn is used to calculate the organizational Defects Per Unit (DPU). Organizational DPU is then converted to organizational Defects Per Million Opportunities (DPMO). The calculation formula is:

$$\text{DPMO} = \frac{\text{Corporate DPU} * 1{,}000{,}000}{\text{Number of Top Executives}}$$

This formula brings the decision-making of top executives responsible for the organization's successes and failures at the highest level into the equation. Finally, the organizational DPMO is converted to a sigma value using the standard table. This Corporate Sigma level is a measure of the organization's variability in performance on its critical measurements.

Final Thoughts on the Six Sigma Business Scorecard

The Six Sigma Business Scorecard combines the Six Sigma methodology for dramatically improving customers' delight with proven methods for achieving financial objectives. The Scorecard was developed to help measure the achievement of better performance across the corporation. It:

- provides a new model for defining a corporate sigma level
- aligns with the business' organizational structure
- maintains visibility of cost, revenue, and profitability
- includes leadership accountability and rate of improvement

The BPIn can be used as a rough baseline of the corporate performance and as a catalyst to commit to an integrated corporate performance measurement system, the Six Sigma Business Scorecard. The BPIn can be used as a leading indicator of the corporate performance. Using the BPIn creates a strategic intent to implement the Six Sigma Business Scorecard.

This chapter has provided a brief overview of the *Six Sigma Business Scorecard. Much more extensive and in-depth information can be found in Praveen Gupta's Six Sigma Business Scorecard: Creating a Comprehensive Corporate Performance Measurement System* (McGraw-Hill, 2004).

Summary

The leadership team must develop some type of a performance measurement system that is suitable to achieve their process improvement objectives. This chapter described three different ways to track progress of improvement activities through Scorecards.

The primary objective of any measurement system is to facilitate growth and profitability together with ensuring a sound return on investment because of its implementation. The initial challenge to leadership is to select the most relevant and critical measures. Then leadership must devise measurable ways to recognize employees for improvement, determine the rate of improvement for each department, quantify innovation by employees, track new business as a portion of total sales, and measure operational excellence in service or support areas. Traditional measurements, such as total expenses, total sales, suppliers' performance, production performance, and customer satisfaction, are easier to quantify.

As measurement systems and scorecards are implemented, the question often arises about the ongoing efforts required. Continual improvement in the measurement system and scorecard itself should occur, whether the improvement involves setting new goals, easing the scorecard's maintenance, or integrating measurements with technology. Organizations must avoid the trap of making the reporting overly complex.

The value of measurement is in how the data are used. The value increases with timely response to the information gained from the measurements. Timely feedback to process owners enables faster and improved process adjustments. Integrating a tool like the Six Sigma Business Scorecard with the corporate information management system can produce faster response times. Using information technology, the collection, aggregation, analysis, and reporting of a Six Sigma measurement data can be automated.

Automating data collection and analysis gives the leadership more time to respond to the information instead of wasting time worrying about getting the information. Similarly, process owners can spend their limited resources in improving the process instead of supervising data collection. Ultimately, the intent of any measurement system for a Six Sigma campaign is to drive a rapid rate of improvement, experience improved profitability, and realize business growth.

Part

Part Six:

Supplemental Information

This part of the book includes two chapters as supplemental information on two specific topics:

- Chapter 20 discusses the topic of Innovation. Innovation is a key factor in achieving truly breakthrough (rather than merely incremental) improvements. It is a combination of art and science. The chapter includes a survey of some of the research and literature on the topic for consideration when looking to improve individual and organizational innovation.
- Chapter 21 delves deeper into the topic of Measurement System Analysis (MSA) previously covered in Chapter 13. But this chapter reviews more information about MSA specifically within non-manufacturing environments where MSA is often neglected.

Chapter

20

Innovating Breakthrough Solutions

Introduction

Achieving a dramatic improvement is a defining attribute of Six Sigma. Without realizing an innovative or breakthrough solution, one misses the main tenet of Six Sigma and lessens the opportunity to succeed. Six Sigma implies lots of improvement very fast. Incremental improvements are not sufficient to achieve Six Sigma performance. Almost every company has been improving its infrastructure, processes, and products. A common customer dilemma is to choose the right supplier. The obvious choice would be to select a supplier that is improving the most. Therefore, it is not just the current performance; instead, the rate of improvement equally matters.

Often times, Six Sigma teams struggle to produce significant improvement and miss the breakthrough opportunity. Leadership must address the challenge to realize Six Sigma benefits through breakthrough improvements, and accelerate the rate of improvement.

For years, continuous improvement has been the "quality" theme. However, due to competitive pressures and customer demands, corporations must aim at continuous dramatic improvement, which can be realized through innovation. The challenge Black Belts and Master Black Belts face is that an innovative methodology has not been incorporated into the Six Sigma toolbox. Some companies have utilized principles of TRIZ (a Russian name for the innovation methodology) which consists of observations based on the analysis of several hundred thousands patents.

However, one still needs to address how to make individuals think more innovatively.

Lack of innovative thinking at the leadership level and/or the project level can be a contributing factor to the failure of Six Sigma. Gupta, in *The Six Sigma Performance Handbook*, identifies the four basic skills that must be learned by individuals in order to achieve dramatic improvement and Six Sigma goals. They are as follows:

1. Time management
2. Process thinking
3. Statistical thinking
4. Innovative thinking

The absence of *time management* skills inhibit the execution of any planned activity. One must adhere to commitments made to others. Typically, people are not able to complete the action items assigned to them because of lack of prioritization. If an activity is scheduled and commitment is made, time must be found for the activity.

Process thinking alludes to Dr. Shewhart's Plan, Do, Check and Act (PDCA) cycle. The "Plan" consists of preparation for service processes, or set up for production processes. The "Do" includes steps to perform tasks involved in a process or projects. The "Check" represents the verification of the output of process activities. The "Act" implies correction of unacceptable process outputs.

Statistical thinking requires understanding of random and assignable variation. Random variation is variation inherent to the process design, while special cause variation occurs because of specific action. Statistical thinking allows one to make decision based on the understanding that variation could be due to natural, inherent, or specific causes. Once a person masters statistical thinking, it becomes easier to detect patterns and related causes, and make decisions accordingly, instead of over-reacting or excessively correcting a process. In other words, one develops a better sense of expectations and predictability of the process output.

Innovative thinking has been used predominately by a selected few who either were in R&D or were self-motivated to do something differently. If an organization plans to benefit from the Six Sigma initiative over a period of time, its leadership must institutionalize innovative thinking throughout the organization.

Understanding Creativity

In his book, *Cracking Creativity*, Michael Michalko indicates studies have shown that for most geniuses, there was no direct correlation between their genius and certain attributes. Even a 1600 score on the SAT, mastering languages, or high IQ scores do not guarantee a person will be a genius. There are some people that are more intelligent and less creative, as well some that are more creative and less intelligent. Geniuses, in general, think of many possible ways of solving a problem, some of which can be absurd, unconventional, or impractical. Richard Feynman believed that people comfortable with numbers tend to be more inventive when thinking of different ways for solving a problem. He suggested that one should learn in productive thinking instead of reproductive thinking (i.e. instead of reinforcing a rigid construct of problem-solving, one should learn a loose way of looking at things). Productive thinking leads to solution while the reproductive thinking leads to economic implementation. Therefore, develop *innovative thinking* by thinking and seeing in a way that no one else is thinking or doing. In other words, challenge the given and question the obvious. In order to solve a problem, one must be able to think of creating a problem in many different ways. Each way of creating a problem would isolate that way from solving the problem, and therefore, get closer to the ideal. Einstein's famous quotation, "Imagination is better than knowledge," would apply here in a sense that one must be able to see beyond what is visible. One must be able to see and think about different sides of the issue or a problem.

Some of the common steps to innovative thinking consist of the following:

- Visualize the problem in different ways, or from different angles
- Represent your thoughts in visuals
- Think fast and frequently
- Try different combinations
- Investigate the opposite side

- Think outside the visible space
- Look for disconnects
- Look for ignorance
- Think in teams, build on others' ideas

Learning *how* to think becomes more critical than learning *what* to think. Our current education and training process emphasizes to think in an established framework and application of it, instead of providing a framework to think and then apply differently. Many Six Sigma Black Belt training emphasizes the construct of DMAIC and associated tools. Black Belts apply these tools in a rote manner, instead of applying the tools to come up with a unique solution to a problem. They emphasize application of tools versus solving the problem using the tools as appropriate. One needs to provide the freedom of thoughts, and tools should expand the space of thoughts and facilitate imagination beyond the visible. Figure 20-1 shows creativity relationship between what and how to think.

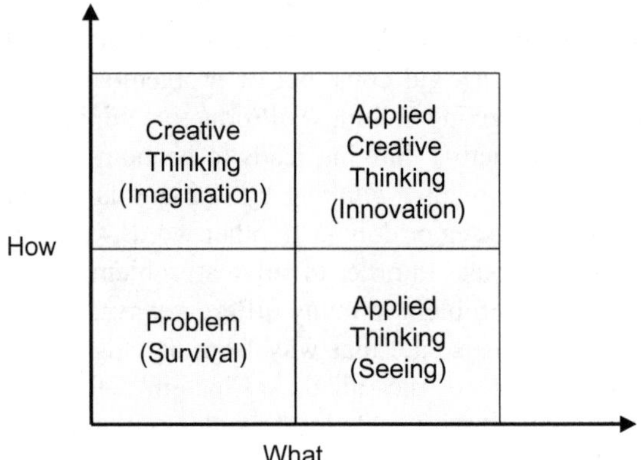

Figure 20-1 What Versus How to Think

The figure shows that when one thinks about what is needed, the focus is a quick fix and ignorant of its adverse side effects. The outcome is from adverse to incremental. When one is highly creative, application of creativity is overlooked, sometimes deliberately, to facilitate the necessary creative process for exploring innovative solutions. In this case, one is exploiting and expanding the boundaries of potential solutions. Once the creative thinking has been fully exploited, one can look at application of the creative thoughts towards solving similar problems. At this point, the breakthrough solutions may occur and lead to significant improvement in more than one areas.

Lynne C. Levesque, in her book *Breakthrough Creativity*, has analyzed combinations of two attitudes (introvert and extravert) and four mental functions (sensation, intuition, thinking, and feeling). Accordingly, she has classified people in eight categories: Adventurer, Navigator, Explorer, Visionary, Pilot, Inventor, Harmonizer and Poet. There are interdependencies among these eight classes of talents - for example, the Inventor combines the characteristics of Adventurer and Explorer.

The innovator's desire to innovate manifests through the questioning and penetrating thinking - delving deep into a subject very quickly. These people ignore the present status, or the current process, think of an ideal scenario, then logically move from a larger space of problem and solution to the solution in a detached and objective manner. The inventors analyze a situation internally through introspection and reasoning. They have developed a personal approach to putting a puzzle together based on their experience, categorizing their ideas, and clarifying and validating them using various events and experience. These people keenly observe everything that goes on around them, experience as many fields of life as they can, capitalize any opportunity to experience and learn from a different situation, and try to reflect on their lessens by combining various situations or variables. In other words, they continually simulate application of their newly gained knowledge until the ideas are clarified, and converged to a point of an application, whether solving a problem, or developing a new product or service. Learning is their motive and fun is critical for them. Observing various individuals in the industry who have been highly innovative, the author believes that for one to be creative, one must be sort of crazy to do different things beyond specified normal behaviors (not destructive), be able to fantasize or imagine as far as possible throughout the universe.

An innovator appears to show some of the following attributes:

- A quick system-level understanding to speed up the creativity process
- Unique and thoughtful ways to overcome any obstacles or get around constraints
- Focus on optimized solutions exploiting many constraints
- Ability to change paradigm
- Indecisiveness due to lack of information

To sustain the innovative thinking, one must continue to do the following:

- Make innovation the purpose of your life
- Identify your primary and auxiliary talents for creativity
- Learn new aspects of life, read books about different topics (e.g. art, science, math, fuzzy logic, universe, earth, nature, visit museums, study noble laureates)
- Learn new tools to use, including creativity tools
- Search for new ways to generate new ideas
- Learn to express ideas through verbal, written or visual communication
- Correlate the outcome with the reality
- Manage information to create more information

Discovering Personal Creativity

Everyone on this earth is born creative. Some people show creativity at an early age, while others manifest creativity at a later stage in life. Gail McMeeken, in her book 12 *Secrets of Highly Creative Women*, emphasizes expansion of personal creativity through results-driven engagement of creative traits. To engage one's personal creativity, one must acknowledge one's creative self through recognizing creative potential, experimenting, modeling successful application of creativity, and learning from other creative people. Creativity is not a geographically, personally or otherwise biased suite of skills that only a few possess. One can see creativity across the world, east, middle-east, south, north, and west, or all continents, throughout the history. Creative people are innovators and they sometimes come from nowhere. As George Bernard Shaw said, all progress comes from unreasonable people, so one cannot rationalize one's creativity. Instead, one must always seek to maximize personal creativity.

One of the ways to become more creative is to engage in many aspects of life around you. Whether reading junk mail, walking in the woods, playing with kids, cooking, sewing, rowing, laying on the ground and watching patterns in the clouds, swinging on trees in the wind, riding waves in the water, reading books on different subjects, listening to music rock to classic, staring at plants, worshipping in a church or temple, or roaming in museums, one can experience creativity at its best. One must decide to have an attitude that everyone is good, and wonder at everything around you. One must first admire if one sees something different, express a personal 'WOW,' and then ask the question how that special thing happened. What is thc creative thought behind the creative expression of an idea? Ultimately, one must see or experience so much that it starts generating new ideas. For an inexperienced person, new ideas come from being busy at different or multiple things. For an experienced person, new ideas come from reflecting on the experience already gained by imagining a combination of different experiences into a new experience. However, either path to creativity requires first, a decision to 'do things differently' and second, to 'just do something' or put energy into it, whether mental or physical. One must question the current and imagine the future, not to plan or panic for the future.

One way to expand one's universe of ideas is to absorb ideas behind the successful products, processes, or success stories. To be creative, one needs not be thinking big. Actually, one can be creative in thinking about creativity - learn from others' success stories by understanding the thought behind the success story and by asking the question "why." The success stories can be from peers, from superiors, or from strangers.

Ideas can be generated by following these suggestions:

- Look around for ideas continually
- Never criticize, wonder a lot
- Imagine the farthest, including uncharted, territories
- Roam around the world, all in the mind
- Visualize situations
- Handle multiple variables
- Competitively think of various combinations

The "never" with criticism is important as ideas are generated out of positive energy. Whenever one criticizes a situation or a person, it creates negative energy and mental stress. Criticism closes the mind, while support opens the mind. Therefore, while in the process of generating ideas, criticism must be minimized.

Great Innovative Minds

Michael Gelb, in his book *Discover Your Genius*, has identified several great people for their accomplishment and genius. Gelb studied great minds like Plato, Columbus, Copernicus, Einstein, Gandhi, Shakespeare, Darwin, Jefferson, Brunelleschi, and Elizabeth. According to Gelb, getting agood idea or discovering your genius requires manifestation of insight, intuition or inspiration that allows personal senses to experience some object from their unique perspective. Every baby demonstrates a spark of genius when born which fades away over time because of the environment and resulting expectations.

To exploit one's full genius potential, one must work hard with a framework for personal development. Most geniuses have prioritized learning to focus, and strategize passionately. To be passionately committed to innovation, one must become obsessed and restless. The ten great people studied in the Gelb's book had demonstrated some unique qualities. For example, Plato explored the questions of learning and making society a better place to live. Christopher Columbus ventured into unknown territories with a compelling vision, optimism, and courage. Copernicus became known for changing the paradigm of the sun revolving around the earth to the earth revolving around the sun by carefully presenting his theory. His approach of addressing a paradigm shift in today's rapidly changing environment could be useful. Einstein was known for his imaginative, childlike way to look at objects.

Cultivating Creativity

Denise Shekerjian, in her book *Uncommon Genius*, analyzes 40 MacArthur Award winners for the sources of their creative impulse. According to Shekerjian, every creative moment occurs due to the conscious application of raw talent, implying creativity or innovation does not occur by luck alone. There is a process of applying innate talents towards a personal vision by taking risks, staying loose, setting up the right environment, and learning by doing. Creativity is triggered by adverse conditions - people become creative because they are just bored, frustrated with something, mad at something, or challenged.

Cultivating an aptitude to be creative for breakthrough solutions is far more important than just inspiration for creativity, or luck. This requires identifying one's innate or peculiar talent, and honing it for good. It has been recognized that to identify one's particular talent in adulthood, one must go back and revisit their childhood and figure out what one loved to do or play with when their view of the world and themselves was not biased and they were not boxed into a framework. This must happen first to maximize creativity.

Creativity requires risk taking, i.e., no pain no gain. The risk of criticism, rejection, or failures by sharing ideas or activities with others is unavoidable for being creative. When someone steps outside the normal process or path to be creative, they take risks with a little chance of being right, at least initially. Innovative companies encourage risk-taking. Therefore, innovators take risks, and they have no fear of failure or losing their work. It has been observed that when people have nothing to lose, they become open to do anything. Many good ideas come when we are not looking for them. When we want ideas, none come to mind.

Staying loose requires that we create conditions or scenarios for doing that. For example, there are many rooms in a house. There is bathroom, kitchen, bedroom, or study room. When we go to the kitchen, we think about food. When we go to bedroom, we get ideas about sleeping. When we go to the bathroom, we have a purpose. When we want to study, we go to the study room. So why not create a creativity or playroom for everyone? The creativity or playroom stimulates our thinking and enable us to generate new ideas about different things. The creativity room may consist of contrasts such as a variety of technologies or equipment, different materials, lighting conditions, juxtaposition of art and science. This allows the thinking process to stretch beyond the norm. In a business environment, the creativity room becomes the meeting room. Every conference or meeting room should be full of creativity symbols. People should sense and feel creativity. Creativity, once triggered, is contagious. People build on each other's creativity. Then, there is no limit to creativity and no shortage of ideas.

People learn a lot more by playing or doing things. The creative room can work like a lab for people to try out new ideas. Even better, the entire facility can be creatively positioned or designed to create a sense of creativity all the time. In order for one to be creative, it is not limited to a room, or moment. Creativity is a continuous process. Experience is a name given to mistakes made. In order to be serially innovative, one must set a direction, even define it as the purpose of one's life. To improve one's luck, one must continually be looking for the opportunities and odd things, noticing details with curiosity or wonder.. By playing with different things, learning from things one sees, eventually, one gets the revelations, or gets the IDEA! According to Edward de Bono, a psychologist, the play must be purposeless without any intent or instruction. Remembering one's childhood, one is most creative when one plays with nothing.

Differentiating Terms

The following terms are based on variety of definitions available on the internet.

> Intelligence - The ability to learn, understand, comprehend, acquire, interpret, reflect and apply knowledge or skills, and improve based on the experience gained. Conventionally, intelligence has been measured using standard tests.
>
> Creativity - A person's ability to generate new ideas or new perspectives, not necessarily practical ones, of existing situations by imagining beyond the predetermined space of vision, or imagining an existing condition in a new way. Creativity manifests itself as an astonishment, revelation, novelty, freeing delight, originality, or extra-ordinary.
>
> Invention - A discovery or creation of an original or new, accidental or deliberate, concept, knowledge, process, apparatus, matter, or living organisms, or device, in mind, or through study, research, or experimentation that solves an existing problem or reshapes an existing thinking or culture.
>
> Genius - An individual with an exceptional intellectual or creative ability, talent, and originality to see things differently than others are seeing it.

> Innovation - A result of an intellectual initiative to introduce new elements through study, experimentation, discovery, or invention to produce a new process or product for the first time, or improve existing process or product that are significantly different or create significant value.

Figure 20-2 shows how a genius may use ones intelligence, creativity, and inventive instincts to generate new ideas, and develop innovative products or services.

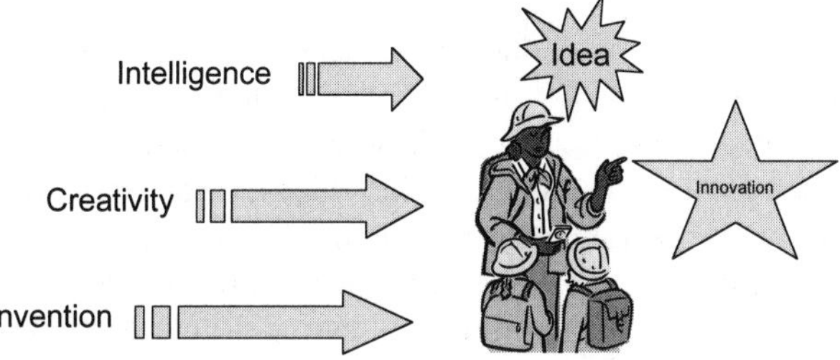

Figure 20-2 Path to Innovation

Innovating for Six Sigma

Six Sigma requires over 20,000 times improvement to advance from a sigma level of three to six sigma level of performance. Such an improvement, at least in terms of numbers, is unheard of in a normal business paradigm. To achieve an aggressive rate of improvement, one must plan to be creative and innovative. Leadership must promote employee involvement and innovation to achieve Six Sigma objectives. Without intellectual involvement of people, realizing breakthrough results would be impossible.

Clayton M. Christensen and Michael R. Raynor, in their book *The Innovator's Solution*, have emphasized sustained innovation in achieving corporate business growth. They indicate a successful era of superior performance in the life of a corporation occurs due to some innovative disruption. Sustaining innovation requires not just the ideas but rather the packaging of ideas for growth opportunities. In the Six Sigma world, breakthrough improvement levels are emphasized. However, methods to produce breakthrough solutions for a project have not been developed very well. The tools included in the DMAIC model allow practitioners to make fact-based decisions, or identify causes of problems that exist. The missing link to achieve sustained breakthrough solutions is the innovation methods, and even quantifying how much improvement would be called a breakthrough improvement. In other words, how to recognize innovation is occurring or just simply continual and incremental process improvement is taking place.

TRIZ

TRIZ, a Russian theory of innovation, is a systematic approach to innovation based on analysis of hundreds of thousands of patents. TRIZ was invented and structured by Genrich Altshuller, a patent examiner for the Russian Navy. Among hundreds of thousands of patents, he observed and compiled some repeating patterns. TRIZ provides the first understanding of the trends, or patterns, of evolution for technical systems. TRIZ captures patterns of creativity without addressing the creativity process itself. TRIZ can be considered an inventive problem solving technique.

Altshuller found that four major technical areas of innovation are Mechanical, Electromagnetic, Chemical and Thermodynamics. If someone is thinking creatively in certain areas, one can generate ideas in these technical areas. In order to innovate on demand, one must recognize the opportunity for innovation. If someone is struggling in making decisions due to more than one situation, that means there is some contradiction. TRIZ classifies contradictions in two categories:

Technical contradiction is when two competing technologies are involved. When trying to improve one characteristic of a technical system, the other characteristic must be compromised.

Physical contradiction occurs when two opposite properties are required from the same element of a technical system or from the technical system itself.

1. Separation in time
2. Separation in space

3. Separation in state of the matter
4. Separation between the whole system and its parts.

The TRIZ methodology consists of describing a specific problem, generalizing it, defining the ideal solution, then applying it to a specific problem. This is shown in Figure 20-3. To generalize a problem, one must go a level or two up to understand a problem at a system level. At a higher level, there are more opportunities with a bigger impact, therefore, there are more opportunities for innovation. At the higher level, the ideal solution is designed first. The ideal system must be more reliable, simple, and effective. TRIZ prompts a person to solve a problem without adding complexity to the system. When inventing a solution, one reduces barriers to a system achieving its ideal state.

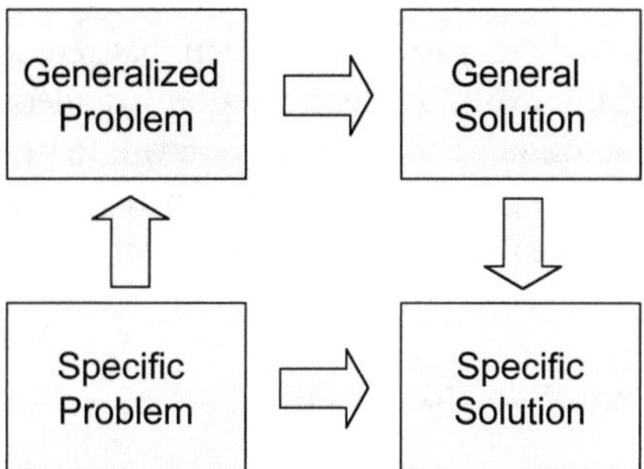

Figure 20-3 TRIZ Approach to Inventive Problem-Solving

Altshuller felt that main objectives of an inventive theory should satisfy the following conditions:

1. Be a systematic procedure
2. Guide through a broad solution space to the ideal solution
3. Be repeatable and reliable
4. Be familiar to inventors

Key TRIZ Principles

In reviewing the thousands of patents, Altshuller distinguished between incremental and breakthrough inventions. The breakthrough inventions were used to develop 40 Inventive Principles. When a contradiction is resolved, an innovation may occur. Key TRIZ principles include the following:

- Do it inversely
- Do it in advance
- Do a little less
- Save space
- Remove contradiction in time or space
- Fragment and consolidate
- Be dynamic

When solving a problem, one must ask why the problem occurred in the first place. Understanding the reason for the problem and thinking about it opens our mind to accept the current situation as it is. Then, one should look into what is the real problem. In other words, what is not happening as desired? This will get to the core of the problem. Sometimes, people try to solve a problem that does not exist and become frustrated. Recognizing a contradiction is a critical step that requires critical thinking. Finally, once the reason or contradiction is understood, one should imagine the ideal solution. The ideal solution would be the perfect world that one must visualize. Then, we channel our resources in resolving the contradiction to achieve the ideal state and investigate many possibilities.

Building an Innovative Corporation

Leadership for Innovation

Successful leaders recognize significance of innovation and the required leadership. The leader must believe and understand that the role an innovative culture can play in the growth of a corporation in future years. They consider innovation in all aspects of business, and consider all aspects of business in creating a culture of innovation. To lead an organization towards a learning and innovating entity, the organizational environment must influence thoughts, planning, and acts. Johnson Controls, an organization that has been existence for more than a century, has recognized role of innovation as stated in its Values as "responding to its customer needs through improvement and innovation."

To launch or sustain the innovation initiative, the leader must commit to recognize intellectual involvement of all employees, value of all information available, and evolution of all employees and processes. Today's competitive marketplace has moved beyond the perfection in products and services. Instead, customers are expecting value through innovation and flexibility. Therefore, leadership must make a strategic decision to explicitly commit and institutionalize innovation. The innovation must be incorporated in their communication, expectations, profitability, growth, and recognition. The expectation that inspires to innovate at all levels, from cleaning service to leadership, including the R&D personnel.

It has been known that in order for a person to be innovative, that person does not have to be intelligent according to some standards. However, the person is knowledgeable in what he or she does. Corporations that have been considered widely innovative include 3M, Rubbermaid, Microsoft, Chrysler, Sony and Boeing, implying that innovation is not limited to expensive products only. One can be innovative in designing a $0.39 container. An innovative organization must enforce innovation at the activity level. One needs to become innovative in everything one does, not every thing one produces.

The leader of an organization sets beliefs, initiative, and environment for innovation. The visionary leadership develops a corporate meaning of innovation in the organizational context, and develops a corporate strategy to learning and innovating success. The leadership establishes expectation and recognition for innovation from employees at all levels. The strategy involves training, internal promotion, innovation expectation and objectives, roles of executives, managers and employees, intellectual property aspects, and transformation from innovation to product or service for realizing economic benefits. The leadership, executives, and managers can set example through their own behaviors, attitude, innovative thinking and actions, and support for innovation. Eventually, it must pay to innovate.

Organization for Innovation

To prepare an organization for innovation, the organization must be appealling to innovators for its appearance, policy and procedures. One cannot practice innovation with lots of limits.

The most significant barrier to innovation, creativity or simply thinking, is fear of failure or punishment. The leadership must demonstrate empathy and practice participation, and the organizational policies must clearly reflect support for thinking and room for creativity. The organizational objectives must include intellectual engagement at all times. The measures must include not just the productivity, but instead, intellectual or innovative output. The Chief Information Officer may even like to become the Chief Innovation Office with the goal of transforming information into innovation.

In order to integrate and promote innovation in our normal daily activities, the organization must create the innovation model with resources allocated to make it work. In other words, the innovation must become an element of the profitability and growth streams. The innovation begins with ideas, i.e., there must be a mechanism to generate ideas from all employees of the company. The ideas are reviewed for relevance and applicability. In either case, the supportive feedback must be given for each idea. Employees must be encouraged to think and reflect on their experience and look into future with new ideas or innovative products, services or solutions. Innovative or learning organizations support some kind of in-house library where employees could browse through some books or literature to get re-energized intellectually.

A recommended organizational structure could incorporate the elements as shown in Figure 20-5 which shows that the culture of creativity cultivates ideas, ideas are managed into innovations, and innovations are transformed into products or services for economic gain.

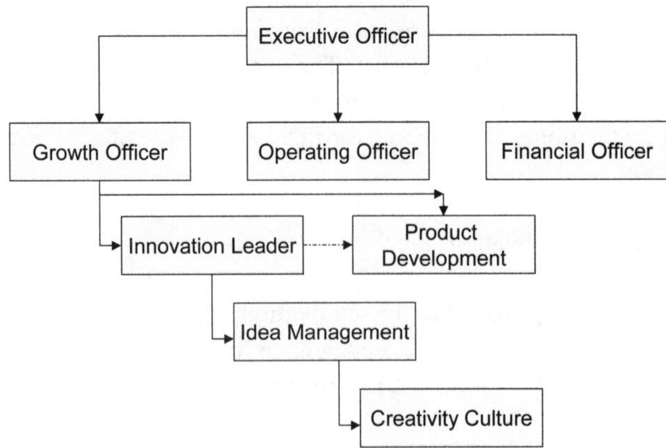

Figure 20-4 Organizing for Innovation

Culture for Innovation

If an organization likes to have ideas from its employees, it needs to create an environment, or even a room for creative ideas. Janine Benyus has said in her book *Biomimicry* that innovation can occur from a purposeful effort, can be "kissed by the muses," or as a result of a "flash of genius." As one creates a culture of innovation, the leadership must increase odds of innovation happening in either a purposeful manner or by bringing the genius out of its employees. According to Janine, there are three conditions for a successful innovation that must be met. They are:

1. innovation is work
2. innovators build on their strengths, and
3. innovation must have an economic impact.

Innovation occurs from hard, focused, and purposeful work demanding persistence, diligence, and commitment.

The purposeful innovative work takes place in the research and development departments in an organization. However, that may not be sufficient to maintain the competitive edge. Besides, the purposeful work tends to be driven towards the longer-term business objectives. Much of the competitive advantage is gained through innovation at work all the time at the activity level. For that to occur, the organization must provide the creative space. The author has seen companies with a creativity room. A creativity room allows thinking without limits, subconscious thinking, extracting concepts, and playing with toys or doing something to unclutter one's mind.

To prepare such an environment, there may be visuals throughout the facility reminding of creativity. The company can provide training to teach tools of creativity, or allow practice to think productively, instead of reproductively. The company can even have successful innovators share their experience with employees, or have a competition for innovative solutions (already defined in statistical terms as greater than 47.5% improvement). The author has found that having a library in-house on industry related books with some books in basics help tremendously in generating ideas. Even if employees go to the library for five minutes, walk around the library, and read titles of the books, they can come up with lots of innovative ideas. Because various book titles stretch their imagination in different direction. However, employees must be there to think of some ideas to do something differently. Going there to read newspaper or sports sometimes may not be as productive. However, reading sports in newspaper does teach us how the best-in-class sports work as a team to achieve excellence.

The creativity room must have an environment that forces minds to wander around. Doug Hall in his book *Jump Start Your Brain* has used the "Eureka!" method where participants receive stimuli through interacting with a variety of objects or ideas to create new ideas utilizing one's imagination and intellect. He has basically even created a Eureka Institute where participants attend the session to train their brains for creativity and innovation.

To bring that approach in-house, each company must create its own Eureka room where four fields of innovation (mechanical, thermal, electrical or magnetic) are explored; where imagination is allowed to expand without boundaries in sensory (colors, taste, feel, or sounds), time, space and applicable industry; where various principles of innovation or inventive problem solving can occur. The purpose is that the creativity room must make a person feel a little different through visuals or experience. The creativity room may have tools and toys to play with to mechanically create new things. In other words, every mind can find some stimuli to get it thinking without being bored. The size or complexity of the creativity room would vary from the corporate needs based on its diversity or complexity of products or services.

Process for Innovation

Everything in life is a process. Therefore, applying Shewhart's PDCA (Plan, Do, Check, and Act) cycle sheds light on its various components. Like any other process, the innovation process requires inputs in terms of machine, material, methods and people (based on Ishikawa's Cause and Effect analysis). One must think creatively about what kind of equipment or tools can be used, material or information, methods, and people who want to innovate.

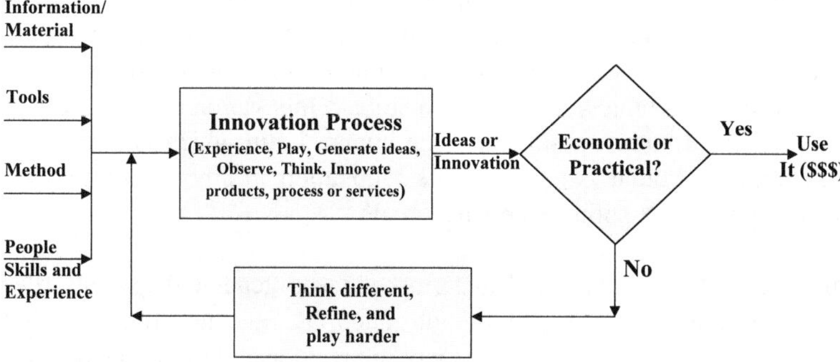

Figure 20-5 Innovation Process

The method of innovation must be loosely defined and difficult to replicate easily. The intent is that within some defined paradigm, one must enjoy the flexibility to try, experiment, fail, learn and innovate. The innovation process must include experiencing a variety of things stretching outside the box or domain of work, creating combinations or associations, mentally validating for 'go' or 'no go' for further play or trials. The mental massage of various concepts or models result in some practical ideas that are explored further for formulation of products, processes or services. Ultimately, the objective of every employee or player is that one must create value through innovation to achieve growth objectives of the business and employees.

If the innovation does turn out to be impractical or not economical for further implementation, one must not become dejected or disappointed. It takes many ideas to convert one into a good (or great) product or service. Therefore, innovation and creativity are a must; idea generation is imperative; and everyone must embrace the process as part of their personal, organizational and societal responsibilities.

Measuring Innovation

To ensure continual innovation and sustained growth, the organization must measure effectiveness of its innovation process, and reinvent the process as needed. Innovation has become a critical success factor for Six Sigma initiatives to be successful, and for organizational growth. The rate of innovation has accelerated with the information age. With all the information available on the Internet, one can quickly answer many unknowns and create new things. People can chat, brainstorm, design 24 hours a day due to globalization. The time to innovate has shrunk as we see more new products on shelves in stores. Innovation utilizes the most under utilized human resources - the human brain which has unlimited potential. The goal is to accelerate innovation in any corporation and measure it against some established goals.

Innovation can be measured in terms of ideas generated, patents filed, engineering awards, new product introductions, revenue from new products, people deployed in innovation, or hours allocated for innovation, etc. Ultimately, measurements must be established to drive innovation in support of organizational improvement and growth. The Six Sigma Business Scorecard (covered in Chapter 19) includes a set of measurements related to the growth and profitability. The ten measurements utilized in determining the Business Performance Index (BPIn) are as follows:

1. Employee recognition by CEO
2. Profitability
3. Rate of improvement
4. Employee recommendations
5. Purchase ($)/Sales($) ratio
6. Suppliers quality
7. Operational Sigma
8. Timeliness
9. New business/Sales ratio
10. Customer satisfaction

Measurements 1, 3, 4, and 9 promote innovation in an organization.

- The employee recognition by CEO is based on innovative solutions with a significant and visible impact on corporate performance.
- The Rate of Improvement measurement requires setting aggressive goals for improvement such that they would mandate employee participation and innovation.
- The Employee Recommendations are measured to ensure ideas continue to flow improving odds of innovation.
- The New Business/Sales ratio could be determined based on the revenue from new products, or revenue from new customers. The intent is to grow the business.

Innovation can be used to achieve lots of improvement very fast, saving lots of resources, and improve profitability. (Refer to Chapter 15 for creativity techniques to use in the Improve phase.) Or, the innovation can be used to create new opportunity for revenue and achieving business growth.

Benefits from Innovation

The sustenance of innovation initiative depends on the continual success of activities, excitement of doing them, and the engagement of everyone. As expectations for innovation in various functions are established, the leadership must verify performance against the expectations, and make appropriate adjustments. If an organization commits to innovation as a key component of its business strategy, it must ensure various phases of the innovation process are executed with excellence. The most critical aspect of the measurement must be the reduction in product or service cost, or impact on the corporate profitability.

Another measure of innovation would be how actively employees are intellectually involved. One can also assess if innovation boosts employee morale, or motivates them to be more productive. Another benefit of implementing the innovation process successfully would be simply learning new skills or ideas.

Recognition and Rewards

In order to promote innovation and creativity, one must have a keen eye to observe successes. Publicizing success is equally critical as ignoring failures. In a corporation, when creativity, innovation, and risk-taking become basic principles, one must establish measures to recognize and reward them. Each success is recognized differently - recognition could be financial, public recognition, or as simple as a 'Thank You' note.

Irrespective of the value or type of recognition or reward, one must recognize specific acts or outcomes of creativity or innovation. The act or outcome could be at the idea level, solution development level, or the outcome of the development process level. Ultimately, innovation must be a rewarding and an enriching experience for everyone involved.

Summary

Innovation is about learning *how* to think rather than *what* to think. Innovation needs to be fostered at the individual and the organizational level. At the individual level, innovation needs to be a personal endeavor that involves experimenting, learning, engaging in many different aspects of life, and establishing a sense of wonder about virtually everything (even the most simple things).

At the organizational level, leadership must understand, believe in and support the role of innovation in the business. The organization needs to develop a process, measurement, and rewards and recognition for innovation.

Chapter

21

Measurement System Analysis in Non-Manufacturing Environments

Introduction

"You get what you measure" is a well-worn axiom, even outside the world of Six Sigma practitioners. Measurements are the foundation of data-based decision-making, and therefore, must be validated before any objective evaluations can be made of current process performance or cause and effect relationships.

The power of Six Sigma lies in its disciplined, structured approach to identifying and solving process issues. The biggest potential pitfall of the Six Sigma approach is failing to adhere to that discipline. It is in the area of Measurement System Analysis (MSA) that discipline often falls by the wayside. When considering measurements for non-manufacturing processes, MSA often becomes the most neglected step. Although manufacturing Six Sigma projects are not immune to this problem, there are far more cases where transactional, business process or service project teams skip the MSA, concluding that it is unnecessary because the data are coming off of "The System." Numbers, survey data and other measurements are often accepted unquestioningly - which is somewhat remarkable in a society where nearly everything is viewed with skepticism. Particularly disturbing is the penchant of many who believe that data retrieved from The System are somehow immune from the pervasive evil of process variation - simply because the process in question is the measurement system.

In this chapter, the impact of measurement system analysis on process improvement will be examined, with attention paid to some of the specific characteristics that make MSA for non-manufacturing processes more challenging.

MSA in the DMAIC Cycle

For any process, there is an output for which a customer expectation and performance standard can be defined - and the customer tells us what represents "good" output versus "bad" output. As the owner of the process supplying the output, the challenge is to be as consistent as possible in delivering what the customer wants. Variation in the process output is, therefore, the enemy we seek to capture and control using the Six Sigma approach.

In its most general sense, a "process" can be depicted by the graphic below: a box into which inputs flow, and out of which an output is produced. The outputs vary because of variation coming from the inputs.

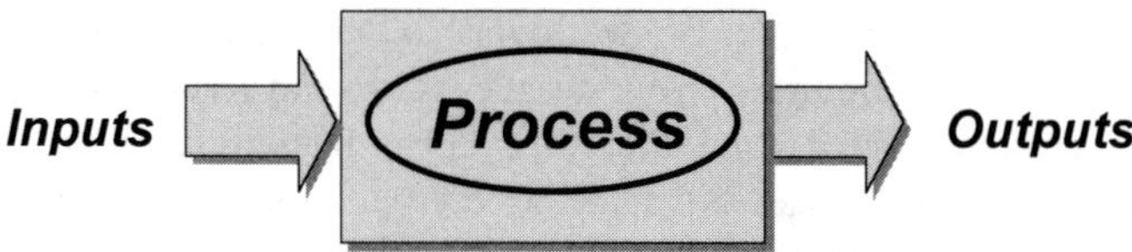

Figure 21-1 A Process

This concept is easily digestible when thinking of parts coming from a machine as outputs, or even when considering invoices as outputs flowing from an Accounts Payable process. The emphasis added by the Six Sigma methodology is considering the measurement system to be a process, with specific inputs that are combined to produce an output: the measurement.

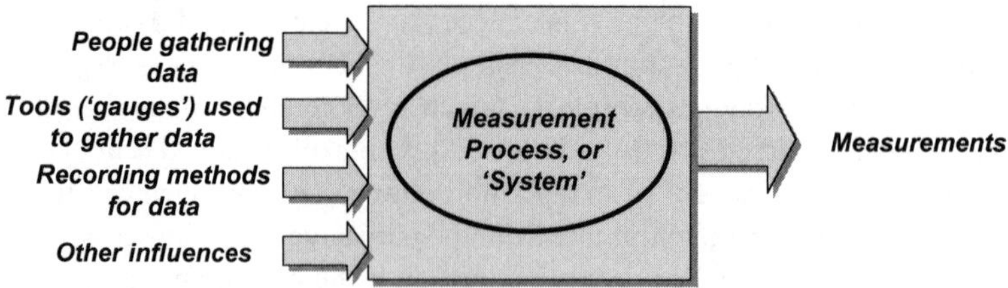

Figure 21-2 A Measurement Process (or System)

Since all processes create variation in the output to some degree, it becomes clear that the measurement system by itself - independent of the process producing the output that the customer sees - can create unacceptable levels of variability. To the person viewing the process delivering the parts or invoices, it is not readily apparent whether the variation in the observed output is coming from the measurement system, or the process itself. In fact, it is not unheard of to find that the measurement system is producing more variation than the process!

Total Observed Variation	**=**	***Variation from the Measurement System***	**+**	***Variation from the Process***

Figure 21-3 Measurement System Contributions to Variation

Through this formula, it becomes obvious that the only way to clearly observe what is really happening in the process is to minimize the variation coming from the measurement system.

MSA Psychology

The theory behind validating the measurement system before taking data is logical. So why does the Six Sigma discipline so often get violated as the project moves into the Measure Phase? Why is it automatically assumed that the measurements coming from the process are okay? There is a psychology at work in non-manufacturing processes that can be blamed for this phenomenon - and although it is far more prevalent in the non-manufacturing world, this psychology can frequently be found in manufacturing as well.

In essence, the problem can be boiled down to a perception issue: when people think of measurements, they think of a gauge or a query from a database rather than the ***process*** that creates the measurements. Consequently, the elements of the measurement process that typically create the bulk of the variation are often overlooked.

The measurement system is not just the gauge or the automatic query - it is also the people that interface with the system, the methods used to record the data and even the environmental conditions as the data are collected. Unless these elements are looked at together as the process that outputs measurement data, it is very tempting to look only at the data gathering tool itself - which is often a digitally-based, no-human-hands-required element - and judge it to be good because of its high level of technology.

The Fast Food Restaurant Example

Consider the drive-through lane of a fast food restaurant that is measuring "order cycle time" as its process output. A sensor in the pavement detects when a customer drives onto the pad with the microphone into which they place their order. The cycle time clock automatically starts. The customer then drives to the pick-up window, where the food is paid for and delivered to the customer. A sensor at that location detects when the customer drives off, thereby ending the order cycle time sequence. The computer subtracts the two time stamps, and an order cycle time measurement is transferred directly into the database - no human intervention whatsoever. Those sensors are pretty reliable, and the algorithm that calculates the time differential is hardly ever wrong . . . so why bother checking the measurement system for this process? It's got to be good.

The fallacy of this argument is the fact that the measurement system is not just the 'gauge' used to measure the process - in this case, the sensors and the computer. The measurement system also includes people. This measurement system can actually be quite variable. When peak times create delays at the pick-up window, customers are often asked to pay for the food and pull off to the side, allowing the order to be brought to them later without obstructing the line. This is not necessarily a deceptive practice on the part of the restaurant staff (especially if it keeps the line moving), but it renders the measurement data in the system for these customers grossly inaccurate for characterizing what is really going on in the order cycle time process.

Other Elements: Why Non-Manufacturing Processes are Different

Although the Six Sigma phases and steps are consistent for any process, whether non-manufacturing or manufacturing, the nature of non-manufacturing processes and the emphasis given to the DMAIC phases are decidedly different. Listed below are some of the primary distinguishing characteristics of typical non-manufacturing processes:

- Cross-functional: processes touch many areas of the business, requiring more extensive teams for improvement project implementation

- People-dependent: processes are not fully automated, so training and employee turnover are consistently large issues
- Undocumented: processes evolve and mutate over time (as people move in and out), and often do not have documented steps
- Significant external "noise": process output is often significantly affected by inputs that may not be easily controlled by the business, such as economic conditions, competitive actions, etc.
- Impossible or difficult to get repeat measures: traditional measurement studies require getting repeat measures on the same unit or process event (e.g. delivery of service) and this can be difficult or impossible for many non-manufacturing situations

From all of these factors, an interesting dynamic emerges: there is a huge improvement opportunity in many non-manufacturing processes, but the execution of projects can be significantly more challenging than in manufacturing applications. Definitions of roles, responsibilities and process steps are often quite nebulous. This means that a significant degree of emphasis during a non-manufacturing Six Sigma project will lie in the Define and Measure phases - prior to taking data for baselining purposes. Some clarity must be obtained up front, and the measurement system is a large part of that effort. In fact, before the classical MSA study begins, there are two prerequisites that must be met to clarify the measurement system:

1. Match the process output metric to the customer metric
2. Create an operational definition for the metric(s)

If a classical MSA study cannot be done because of the last characteristic listed above (inability to get repeat measures), these two prerequisites become in essence the validation of the measurement system and become even more important.

Prerequisite: Matching Metrics

Six Sigma is an outside-in approach to process improvement: it begins with the customer's stated requirements of what is needed from the provider, which is then translated into an internal business process that fulfills the customer need. When it comes to measurement systems, first and foremost is the requirement that the metric used to characterize the internal process output matches (or at least reflects) how the customer is judging the process output. This is true even if the customer is internal - the next downstream user of the process output under consideration.

Case Study: On Time Delivery

A major manufacturer routinely delivered truckloads of their products to a large distributor in their network on a weekly basis. The truck driver would call the warehouse supervisor at the distributor's location on the night before to confirm the scheduled delivery time for the next day. On numerous occasions, the supervisor would request that the delivery be delayed until the following day due to labor issues or backed-up work in their warehouse. The driver would enter the revised delivery date into the manufacturer's system and comply with the new request. However, the distributor's purchasing agent who placed the order was not aware of the revised delivery date since it had been negotiated between the driver and warehouse supervisor. Consequently, Purchasing judged performance of the supplier based on their compliance to original schedule, and the manufacturer judged performance based upon their compliance to the revised schedule. The mismatched measurements were easy to rectify, once the problem was identified; however, up to this point the distributor, had deemed the supplier a poor performer, and the business relationship was quite strained.

Lesson: Be sure that the customer measures the process output the same way as the process owner.

Case Study: On Time Departure (Airline Industry)

A classic case of mismatched measurement systems can be found in the airline industry. Most air passengers, when asked, consider departure time as the time the wheels leave the runway. Most airlines, however, measure departure time through an electronic system that monitors when the aircraft parking brake is released. Therefore, from the airlines' perspective, the flight departs on time when the brake is released at the loading gate and the plane begins to taxi. For those passengers who have experienced a flight that moves away from the loading gate and parks on a remote section of the tarmac for extended time periods, on time departure would not be an accurate characterization of their experience. In an example of unfortunate timing, a news story reported the airline industry's contention that departure and arrival performance had improved - even though customer satisfaction measurements regarding departure/arrival performance had significantly declined during this same time period.

Lesson: this example of mismatched metrics is difficult to correct by itself, since the aircraft electronic monitoring system is also used to calculate pilots' and flight attendants' pay. However, a mitigation technique would be to monitor both operational and customer satisfaction metrics simultaneously, to avoid optimizing one metric and sub-optimizing another.

Prerequisite: Operational Definitions

Once a metric (or metrics) that satisfies both the customer needs and the requirements of the process owner is identified, another pre-requisite to the MSA validation study is the creation of an operational definition for that metric. An operational definition is a specific description of how the process output is to be measured, including by whom, with what measurement tool, and in what data capture format (tick sheets, direct computer entry, etc.). Customer-based specifications must be identified in the operational definition, since these become the standards by which the goodness or badness of the process output are judged. The purpose of the operational definition is to ensure that everyone affiliated with the measurement process has a clear understanding of how the output will be measured and recorded, so that variability from people interfacing with the measurement system is minimized.

Example Operational Definition for On Time Delivery

> The units measured will be the time differential (in minutes) between the customer-requested schedule and the actual arrival of the product at the customer's location. The system will calculate the differential, comparing the time at which the customer signs the electronic ticket in the hand-held unit carried by the driver to the 'requested delivery time' field on the order. The target

value is 0 minutes differential; the customer will accept arrivals within +/- 15 minutes of the target as 'on time'.

MSA: Repeatability & Reproducibility (R &R) Studies

The discipline dictated by the Six Sigma approach requires that measurement systems be validated before collecting process data. Now that the metric has been defined and the operational definition written, the Measurement System Analysis can be performed to quantify the sources of variation in the measurement system.

Since the process data will be used to answer some very important questions such as 'How well is the process performing?' and 'What is driving the process output to vary?', it is critical to ensure that the data produced by the measurement system are "precise" (as consistent as possible.) MSA is, therefore, an important element needed before characterization of the y = f(x) equation can be done.

Just as total observed variation can be broken into two pieces (variation due to the measurement system and variation due to the process), measurement system variation can also be further broken down. Two common sources of variation are called repeatability and reproducibility. A study, known as an R & R study, is conducted to understand these sources. These sources and example studies are described below. However, the first decision before conducting the study is the type of data being used.

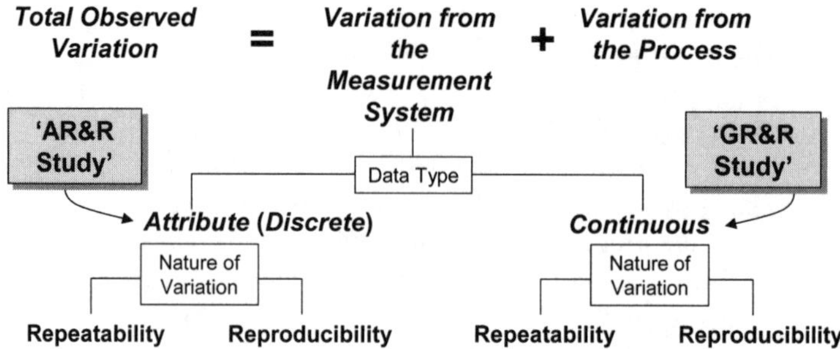

Figure 21-4 Measurement System Sources of Variation

Types of Data

The first task is to identify the type of data being measured. Data that are discrete can be measured as counts of occurrences, categories of results or pass/fail proportions. Discrete data are often described as "classification" data, and are also known as process attributes. Examples of discrete data in non-manufacturing processes include:

- Number of damaged containers
- Customer satisfaction: fully satisfied vs. neutral vs. unsatisfied
- Error-free orders vs. orders requiring re-work

Data that are continuous can be measured by numerical values, typically in which a decimal could make sense. Continuous data, by definition, can theoretically be divided into finer and finer increments of measurement, and so can be used to *quantify* the output. Continuous data are also known as "variable" data. Examples of continuous data in non-manufacturing processes include:

- Cycle time needed to complete a task
- Revenue per square foot of retail floor space
- Costs per transaction

From a process point of view, continuous data are always preferred over discrete data, because they are more efficient (fewer data points are needed to make statistically valid decisions) and they allow the degree of variability in the output to be quantified. For example, it is much more valuable to know how long it actually took to resolve a customer complaint than simply noting whether it was late or not. Measurement system analysis can be performed on processes using either data type - MSA for discrete or attribute data is known as 'AR&R', and MSA for variable data is known as 'GR&R'.

Sampling for MSA Studies

When a measurement study is to be performed, careful consideration must be given to the sample being used for the study. The measured elements of an MSA study should be representative of the full range of variation which could be seen from the process. Thus, it is an engineered sample, containing much more variability for the number of observations gathered than what would typically be seen from an actual process sample. It is very important to realize that the MSA study sample is NOT reflective of actual process performance, and cannot be used to characterize the process output. Its sole intent is to test the measurement system over the full range of output which could occur - including out-of-specification (customer requirement) results. Through the MSA, the precision, or consistency, of the measurement system can be quantified. The MSA sample and the process baseline sample are completely independent of one another.

The size of the MSA sample depends upon the nature of the data being measured. If the data are discrete (attribute), then more observations will be needed to characterize the measurement system than for continuous (variable) data. For example, a continuous study may have a sample size of 30 to 40 while an attribute study may have 100 or more observations.

Types of Measurement Error

Regardless of the nature of the data, the variation (error) from the measurement system can be further subdivided into two sources, known as R & R:

I. Repeatability error - when one operator is using the same gauge to measure the same element multiple times and obtains different results. Repeatability error is often called equipment variation because it is usually caused by a problem with the measuring device itself.
Remedy: modify the gauge, or improve the environmental conditions for reading the gauge (lighting, location, etc.)

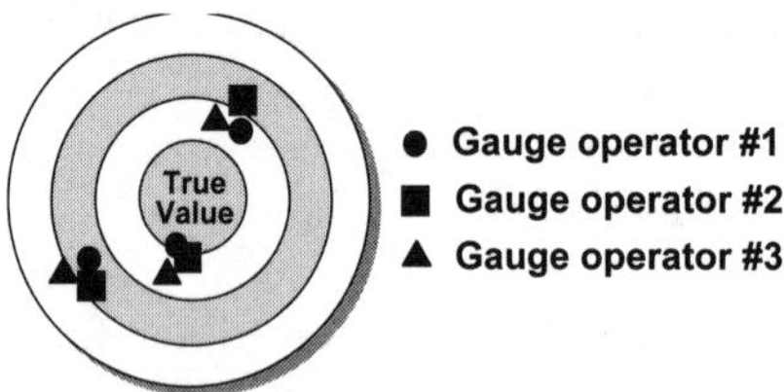

Figure 21-5 Measurement System Variation Due to Repeatability

II. Reproducibility error - when multiple operators are using the same gauge to measure the same element multiple times, and the results are different. Reproducibility error is often called appraiser variation, because it is usually caused by inconsistencies in measurement methods used by the operators. Remedy: train the operators, using the detailed operational definition for the measurement system.

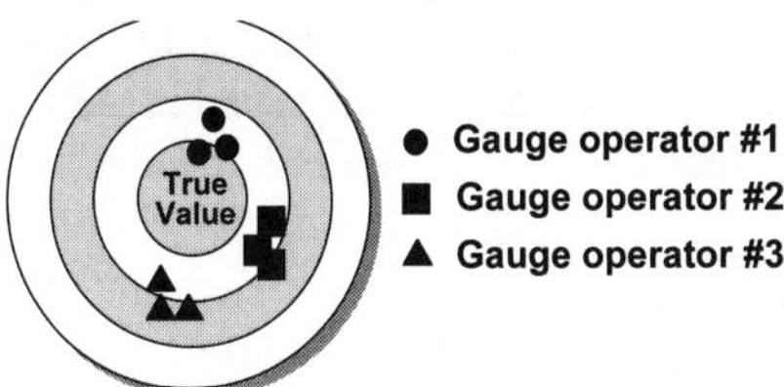

Figure 21-6 Measurement System Variation Due to Reproducibility

Attribute R & R (AR&R)

Attribute gauge studies are typically used when the measurement result is binary, such as defect/no defect or successful/unsuccessful, although rating scales can also be validated with this method. Multiple measurement system operators are chosen to measure a sample set two or more times. In this way, both repeatability (variation within the operator) and reproducibility (variation between the operators) can be quantified.

In an attribute study, a standard can be used for comparison with the results from the measurement system operators. The standard is the 'truth' and any discrepancy from truth due to the measurement system is considered an error (or defect).

For attribute measurements, a rule of thumb is that the agreement with truth should be 99% or better for the measurement system to be considered "good." Agreement of 95% or more is often considered marginal; if the agreement is less than 95%, the measurement system should be considered unacceptable. AR&R studies can be done using statistical software packages which provide graphical output and other summary information; however, they are often done by hand due to the straightforward nature of the calculations.

Example

An attribute measurement instrument with notoriously bad R & R is the "reason code" form. Whether it is reason codes for customer complaints, product returns or help requests, this type of measurement instrument is rarely validated and usually wildly variable. Process associates are asked to categorize events into pre-defined codes, and the meaning of the codes is seldom clear. In one case, a company used reason codes to categorize the reason for product returns. Customer service representatives were required to fill out an on-line form and pick from a pre-determined, numerical code that represented a return reason. When the Six Sigma project reached the MSA phase, they investigated the "goodness" of this measurement system and found that of the 27 codes on the list, only 9 were ever used. They also discovered that there was a common return reason that was apparently not an issue when the reason codes were developed years before, so the representatives just picked a category to label these returns. Through the Six Sigma project, reason codes were streamlined to seven categories, based upon input from the associates. In addition, concise operational definitions were provided on-line to help them choose the appropriate categories.

The team then conducted an AR&R study. They consulted an expert to determine the reason for the product returns. This was considered the "truth." Three service representatives were then asked to determine the reason code for 20 different product returns. Then, on a different occasion, the three representatives were asked to determine the return codes on the same 20 returns. It was a blind study; the representatives did not know they were looking at the same product returns both times. The Six Sigma team was then able to determine repeatability (did the representatives assign each product the same reason code?), reproducibility (were the same reason codes assigned from representative to representative?) and did they assign the right reason code per the expert. The results of the study were 100% across all categories. The team felt confident in their measurement system.

Gauge R & R (GR&R)

GR&R studies are performed on continuous output data. As in the AR&R study, multiple measurement system operators are used to measure sample data multiple times in order to quantify repeatability and reproducibility. Since the data are continuous though, the variation in the measurement system is characterized by standard deviation calculations that are best left to statistical software packages. GR&R results are usually expressed as the % of total variability coming from the gauge, or measurement system. Since variability from the measurement system should be small relative to the total variability, a typical criterion for acceptability is 10% or less. If the measurement system contributes between 10% and 30% of the total variability, it may be considered conditionally acceptable.

There are three approaches that can be used to calculate the percentage of variability attributable to the measurement system:

1. % of Total Variability (%R&R) - used to determine if the measurement system variability is small enough to understand process variability
2. % of Tolerance (%P/T) - used when the customer specification range is given (Upper Specification Limit - Lower Specification Limit). The criterion for acceptability for GR&R as a % of tolerance is the same as for % of Study: 10% or less is acceptable, and 10-30% is marginally acceptable.
3. % of Contribution - a special case of the variability calculation, which uses the statistical variation rather than the standard deviation (which allows the contributions to be arithmetically added). The criterion for acceptability is: < 2% variation is acceptable, 2-9% is conditionally acceptable.

Statistical software packages often provide the added benefit of quantifying the discrimination of the continuous measurement system. This can often be seen as the number of distinct categories in the analysis output. When the number of distinct categories is greater than five, the measurement system can discern over five groups within the data range and is often considered to have acceptable discrimination.

A summary of typical acceptance criteria is shown in the chart below:

	AR&R	GR%R %Total Variability	GR&R % Tolerance	GR&R %Contribution	# of Distinct Categories
Acceptable	99%+ Agreement	< 10%	< 10%	< 2%	> 10
Marginal	95% -99% Agreement	10% - 30 %	10% - 30 %	2% - 9%	5 - 10
Unacceptable	< 95 % Agreement	> 30%	> 30%	> 9%	< 5

Figure 21-7 R & R Acceptance Criteria

Continuous gauge studies occur less frequently in non-manufacturing environments. The following gives an example where a team decided to use technology to get repeat measures.

Example

A team was going to use cycle time as their 'y' for their project. They were studying a customer order process. It was a high-volume area and the team was concerned that having the representatives in the process collect the data on cycle times may **change** the cycle times. Each order took about ten minutes.

Cycle time had never been collected for this process before and the team was concerned about how well they could collect this type of data. They decided to have team members do the collection for the project but they were not trained analysts.

To conduct a measurement system study, the team videotaped 10 different customer orders being handled through process. Before videotaping the orders flowing through the process, the team explained the purpose of the taping to the representatives. The purpose was to assess the ability to get valid cycle times - not to critique the representatives in the process.

They had three team members measure the cycle times using the video-tapes. After some time had passed, they had the team members look at the videos again and re-assess the cycle times. From this study, they determined the repeatability and reproducibility.

They found the repeatability (a person's ability to get the same value on the same unit) to be relatively small. The reproducibility (differences from person to person) was much larger, not surprisingly.

Upon investigation, they found some differences in how the team members timed the processes:

1. Two team members considered the time it took to get the order from the fax machine as part of the process; one member didn't start the clock until the person had the order at their desk.
2. One team member stopped the clock when a person in the process had to answer the phone or leave the process for a moment. The other two didn't.
3. In one videotape, a person needed help and advice from another representative. Two team members considered this as part of the process and one team member didn't.

After reviewing the differences in the times, the team saw the need for more clarification on what they meant by cycle time. They had previously developed an operational definition but without testing this definition there were many situations that they hadn't addressed.

Comparison of MSA Acceptance Criteria

Those new to MSA are often surprised to find that the criteria for good attribute data measurement systems are more stringent than those for continuous data measurement systems. It is the nature of the measurement that drives the difference in standards:

- For discrete (attribute) data, an incorrect measurement is not just "off," it is dead *wrong*; therefore, there must be high agreement between "truth" and the measurements, or the measurement system will be rendered completely useless for decision-making.
- For continuous (variable) data, the standards are relaxed somewhat because a slight error in the measurement result may not be as devastating. For example, if the true value of the measurement is 25.40 and the measurement system outputs 25.35, the difference is small enough that it may not have a deleterious effect on the deci-

sions that are made from that data (unless the customer specifications are in thousandths)

MSA for Surveys

Non-manufacturing processes often utilize surveys to obtain process measurement data, the most common being customer satisfaction surveys. Unfortunately, the tendency towards MSA avoidance becomes even more evident when surveys are the measurement tools, and the risk of excessive measurement system variability is even greater in these cases.

Validation of survey instruments is somewhat different from the R & R approaches described previously. MSA may be done by piloting the survey prior to broad issuance, and confirming that the meaning of the questions (or statements) on the survey is clearly understood. One approach is to have the survey author list what was intended by the question - two to three bullets for each question. Then, the survey is given to the pilot study group (usually 10 to 20 people) as a 'pre-survey' - not for them to take the survey, but to list for each question what ***they*** think the survey is asking.

As in an R & R study, the sample of people evaluating the survey instrument should be engineered - customers, people inside the process, and people who are completely independent of the process should be included. A comparison is then made of what the pre-survey recipients thought the questions meant to the meanings defined by the survey author. If discrepancies are found, the questions are re-worded and another pre-survey is piloted.

Performing an MSA on a survey instrument can be very enlightening, especially in the global business community. Many surveys that go out without validation return surprising results because of the language differences and the difficulties in interpreting colloquialisms and slang. Creating a common understanding of the elements on a survey form is the best way to minimize variation in the measurement results.

Near Misses: Other Non-manufacturing MSA Case Studies

In many non-manufacturing processes, the measurement study is bypassed, for all of reasons previously mentioned. Hopefully, the mistake is caught in time and the poor measurement system is uncovered and corrected. In these situations, there is usually a period of shock as the team realizes how close they came to disaster - basing high cost, high impact decisions on bogus data. The two case studies below highlight processes in which this nearly occurred.

Case Study: Order Placement Process

A high-volume call center for wholesale and retail products handled hundreds of orders per day. The order-taking system used had been in place for enough time that all of the associates felt very comfortable with it. Orders from customers could be submitted in any of four ways: by phone, by fax, by mail and by a private internet-based order form. There was a big push from corporate to increase the number of orders placed over the internet, since order service associates would not have to key in any of the order data (saving time and improving productivity). A Six Sigma project was initiated to uncover the key drivers of orders placed by internet within a specific sales channel with the goal of increasing the percentage of orders placed via internet. The team decided on a data collection plan that involved querying an extensive historical database, triggering on an order form field called "Placed By." On incoming orders, the associates coded the field according to the method of placement: "T" for telephone, "F" for fax, "M" for mail. The "I" for internet was automatically populated in the field. Given the maturity of the system and the simplicity of the field in question, a measurement system study was not done. When the first query was returned and studied, the team was surprised to find a 98.2% order placement by telephone. This was particularly surprising given the bank of fax machines that ran almost constantly, and the team of associates who did nothing but enter faxed-in orders into the system. A study was initiated to validate the measurement system. After some discussion, the team decided to create a sample of orders from faxes pulled randomly from the fax machine over several days. Information from 50 faxed orders was recorded, then the faxes were sent back through the system normally. After three days, the team queried the database to find the electronic version of the orders in their sample - 75% of the orders were identified in the "Placed By" field with a "T" even though the team knew the orders came in by fax. How could this have happened? It turns out that the "Placed By" field was defaulted to "T" and associates would have had to manually change it as they entered the orders. This field had not generated much interest at corporate before this project, so the attention to that detail evaporated over time as associates moved in and out of their high-turnover jobs. The repair was relatively painless, and good data began flowing from the database once the "Placed By" field was changed to a non-default, required entry.

Case Study: Sales Effectiveness Survey

The sales arm of a major manufacturer was interested in improving the effectiveness of their sales force that routinely went out and met with distributors and retailers of their products. A Six Sigma project to improve sales force effectiveness was launched with the aim of identifying statistically those elements that customers are looking for in a salesperson. A focus group was used to develop the categories on the survey form that was going to be used to solicit feedback from the customers, and a 9-statement form was developed from their input. The regional manager had already sent out five of the survey forms to customers in his vicinity. During a Six Sigma project review, the topic of MSA came up, and the regional manager realized that he had not validated the content of the statements on his form. He recovered the forms that had been sent already, then created a list of what he meant by each of the statements. He went to seven colleagues (customers, peers and his own team members) and asked them to list what they thought was meant by each of the statements. An excerpt of the statements is shown below. Each statement was to be ranked on a 1 to 5 rating scale, with 1 meaning "strongly agree" and 5 meaning "strongly disagree."

Sample survey statements:

- My salesperson is knowledgeable about current sales events
- My salesperson keeps me up-to-date on new products
- My salesperson can respond to technical questions in a timely manner

The first shock came on the statement about "keeping up-to-date on new products." Four of the seven colleagues thought that meant that the salesperson went to the customer's location and replaced pages in the customer's sales binder - a customer requirement of which the sales force was not even aware! The second surprise was the number of people who misunderstood the rating scales - transposing a 5 for strongly agree, and a 1 for strongly disagree. The regional manager reworded the statement to read: "My salesperson discusses new products with me within 2 weeks of product launch," and changed the rating scale to the more traditional 5 for strongly agree (with clearer labels for those taking the survey). In addition, he and the sales force held expectation-setting meetings with their customers over the subsequent months in order to be sure that everyone understood the requirements of their relationship (a bonus takeaway, outside the direct scope of the Six Sigma project).

Summary

A basic rule of thumb for any Six Sigma project is that the measurement system is probably not as good as you think it is, and if the measurement system is not validated, then any data coming from that system are suspect - introducing significant risk into the subsequent decision-making regarding process improvements.

The measurement system is a process with inputs and outputs, and those inputs go beyond just the gauge used. Since the measurement process variation adds directly to the actual process variation, it can create a very distorted view of the real process and must, therefore, be minimized.

Measurement system validation is not a step that can be skipped. If "you get what you measure," then the measurement system must provide a correct picture of what you are getting or the benefits of measurement are lost. In transactional processes, business processes and service processes, a significant portion of the variation from the measurement system can be attributed to mismatched measurements and/or poorly written (or non-existent) operational definitions. Even if the formal R & R study is not performed, matching of measurement systems and construction of operational definitions should always be done to validate the measurements before collecting actual process data.

Index

ABOUT THE AUTHORS

TOM MCCARTY is the Director of Consulting Services for Motorola University. Tom and the Consulting Services team are dedicated to improving the business performance of Motorola's suppliers, channel partners, customers, and business alliances through the full implementation of Six Sigma Business Improvement Campaigns. For the past five years, Tom has worked with numerous executive teams to help them formulate strategic plans, balanced scorecards, and focused Six Sigma implementation plans. Tom is trained as a Six Sigma Black Belt, has 26 years of Motorola management experience, and has successfully led Six Sigma implementations for a variety of clients, across a variety of industries. Tom is a contributing author to *The Corporate University Handbook* and a co-author of *The New Six Sigma*.

LORRAINE DANIELS is a Master Black Belt working as a Six Sigma Senior Consultant and Master Instructor worldwide for Motorola University. She is a doctoral candidate in industrial engineering at Arizona State University. She lives in Tempe, Arizona.

MICHAEL BREMER is the President of the Cumberland Group, a consulting firm that specializes in process and business performance improvement. A certified Six Sigma Black Belt from Motorola, he lives in Hinsdale, Illinois.

PRAVEEN GUPTA is a PE, Master Black Belt, and ASQ Fellow. Praveen, who was there at time and place of the birth of Six Sigma, has taught Six Sigma at Motorola University for over ten years. His first Six Sigma project was completed in 1988. He lives in Schaumburg, Illinois.